Algebra and Operator Theory

Algebra and Operator Theory

Proceedings of the Colloquium in Tashkent, 1997

edited by

Yusupdjan Khakimdjanov

Institute of Mathematics,
Uzbekistan Academy of Sciences,
Tashkent, Uzbekistan
and
Department of Mathematics,
Université de Haute Alsace,
Mulhouse-Colmar, France

Michel Goze

Department of Mathematics,
Université de Haute Alsace,
Mulhouse-Colmar, France

and

Shavkat A. Ayupov

Institute of Mathematics,
Uzbekistan Academy of Sciences,
Tashkent, Uzbekistan

SPRINGER SCIENCE+BUSINESS MEDIA, B.V.

A C.I.P. Catalogue record for this book is available from the Library of Congress.

ISBN 978-94-010-6130-8 ISBN 978-94-011-5072-9 (eBook)
DOI 10.1007/978-94-011-5072-9

Printed on acid-free paper

TABLE OF CONTENTS

PREFACE

The first Franco-Uzbek colloquium in Mathematics took place at the University of Haute Alsace in May 1995. Thanks to the support of the French Embassy in Uzbekistan, of the "Conseil Général du Haut Rhin", and the "Villes de Mulhouse et Colmar", a group of Uzbek mathematicians could come to Alsace in order to present their research works. Everyone will understand the interest of such a colloquium for some mathematicians, somewhat isolated from the rest of the world. In order to reinforce these ties and to put in concrete form a co-operation between the laboratory of Mathematics of the "Université de Haute Alsace" and the mathematical Institute of the Academy of Sciences of Uzbekistan, a second colloquium in mathematics had simply to follow. Also, this had allowed to widen the relations with other French and Uzbek laboratories.

This volume presents the lectures given during this symposium which took place in Tashkent from 29 September to October 5, 1997 at the Mathematical Institute of Uzbekistan Academy of Sciences.

The chosen themes deal with Algebra and the Theory of Operators. This choice was dictated mainly by the mathematical relations between the two countries; they focus essentially on these two themes. Some works signed together by some Uzbek and European mathematicians have to do with the theory of deformations of the Lie algebras, the study of the varieties of laws of algebras, the study of the representations of the Lie algebras. So, a certain number of articles concern the study of Lie algebras, Leibniz algebras or superalgebras. The contributions of European mathematicians also treat of Hochschild cohomology for Klein surfaces, real K-theory, Euler equations on Lie algebras. This volume also contains a detailed introduction to the theory of quantum groups, which had been presented at the first colloquium. As far the theory of operators is concerned the team of Professor Ayupov, member of the Uzbekistan Academy of Sciences, has specialized for a long time in this domain. The works of the Uzbek mathematicians to a great extent center around this topic and essentially concern the study of certain type of Operator Algebras. A more geometric presentation related to a problem of Elie Cartan was also given. This reflects thus an other branch of research carried through in the Mulhouse laboratory.

The organizers want to thank the French Embassy in Tashkent which awarded some grants permitting a certain number of French mathematicians to participate in the colloquium and discuss with their Uzbek colleagues; the cities of Mulhouse and Colmar for their contribution to the publication of this proceeding and the University of Haute Alsace for its material and financial support.

ON LEIBNIZ ALGEBRAS

Sh.A. AYUPOV, B.A. OMIROV

Institute of Mathematics,
Uzbekistan Academy of Sciences
F. Hodjaev str., 29, 700143 Tashkent, Uzbekistan.

This work is devoted to study of comparatively new algebraic object - Leibniz algebras, introduced by Loday [1], as a "non commutative" analogue of Lie algebras.

For Leibniz algebras naturally the question arises - to prove analogues of results from the theory of Lie algebras. Specifically, in this work for Leibniz algebras, satisfying Engel's condition we prove basic theorems, which were known for Malcev algebras, also presenting a generalization of Lie algebras.

In this work we define the following correspondence: for each Leibniz algebra L corresponds the Lie algebra $R(L)$, the set of right multiplications.

We also introduce the notion of right annihilator, which is the kernel of the map $\pi : L \to R(L)$, in order to describe Leibniz algebras of dimension three over a field of zero characteristic.

Relation of Leibniz algebra with its factor algebra by right annulator is considered and sufficient conditions are found for a Leibniz algebra to become a Lie algebra.

1. Nilpotency properties.

A Leibniz algebra L over a field F is an F-module equipped with a bilinear map called bracket: $[-,-] : LxL \to L$, satisfying the Leibniz identity:

$$[x,[y,z]] = [[x,y],z] - [[x,z],y] \tag{1}$$

for all x, y, z in L.

Y. Khakimdjanov et al (eds.), Algebra and Operator Theory, 1-12.

Observe, that if the identity $[x,x] = 0$ is also valid in L, then the Leibniz identity coincides with the Jacobi identity. So, Leibniz algebras are "non commutative" analogues of Lie algebras.

For a Leibniz algebra L define the following well-known sequences:

a) $L^{<1>} = L, \quad L^{<n+1>} = [L^{<n>}, L]$;

b)$L^{[1]} = L, \quad L^{[n+1]} = [L^{[n]}, L^{[n]}]$;

c)$L^1 = L, \quad L^{n+1} = [L^1, L^n] + [L^2, L^{n-1}] + ... + [L^{n-1}, L^2] + [L^n, L^1]$:

The algebra L is called *nilpotent*, if there exists $n \in N$ such that $L^n = 0$. Algebra L is called *solvable*, if there exists $m \in N$ such that $L^{[m]} = 0$.

Algebra L is called *right nilpotent* if $L^{<n>} = 0$ for an appropriate $s \in N$. For these sequences the following two lemmas are true

Lemma 1 *For any $i, j \in N$ the relation $[L^{<i>}, L^{<j>}] \subseteq L^{<i+j>}$ is valid.*
Proof Proof is produced by induction on i for each j. The base of induction is obvious. Suppose, that inclusion $[L^{<i>}, L^{<j>}] \subseteq L^{<i+j>}$ is valid. Then $[L^{<i>}, L^{<j+1>}] = [L^{<i>}, [L^{<j>}, L]] \subseteq [[L^{<i>}, L^{<j>}], L] - -[[L^{<i>}, L], L^{<j>}] \subseteq [L^{<i+1>}, L] - [L^{<i+1>}, L^{<j>}] \subseteq L^{<i+j+1>}$.□

By lemma 1 the following lemma is clear.

Lemma 2 *For any $n \in N$ we have $L^{<n>} = L^n$.*

Thus, the notions of right nilpotent and nilpotent Leibniz algebras coincide.

For any element x of L consider the operator $R_x : L \to L$ of right multiplication, where $R_x(z) = [z, x]$. The set $R(L) = \{R_x : x \in L\}$ of all operators of right multiplication with commutator product is a Lie algebra and the equality:

$$R_x R_y - R_y R_x = R_{[y,x]} \tag{2}$$

is true in $R(L)$. Using the equality (2), it is easy to observe that the Lie algebra $R(L)$ is solvable if and only if the algebra L is solvable.

Further we suppose that the field F has zero characterictic.

A subset B of an associative algebra U over a field F is called *weakly closed* if for any pair (a, b), where $a, b \in B$, there exist $\gamma(a, b) \in F$ such that $ab + \gamma(a, b)ba \in B$.

Theorem 1 *Let B be a weakly closed subset of the associative algebra U of all linear maps on a finite dimensional vector space V over the field F. Suppose that each element of B is associatively nilpotent, i.e. $b^k = 0$ for some $k \in N$. Then the associative subalgebra $B^* = alg < B >$ generated by the system B is nilpotent.*

Proof see [2].

For Leibniz algebras the following analogue of Engel's theorem is true.

Theorem 2 *Let L be a Leibniz algebra with $\dim L = k < \infty$. Then L is a nilpotent algebra, if and only if R_x is nilpotent for any $x \in L$.*

Proof Let L be a nilpotent Leibniz algebra. By lemma 2 there exists $m \in N$, such that $[...[[x_1, x_2], x_3]..., x_m] = 0$ for any $x_i \in L$, $i = \overline{1,m}$. Thus $[...[[y, x], x]..., x] = 0$, where x appears $m-1$ times, i.e. $R_x^{m-1} = 0$.

Conversely, let operator R_x be nilpotent for any $x \in L$. The set $R(L)$ of right multiplications is a Lie algebra of nilpotent linear maps. By theorem 1, the associative algebra $(R(L))^*$ is nilpotent, i.e. there exists $n \in N$, such that $R_{x_1} R_{x_2} ... R_{x_n} = 0$ for any $x_i \in L$. Thus for any $y, x_i \in L$, we have $[[...[y, x_1], x_2]..., x_n] = 0$, i.e. $L^{n+1} = 0$.□

Theorem 3 *Let $\Im$ be a Lie algebra of linear maps on a finite dimensional vector space over a field of zero characteristics and suppose that Θ is a solvable radical of the algebra $\Im$ and $\Re$ is a nilpotent radical of the associative algebra $\Im^*$. Then $[\Theta, \Im] \subseteq \Re$.*

Proof see [2].

Similar to the case of Lie algebras we have the following theorem:

Theorem 4 *Let L be a Leibniz algebra with* $\dim L = k < \infty$. *Then L is solvable if and only if L^2 is a nilpotent algebra.*

Proof The solvability of the algebra L easily follows from the nilpotency of the algebra L^2. Conversely, let L be a solvable algebra, i.e. $L = \Theta(L)$. It is easy to see, that the solvability of the algebra L is equivalent to the solvability of the Lie algebra $R(L)$ and hence by theorem 3, $[R(L), R(L)]$ is contained in the nilpotent radical of the algebra $R(L)^*$. Hence, there exists $m \in N$ such that for any maps of the form $[R_{x_i}, R_{y_i}]$, where $x_i, y_i \in L$, $i = \overline{1,m}$, one has the equation $[R_{x_1}, R_{y_1}]...[R_{x_m}, R_{y_m}] = 0$. Since

$$[R_{x_i}, R_{y_i}] = -R_{[x_i, y_i]}$$

we obtain $(-1)^m R_{[x_1,y_1]} ... R_{[x_m,y_m]} = 0$, i.e.

$$[...[z, [x_1, y_1]], [x_2, y_2], ..., [x_m, y_m]] = 0$$

for all $z \in [L, L]$. Therefore the algebra L^2 is nilpotent. □

Corollary 1 *Let L be a solvable finite dimensional Leibniz algebra over a field of zero characteristic, satisfying the condition $R_{e_i}^{n(e_i)} = 0$ for any element e_i from the basis of L. Then the algebra L is nilpotent.*

Proof It is sufficient to change the notion of centre by the notion of right annihilator in the proof [3.theorem 2]. □

Corollary 2. *Let L be a solvable finite dimensional Leibniz algebra over a field of zero characteristic, satisfying the condition $R^2_{e_i} = 0$ for any element e_i from the basis of L. Then the algebra L is nilpotent.*

Proof In theorem 3 of [3] there is a similar statement for Lie algebras. For the Leibniz algebra L we can pass to the Lie algebra $R(L)$, which is nilpotent and therefore is solvable. Thus L is also solvable. □

An algebra L is called *locally nilpotent* if its any finitely generated subalgebra is nilpotent.

We say, that an algebra L satisfies "n Engel's" condition (condition E_n) if $R^n_x = 0$ for any $x \in L$.

Kostrikin [2] has proved the nilpotency of Lie algebras over a field of zero characteristic, which satisfies the condition E_n .

It is easy to see that if L satisfies the condition E_n, then the Lie algebra $R(L)$ also satisfies the condition E_n , and thus the algebra L with the condition E_n is solvable in the case of zero characteristic.

For finitely generated algebra L with generators $\{x_1, ..., x_k\}$ define the size of the word: $r(R_{x_i}) = 1$ for any $i \in (1, ..., k)$ and $r(R_u * R_v) = d(u) + d(v)$, where $d(u)$ is the length of the word u with respect to variables $\{x_1, ..., x_k\}$.

For algebras L with the condition E_n the following theorem is hold.

Theorem 5 *Let L be a Leibniz algebra, which satisfies E_n . Then L is locally nilpotent.*

Proof: As it was mentioned above the algebra $R(L)$ also satisfies the condition E_n and in accordance with [4] we obtain its nilpotency. Since $R_{[x,y]} = [R_y, R_x]$, there exists $s \in N$ such that $R(L^s) = 0$. Now let L' be the subalgebra generated by finite number of variables $\{x_1, ..., x_k\}$ from the algebra L. The algebra L' obviously satisfies the conditions E_n and $R(L'^s) = 0$. By Leibniz identity it is easy to see, that any word $u \in L'$ can be decomposed into the sum of rightnormed words $((((x_{i1}, x_{i2})x_{i3})x_{i4})...x_{ip})$, where $x_{ip} \in \{x_1, ..., x_k\}$. Let $x_1 < x_2 < ... < x_k$, $B_i := \{$ rightnormed words with the length l on variables $\{x_1, ..., x_k\}\}$ where $l = 1, 2, ..., \quad s-1$, and $B := \bigcup_{l=1}^{s-1} B_l$.

Consider the ordering on the set B, putting $u < v$, if $d(u) < d(v)$. Denote by N the cardinality of the set B. It is clear, that $N < \infty$ Further we will suppose that all words are rightnormed. Consider the notion of regular word: the word $R = R_{v1}R_{v2}R_{v3}...$ called regular if $v_1 \leq v_2 \leq v_3 \leq ....$ By induction on the length of words let us show that any word R can be represented as the sum of regular words of the same size as R. For $k = 2$ $R = R_{v1}R_{v2}$ is either regular or $R_{v1}R_{v2} = R_{v2}R_{v1} + R_{[v2,v1]}$, where each term is regular. Suppose that this is true for words of the length $\leq k$. Consider the word $R = R_{v1}R_{v2}...R_{vk}$ and let $v_i > v_{i+1}$. Then

$$R_{v1}R_{v2}...R_{vi}R_{v(i+1)}...R_{vk} = R_{v1}R_{v2}...R_{v(i+1)}R_{vi}...R_{vk}+$$

$$+R_{v1}R_{v2}...R_{[v+i,vi]}...R_{vk}.$$

In the first term the number of order breaks is less that in R. Repeating the same procedure with this term finite number of times, we obtain the sum of regular words of the same size, as R. The length of the second term is less than k and by the assumption of induction it also can be decomposed into the sum of regular words with $r(R)$ size. Consider the word: $R = R_{x_{i1}}R_{x_{i2}}R_{x_{i3}}...R_{x_{iM}}R_{x_{i(M+1)}}$, where $M = N(n-1)$. It is decomposed into the sum of regular words with $(M+1)$ size, but the maximal regular word different from zero has M size. Thus $L^{\prime<M+2>} = 0.\square$

Remark. In theorem 3 it is not clear whether L is globally nilpotent. But for $n = 2$ we can prove global nilpotency.

It is easy to see that in a Leibniz algebra L, which satisfies the condition E_n there exists a maximal locally nilpotent ideal and it is equal to the maximal locally solvable ideal. This maximal locally nilpotent ideal is called *a radical* and denoted by R. The following theorem shows that the ideal R is a radical in the sense of Kurosh.

Corollary *Let L be a Leibniz algebra with the condition E_n. Then the algebra L/R is semisimple.*

2. The description of three dimensional Leibniz algebra.

Consider the set $Z(L) = \{x \in L : [y,x] = 0$ for any $y \in L\}$ of L. It is easy to see that $Z(L)$ is an ideal of L and contains elements of the form

$$[x,y] + [y,x], [x,x], [[x,y],[y,x]]$$

for any $x, y \in L$.

Therefore, in the Leibniz algebra $L/Z(L)$ the identity $[\overline{x},\overline{x}] = 0$ is valid and so it is a Lie algebra.

Lemma 3 *Let L be a Leibniz algebra, which is not a Lie algebra. Then $Z(L) \neq 0$ and $Z(L) \neq L$.*

Let L be a Leibniz algebra of dimension 3, which is not a Lie algebra, and suppose that dim $Z(L) = 1$. Let x, y, z be basis vectors of L and $x \in Z(L)$. Then taking into account that $Z(L)$ is an ideal of L we obtain the following multiplication in L :

$[x,x] = 0;\quad [y,y] = a_1x;\quad [x,z] = a_4x;$

$[z,x] = 0;\quad [z,z] = a_2x;\quad [z,y] = \beta_1x + \beta_2y + \beta_3z;$

$[z,x]=0;\ \ [x,y]=a_3x;\ \ [y,z]=\gamma_1x-\beta_2y-\beta_3z;$
where $a_1^2+a_2^2\neq 0$.
Denoting $\overline{y}:=\beta_2y+\beta_3z$, and omitting the bars we obtain

$$\begin{aligned}
&[x,x]=0;\ \ [\overline{y},\overline{y}]=a_1'x;\ \ [x,z]=a_4'x;\\
&[z,x]=0;\ \ [z,z]=a_2'x;\ \ [z,\overline{y}]=\beta_1'x+\beta_2'\overline{y};\\
&[\overline{y},z]=0;\ \ [x,\overline{y}]=\alpha_3'x;\ \ [\overline{y},z]=\gamma_1'x-\beta_2'\overline{y};
\end{aligned}$$

where ${\alpha'_1}^2+{\alpha'_1}^2\neq 0$.

Identities for the structure constants in this case are the following:

$$\left[\begin{array}{c}
2\beta_2\alpha_1=\gamma_1\alpha_3-\alpha_1\alpha_4\\
\beta_2\alpha_3=0\\
\alpha_2\alpha_3=\beta_2\beta_1+\beta_1\alpha_4+\beta_2\gamma_1
\end{array}\right. \tag{3}$$

a) Let $\beta_2=0$, then we obtain the following multiplication:

$[y,y]=\alpha_1x;\ \ [x,y]=\alpha_3x;\ \ [z,y]=\beta_1x;\ \ [z,z]=\alpha_2x;$
$[x,z]=\alpha_4x;\ \ [y,z]=\gamma_1x;\ \ [x,x]=[z,x]=[y,x]=0.$
From Leibniz identity we obtain the conditions:

$$\left[\begin{array}{l}
\gamma_1\alpha_3=\alpha_1\alpha_4\\
\beta_1\alpha_4=\alpha_2\alpha_3
\end{array}\right. \tag{4}$$

I) a') Suppose $\alpha_1\neq 0$ and $\alpha_3\neq 0$ then from (4) we have $\alpha_4=\frac{\gamma_1\alpha_3}{\alpha_1};\ \ \alpha_2=\frac{\beta_1\gamma_1}{\alpha_3}$.
Denoting $\overline{y}:=\frac{y}{\sqrt{\alpha_1}}$ we have the following:

$[\overline{y},\overline{y}]=x;\ \ [x,\overline{y}]=\alpha_3'x;\ \ [z,\overline{y}]=\beta_1'x;$
$[z,z]=\beta_1'\gamma_1'x;\ \ [x,z]=\alpha_3'\gamma_1'x;\ \ [\overline{y},z]=\gamma_1'x;$

where $\beta_1'=\frac{\beta_1}{\sqrt{\alpha_1}},\ \ \gamma_1'=\frac{\gamma_1}{\sqrt{\alpha_1}},\ \ \alpha_3'=\frac{\alpha_3}{\sqrt{\alpha_1}};$

Since $dimZ(L)=1$, it follows that $\gamma_1'\neq 0$ Denote $\overline{z}:=-\frac{z}{\gamma_1'}$, then multiplication will have the following form:
$[\overline{y},\overline{y}]=x;\ \ [x,\overline{y}]=\alpha_3'x;\ \ [\overline{z},\overline{y}]=\beta"_1x;$
$[\overline{z},\overline{z}]=\beta"_1x;\ \ [x,\overline{z}]=\alpha_3'x;\ \ [\overline{y},\overline{z}]=x;$
where $\beta"_1=-\frac{\beta_1'}{\gamma_1'}$

b') Suppose now $\alpha_1 \neq 0$ and $\alpha_3 = 0$ then $\alpha_4 = 0$. Putting $\overline{y} := \frac{y}{\sqrt{\alpha_1}}$ we obtain

$$[\overline{y}, \overline{y}] = x; \quad [x, \overline{y}] = 0; \quad [z, \overline{y}] = \beta_1' x;$$

$$[z, z] = \alpha_2 x; \quad [x, z] = 0; \quad [\overline{y}, z] = \gamma_1' x;$$

where $\beta_1' = \frac{\beta_1}{\sqrt{\alpha_1}}$; $\gamma_1' = \frac{\gamma_1}{\sqrt{\alpha_1}}$;

In the case $\gamma_1' \neq 0$ denoting $\overline{z} := \frac{z}{\gamma_1'}$, we obtain:

$$[\overline{y}, \overline{y}] = x; \quad [x, \overline{y}] = 0; \quad [\overline{z}, \overline{y}] = b"_1 x;$$

$$[\overline{z}, \overline{z}] = a_2' x; \quad [x, \overline{z}] = 0; \quad [\overline{y}, \overline{z}] = x;$$

where $\beta"_1 = \frac{\beta_1'}{\gamma_1'}$;

But if $\gamma_1' = 0$, then $\alpha_2 \neq 0$ (otherwise $z \in Z(L)$) and denoting $\overline{z} := \frac{z}{\sqrt{\alpha_2}}$, we obtain:

$$[\overline{y}, \overline{y}] = x; \quad [x, \overline{y}] = 0; \quad [\overline{z}, \overline{y}] = \beta"_1 x;$$

$$[\overline{z}, \overline{z}] = x; \quad [x, \overline{z}] = 0; \quad [\overline{y}, \overline{z}] = 0;$$

where $\beta"_1 = \frac{\beta_1'}{\sqrt{\alpha_2}}$;

Thus under the assumption $\alpha_1 \neq 0$ we have 3 classes of algebras:

$[y, y] = x$; $[x, y] = \alpha_3 x$; $[z, y] = \beta_1 x$;
$[z, z] = \beta_1 x$; $[x, z] = \alpha_3 x$; $[y, z] = x$; $(\alpha_3 \neq 0)$

$[y, y] = x$; $[x, y] = 0$; $[z, y] = \beta_1 x$;
$[z, z] = \alpha_2 x$; $[x, z] = 0$; $[y, z] = x$;

$[y, y] = x$; $[x, y] = 0$; $[z, y] = \beta_2 x$;
$[z, z] = x$; $[x, z] = 0$; $[y, z] = 0$;

II) Now let $\alpha_1 = 0$, then $\alpha_2 \neq 0$ and from (4) we obtain: $\alpha_3 = \frac{\beta_1 \alpha_4}{\alpha_2}$ and $\gamma_1 \beta_1 \alpha_4 = 0$. Note, that $\beta_1 \neq 0$ otherwise $y \in Z(L)$, i.e. $\gamma_1 \alpha_4 = 0$.

a') Let $\gamma_1 = 0$ then putting $\overline{z} = \frac{z}{\sqrt{\alpha_2}}$, we have the following multiplication:
$[y, y] = 0$; $[x, y] = \beta_1' \alpha_4' x$; $[\overline{z}, y] = \beta_1' x$;
$[\overline{z}, \overline{z}] = x$; $[x, \overline{z}] = \alpha_4' x$; $[y, \overline{z}] = 0$;

where $\beta_1' = \frac{\beta_1}{\sqrt{\alpha_2}}$; and $\alpha_4' = \frac{\alpha_4}{\sqrt{\alpha_2}}$;

b') Now let $\gamma_1 \neq 0$, then $\alpha_3 = \alpha_4 = 0$ and again denoting $\overline{z} = \frac{z}{\sqrt{\alpha_2}}$, we have:

$$[y, y] = 0; \quad [x, y] = 0; \quad [\overline{z}, y] = \beta_1' x; \quad [\overline{z}, \overline{z}] = x; \quad [x, \overline{z}] = 0; \quad [y, \overline{z}] = \gamma_1' x;$$

where $\beta_1' = \frac{\beta_1}{\sqrt{\alpha_2}}$; and $\gamma_1' = \frac{\gamma_1}{\sqrt{\alpha_2}}$;

In the cases II.a'and II.b', by denoting $\overline{y} := \frac{y}{\beta_1'}$, we obtain that under the assumption $\alpha_1 = 0$ there are only 2 classes of algebras:

$$[\overline{y},\overline{y}] = 0;\quad [x,\overline{y}] = \alpha_4' x;\quad [\overline{z},\overline{y}] = x;\quad [\overline{y},\overline{y}] = 0;\quad [x,\overline{y}] = 0;\quad [\overline{z},\overline{y}] = x;$$

$$[\overline{z},\overline{z}] = x;\quad [x,\overline{z}] = \alpha_4' x;\quad [\overline{y},\overline{z}] = 0;\quad [\overline{z},\overline{z}] = x;$$

$$[x,\overline{z}] = 0;\quad [\overline{y},\overline{z}] = \gamma"_1 x;\quad (\gamma"_1 \neq 0)$$

b) Now let $\beta_2 \neq 0$ then the multiplication will be the following:

$[y,y] = \alpha_1 x;\quad [x,y] = \alpha_3 x;\quad [z,y] = \beta_1 x + \beta_2 y;$
$[z,z] = \alpha_2 x;\quad [x,z] = \alpha_4 x;\quad [y,z] = \gamma_1 x - \beta_2 y;$

We can assume that $\beta_1 = 1$ (putting $\overline{z} = \frac{z}{\beta_2}$ and thus the equalities will have the following form:

$$\left[\begin{array}{l} \alpha_3 = 0 \\ 2\alpha_1 + \alpha_1\alpha_4 = 0 \\ \beta_1\alpha_4 = \gamma_1 + \beta_1 = 0 \end{array}\right. \tag{5}$$

I) a') Let $\alpha_1 = 0$, then $\alpha_2 \neq 0$ (otherwise L is a Lie algebra). In view of (0.3) we obtain:

$[y,y] = 0;\quad [x,y] = 0;\quad [z,y] = \beta_1 x + y;\quad [z,z] = \alpha_2 x;$
$[x,z] = \alpha_4 x;\quad [y,z] = -\beta_1(1+\alpha_4)x - y;$

b') Now let $\alpha_1 \neq 0$ then (5) implies $\alpha_4 = -2$ and we obtain the following multiplication:
$[y,y] = \alpha_1 x;\quad [x,y] = 0;\quad [z,y] = \beta_1 x + y;\quad [z,z] = \alpha_2 x;$
$[x,z] = -2x;\quad [y,z] = -\beta_1 x - y;$

Now consider the case, when $dimZ(L) = 2$. Let $x, y \in Z(L)$, then the multiplication in this case will be the following:

$$[z,z] = \alpha_1 x + \alpha_2 y;\quad [x,z] = \gamma_1 x + \gamma_2 y;\quad [y,z] = \beta_1 x + \beta_2 y;$$

(the rest products are equal to zero).

a) Let $dim[L,L] = 1$. Putting $\overline{x} = \alpha_1 x + \alpha_2 y$, we obtain

$$[z,z] = \overline{x};\quad [\overline{x},z] = k_1\overline{x};\quad [y,z] = k_2\overline{x};$$

b) Let $dim[L,L] = 2$. Putting $\overline{x} = \alpha_1 x + \alpha_2 y$, we have
$[\overline{x},z] = \alpha_1[x,z] + \alpha_2[y,z] = \alpha_1(\gamma_1 x + \gamma_2 y) + \alpha_2(\beta_1 x + \beta_2 y) =$

$= x(\alpha_1\gamma_1 + \alpha_2\beta_1) + y(\alpha_1\gamma_2 + \alpha_2\beta_2)$.

Consider the following matrix: A

$$\begin{pmatrix} \alpha_1, & \alpha_2 \\ (\alpha_1\gamma_1 + \alpha_2\beta_1), & (\alpha_1\gamma_2 + \alpha_2\beta_2) \end{pmatrix},$$

If $detA \neq 0$, then denoting $\overline{y} := x(\alpha_1\gamma_1 + \alpha_2\beta_1) + y(\alpha_1\gamma_2 + \alpha_2\beta_2)$ we have:

$$[\overline{x}, z] = \overline{y}; \quad [z, z] = \overline{x}; \quad [\overline{y}, z] = k_1'\overline{x} + k_2'\overline{y}.$$

If $detA = 0$ then $[z, z] = \overline{x}$; $[x, z] = k_1'\overline{x}$. Denote $\overline{y} := \beta_1 x + \beta_2 y$. Notice that the determinant of the matrix

$$\begin{pmatrix} \alpha_1 & \alpha_2 \\ \beta_1 & \beta_2 \end{pmatrix},$$

differs from zero, because otherwise $dim[L, L] = 1$, but this case is already considered. Hence

$$[\overline{y}, z] = \beta_1[x, z] + \beta_2[y, z] = \lambda_1\overline{x} + \lambda_2\overline{y},$$

where $\lambda_2 \neq 0$.

Thus in the case $dim[L, L] = 2$ and $dimZ(L) = 2$, we have two Leibniz algebras:

$[x, z] = y; \quad [z, z] = x; \quad [y, z] = k_1 x + k_2 y$

$[x, z] = k_1' x; \quad [z, z] = x; \quad [y, z] = \lambda_1 x + \lambda_2 y \quad (\lambda_2 \neq 0)$

In the notations $\overline{z} = \lambda_2 z$ and $\overline{x}' = \lambda_2^2\overline{x}$ the second algebra can be represented in the following form:

$$[x, z] = k"_1 x; \quad [z, z] = x; \quad [y, z] = k"_1 x + y.$$

The following table represents all Leibniz algebras of dimension 3, which are not Lie, and shows when they are nilpotent or solvable.

	$dim Z(L) = 2$	$dim Z(L) = 1$
Solvable	$[y, z] = \alpha_1 x; [x, z] = \alpha_2 x$ $[z, z] = x;$	$[y, y] = x, [z, z] = \beta x, [x, y] = \alpha x,$ $[x, z] = \alpha x, [z, y] = \beta x; [y, z] = x, \quad \alpha \neq 0$
	$[z, z] = x, \quad [x, z] = y,$ $[y, z] = \alpha_1 x + \alpha_2 y;$	$[x, y] = \alpha x, [x, z] = \alpha x, [z, y] = x,$
	$[y, z] = \alpha_2 x + y, [z, \dot{z}] = x;$ $[x, z] = \alpha_1 x;$	$[x, z] = \alpha x, [z, y] = \beta x + y,$ $[y, z] = -\beta(1 + \alpha)x - y,$
		$[y, y] = x, [z, z] = \alpha x,$ $[x, z] = -2x, [z, y] = \beta x + y, [y, z] = \beta x - y,$
nilpotent		$[y, y] = x, \quad [z, z] = \alpha x,$ $[z, y] = \beta x, \quad [y, z] = x,$
		$[y, y] = [z, z] = x, \quad [z, y] = \beta x,$
		$[z, z] = [z, y] = x, \quad [y, z] = \alpha x, \quad \alpha \neq 0$

So we obtain 10 classes of Leibniz algebras, which are not Lie algebras. Description of all three dimensional Lie algebras is given in [2]: there are 6 such classes. Therefore we have 16 classes of Leibniz algebras of dimension 3 over the field of zero characteristic.

3. On a case of degeneration of Leibniz algebras

Theorem 6. *Let L be a finite dimensional Leibniz algebra over a field F of zero characteristic, which satisfies the following conditions:*

a) $[Z(L), L] = 0$;

b) $L/Z(L)$*- one of the classical Lie algebras:* $A_l, \quad B_l, \quad C_l, \quad D_l$.

Then L is a Lie algebras.

Proof Let the factor algebra $L/Z(L)$ be isomorphic to the algebra A_l. In the work [2] the matrix basis $e_{i,j} (i \neq j)$, where $i, j \in \{1, ..., l+1\}$; $h_k = e_{kk} - e_{l+1,l+1}$, where $k \leq l$, was constructed in A_l. Using this basis it is easy to describe the multiplication in $L/Z(L)$. In particular,

$$[h_i, e_{ij}] = e_{ij}; \text{ for } j \neq l+1, \quad i \neq l+1; \quad i \neq j$$
$$[h_i, e_{i,l+1}] = 2e_{i,l+1}; \text{ for } i \neq l+1, \quad i \neq j$$
$$[e_{l+1,i} e_{ji}] = e_{l+1,i}; \text{ for } j \neq l+1, \quad i \neq j$$
$$[e_{l+1,k}, e_{k,l+1}] = -h_k; \text{ for } k \neq l+1$$
$$[e_{l+1,i}, h_i] = 2e_{l+1,i} \text{ for } i \neq l+1$$

Here $[.,.]$ is the commutator.

Let $e'_{ij}, h'_k, x_1, ..., x_m$ form a basis of the algebra L and $\{x_1, ..., x_m\} \in Z(L)$, where e'_{ij}, h'_k are the equivalence classes for the elements e_{ij}, h_k respectively, i.e. $e'_{ij} = e_{ij} + Z(L); \quad h'_k = h_k + Z(L)$.

Consider the multiplication in L:

$$[e'_{ij}, [h'_i, e'_{ij}]] = [e'_{ij}, e'_{ij} + Z(L)] = {e'}_{ij}^{2}.$$

Using Leibniz identity, we have:

$$[e'_{ij}, [h'_i, e'_{ij}]] = [[e'_{ij}, h'_i], e'_{ij}] - [[e'_{ij}, e'_{ij}], h'_i] = [[-e'_{ij} + Z(L], e'_{ij}] = -{e'}_{ij}^{2}.$$

Since ${e'}_{ij}^{2} \in Z(L)$, the condition a) implies $[[e'_{ij}, e'_{ij}], h'_i] = 0$. Comparing the above equalities and since the field has zero characteristic we obtain: ${e'}_{ij}^{2} = 0$ for $j \neq l+1, \quad i \neq l+1, \quad i \neq j$.

Consider the multiplication in L :

$$[e'_{i,l+1}, [h'_i, e'_{i,l+1}]] = [e'_{i,l+1}, 2e'_{i,l+1} + Z(L)] = {e'}_{i,l+1}^{2};$$

Using Leibniz identity, we obtain:

$$[e'_{i,l+1}, [h'_i, e'_{i,l+1}]] = [[e'_{i,l+1}, h'_i], e'_{i,l+1}] - [[e'_{i,l+1}, e'_{i,l+1}], h'_i] =$$

$$= [[-2e'_{i,l+1} + Z(L)], e'_{i,l+1}] = -2{e'}_{i,l+1}^{2}.$$

because ${e'}_{i,l+1}^{2} \in Z(L)$, then $[[e'_{i,l+1}, e'_{i,l+1}], h'_i] = 0$

Comparing these equalities we obtain: ${e'}_{i,l+1}^{2} = 0$.

Consider the multiplication in L:

$$[e'_{l+1,i}, [h'_i, e'_{i,l+1}]] = [e'_{l+1,i}, 2e'_{i,l+1} + Z(L)] = [2e'_{l+1,i}, e'_{i,l+1}];$$

From Leibniz identity, it follows:

$$[e'_{l+1,i}, [h'_i, e'_{i,l+1}]] = [[e'_{l+1,i}, h'_i], e'_{i,l+1}] -$$

$$-[[e'_{l+1,i}, e'_{i,l+1}], h'_i] = [2e'_{l+1,i}, e'_{i,l+1}] + {h'}_i^{2};$$

Considering as above we obtain, that ${h'}_i^{2} = 0$.

Consider $[e'_{l+1,i}, [e'_{l+1,j}, e'_{j,i}]] = {e'}_{l+1,i}^{2}$, where $i \neq j$ and $i \neq l+1$. From $[e'_{l+1,i}, e'_{l+1,j}] \in Z(L)$ and $[e'_{l+1,i}, e'_{j,i}] \in Z(L)$ for $i \neq j, \quad i \neq l+1$ and Leibniz identity, we have:

$$[e'_{l+1,i}, e'_{l+1,j}, e'_{j,i}]] = [[e'_{l+1,i}, e'_{l+1,j}], e'_{j,i}] -$$

$$-[[e'_{l+1,i}, e'_{j,i}], e'_{l+1,j} = 0.$$

Hence ${e'}_{l+1,i}^{2} = 0$.

Thus, we have showed that in the case when the algebra $L/Z(L)$ is isomorphic to the algebra A_l and L satisfies the condition a), one has $[x, x] = 0$ for any $x \in L$, i.e. L is Lie algebra.

The remaining cases can be proved in a similar way. □.

Remark. There are Leibniz algebras, which satisfy the condition b), but does not satisfy the condition a) and which are not Lie algebras.

Corollary. *Let L be a finite dimensional Leibniz algebra over a field F of zero characteristic which satisfies the following conditions:*

a) $DimZ(L) = 1$;

b) $L/Z(L)$ is one of classical Lie algebras: A_l, B_l, C_l, D_l. Then L is a Lie algebra.

Proof Let $\{x_1, ..., x_n, d\}$ be a basis of the algebra L and $d \in Z(L)$. Since the algebra $L' = L/Z(L)$ is simple, $L'^2 = L'$. It means, that for any $x \in L$ there exist $a_i, b_i \in L$ and $\lambda, \alpha_i \in F$ such, that $x = \sum \alpha_l[a_i, b_i] + \lambda d$. Therefore

$$[d, x] = [d, \sum \alpha_i[a_i, b_i] + \alpha_i d] = \sum \alpha_i[d, [a_i, b_i]] =$$

$$= \sum \alpha_i([[d, a_i], b_i] - [[d, b_i], a_i]) = \sum \alpha_i(f(a_i)[d, b_i] -$$

$$-f(b_i)[d, a_i]) = \sum \alpha_i(f(a_i)f(b_i)d - f(b_i)f(a_i)d) = 0,$$

where $f(a_i), f(b_i) \in L$. Thus $[d, L] = 0$, i.e. $[Z(L), L] = 0.\square$

References

[1] Loday J.L. Une version non commutative des algèbres de Lie: les algèbres de Leibniz, *L'Enseignement Mathematique* 39 (1993), 269-293.

[2] Jacobson N. *Lie algebras* M: Nauka, 1984.

[3] Latyshev V.N. On zero divisors and nilelements in Lie algebras, *Syberian Math.J.*,4 (1963-4), 830-836.

[4] Kostrikin A.I. *Around Burnside* M: Nauka, 1986.

[5] Kuzmin E.N. The locally nilpotent radical of Mal'cev algebras, which satisfies the condition of Engel. *Dokl.Akad.Nauk SSSR*, 177,3(1967)

E-Mail : ayupov @ im.tashkent.su, shayupov @ hal.freenet.uz

A MODULI PROBLEM RELATED TO COMPLEX SUPERMANIFOLDS [1]

A.L. ONISHCHIK

Kolomenskaia nab. 10-153,
115142 Moscow

Abstract. We study the problem of classifying complex analytic supermanifolds that have a given retract, i.e., the associated split supermanifold. The problem is solved in the case of the split supermanifold (M, Ω), where M is a simply connected irreducible compact Hermitian symmetric space and Ω the sheaf of holomorphic forms on M. We also give a construction of a family of non-split supermanifolds with retract (M, Ω) for a general complex manifold M. This family is non-empty whenever M is a compact Kähler manifold.

1 Preliminaries

Many moduli problems appearing in complex analysis and geometry can be expressed in terms of the 1-cohomology with values in a sheaf of non-abelian groups. Let, e.g., M be a complex analytic manifold. Then the problem of classifying holomorphic vector bundles E of rank m over M is equivalent to that of calculating the cohomology set $H^1(M, \mathcal{F}_{\mathrm{GL}_m(\mathbf{C})})$, where $\mathcal{F}_{\mathrm{GL}_m(\mathbf{C})}$ is the sheaf of holomorphic functions on M with values in the group $\mathrm{GL}_m(\mathbf{C})$. In other cases one has to deal with the cohomology set with values in a sheaf of automorphisms. We will study here a problem of this latter sort.

More precisely, let us fix a holomorphic vector bundle E of rank m over

[1]Work supported in part by Russian Foundation of Fundamental Research,

Y. Khakimdjanov et al (eds.), Algebra and Operator Theory, 13-24.

a complex manifold M and let $\mathcal{F}$ and $\mathcal{E}$ denote the sheaves of holomorphic functions on M and of holomorphic sections of E, respectively. Consider the sheaf

$$\mathcal{O} = \bigwedge_{\mathcal{F}} \mathcal{E} = \bigwedge \mathcal{E} \tag{1}$$

of Grassmann algebras over $\mathcal{F}$ generated by $\mathcal{E}$. It possesses the natural $\mathbf{Z}$-grading by the exterior powers $\mathcal{O}_p = \bigwedge^p \mathcal{E}$, and hence the $\mathbf{Z}_2$-grading $\mathcal{O} = \mathcal{O}_{\bar{0}} \oplus \mathcal{O}_{\bar{1}}$, where

$$\begin{aligned} \mathcal{O}_{\bar{0}} &= \bigoplus_{p\geq 0} \bigwedge^{2p} \mathcal{E}, \\ \mathcal{O}_{\bar{1}} &= \bigoplus_{p\geq 0} \bigwedge^{2p+1} \mathcal{E}. \end{aligned}$$

Thus, $\mathcal{O}$ has a structure of the sheaf of $\mathbf{Z}_2$-graded algebras (or *superalgebras*). Denote by $\mathcal{A}ut\ \mathcal{O}$ the sheaf of automorphisms of this sheaf of superalgebras (as usually, any automorphism is even and maps each stalk $\mathcal{O}_x$, $x \in M$, onto itself). This is a sheaf of groups. Consider the filtration

$$\mathcal{A}ut\ \mathcal{O} = \mathcal{A}ut_{(0)}\mathcal{O} \supset \mathcal{A}ut_{(2)}\mathcal{O} \supset \dots, \tag{2}$$

where

$$\mathcal{A}ut_{(2p)}\mathcal{O} = \{a \in \mathcal{A}ut\ \mathcal{O} \mid a(u) - u \in \bigoplus_{q\geq 2p} \mathcal{O}_q \text{ for all } u \in \mathcal{O}\}. \tag{3}$$

Clearly, $\mathcal{A}ut_{(2p)}\mathcal{O}$ are subsheaves of normal subgroups in $\mathcal{A}ut\ \mathcal{O}$. Our aim is to study the 1-cohomology set $H^1(M, \mathcal{A}ut_{(2)}\mathcal{O})$. Let us now describe its geometrical interpretation concerning the classification of complex analytic supermanifolds.

We use the standard definition of complex supermanifold of dimension $m|n$ (see [10]) as a $\mathbf{Z}_2$-graded complex ringed space $(M, \mathcal{O})$ locally isomorphic to a model of the form $(U, \bigwedge_{\mathcal{F}}(\xi_1, \dots, \xi_m))$, where U is an open set in $\mathbf{C}^n$ and $\mathcal{F}$ the sheaf of holomorphic functions on U. Any holomorphic vector bundle E over a complex manifold M gives rise to the complex supermanifold $(M, \mathcal{O})$, whose structure sheaf $\mathcal{O}$ is given by (1), its dimension being $m|n$. A supermanifold isomorphic to that of this form is called *split*. For any pair $(m|n)$, $m \geq 2, n \geq 1$, there exist non-split supermanifolds of dimension $m|n$. Now, any complex supermanifold $(M, \mathcal{O})$ determines a split supermanifold $(M, \operatorname{gr} \mathcal{O})$ called its *retract*. The retract is defined in the following way. Denote by $\mathcal{J}$ the subsheaf of ideals of the sheaf $\mathcal{O}$ generated by its odd part $\mathcal{O}_{\bar{1}}$ and consider the filtration

$$\mathcal{O} = \mathcal{J}^0 \supset \mathcal{J} \supset \mathcal{J}^2 \supset \dots.$$

Form the corresponding graded sheaf $\operatorname{gr} \mathcal{O} = \bigoplus_{p\geq 0} \operatorname{gr}_p \mathcal{O}$, where

$$\operatorname{gr}_p \mathcal{O} = \mathcal{J}^p / \mathcal{J}^{p+1}.$$

Denoting $\mathcal{F} = \mathrm{gr}_0\mathcal{O}$ and $\mathcal{E} = \mathrm{gr}_1\mathcal{O}$, we see that $(M, \mathcal{F})$ is a complex manifold, $\mathcal{E}$ is a locally free analytic sheaf on it, and $\mathcal{O} \simeq \bigwedge_{\mathcal{F}} \mathcal{E}$. The following theorem is due to Green ([5]):

Theorem 1 *Let E be a holomorphic vector bundle over a complex manifold M, and denote $\mathcal{O}_{\mathrm{gr}} = \bigwedge \mathcal{E}$. To any supermanifold $(M, \mathcal{O})$ having $(M, \mathcal{O}_{\mathrm{gr}})$ as its retract there corresponds an element of the set $H^1(M, \mathcal{A}ut_{(2)}\mathcal{O}_{\mathrm{gr}})$, this correspondence giving rise to a bijection between the isomorphy classes of supermanifolds satisfying the above condition and the orbits of the group* Aut E *on $H^1(M, \mathcal{A}ut_{(2)}\mathcal{O}_{\mathrm{gr}})$ under a natural action* Int. *The given split supermanifold $(M, \mathcal{O}_{\mathrm{gr}})$ corresponds to the unit element $e \in H^1(M, \mathcal{A}ut_{(2)}\mathcal{O}_{\mathrm{gr}})$.*

Note that any automorphism of E gives rise to an automorphism of $(M, \mathcal{O}_{\mathrm{gr}})$, and thus determines an inner automorphism of the sheaf $\mathcal{O}_{\mathrm{gr}}$. It follows that Aut E acts on the sheaf $\mathcal{A}ut_{(2)}\mathcal{O}_{\mathrm{gr}}$, and hence on its 1-cohomology set, leaving fixed the unit element e. This is the action Int mentioned in Theorem 1.

Let us describe the correspondence mentioned in Theorem 1. Let $(M, \mathcal{O})$ be a supermanifold such that $\mathrm{gr}\,\mathcal{O} = \mathcal{O}_{\mathrm{gr}}$. We can choose an open cover $\mathbf{U} = (U_i)_{i \in I}$ of M such that there exist isomorphisms $h_i : \mathcal{O}|U_i \to \mathcal{O}_{\mathrm{gr}}|U_i$, $i \in I$, inducing the identity isomorphisms of the $\mathbf{Z}$-graded sheaves. Setting $g_{ij} = h_i \circ {h_j}^{-1}$, we obtain a 1-cocycle $g = (g_{ij}) \in Z^1(\mathbf{U}, \mathcal{A}ut_{(2)}\mathcal{O}_{\mathrm{gr}})$. Its cohomology class $\gamma \in H^1(M, \mathcal{A}ut_{(2)}\mathcal{O}_{\mathrm{gr}})$ does not depend of the choice of h_i; this is the desired class.

2 Exponential mapping and its applications

We retain the notation of the previous section. It is in general a difficult problem, to describe the set $H^1(M, \mathcal{A}ut_{(2)}\mathcal{O}_{\mathrm{gr}})$ explicitely. But, as we will see below, this set, under certain strong conditions, coincides with the 1-cohomology group of a locally free analytic sheaf on M. This simple case is sufficient for further applications.

We will use the linearization method proposed in [14]. Let $(M, \mathcal{O})$ be an arbitrary complex supermanifold of dimension $n|m$ and $(M, \mathcal{O}_{\mathrm{gr}})$ its retract. As in the split case (see Sect. 1), we may define the sheaf of automorphisms $\mathcal{A}ut\ \mathcal{O}$ and its filtration by subsheaves $\mathcal{A}ut_{(2p)}\mathcal{O}$ that are defined as in (3), but replacing the subsheaf $\bigoplus_{q \geq 2p} \mathcal{O}_q$ by $\mathcal{J}^{2p}$. For the retract, this filtration coincides with (2). We also consider the *tangent sheaf* $\mathcal{T} = \mathcal{D}er\,\mathcal{O}$, i.e., the $\mathbf{Z}_2$-graded sheaf of derivations of the structure sheaf $\mathcal{O}$. The even part of this sheaf admits the filtration

$$\mathcal{T}_{\bar{0}} \supset \mathcal{T}_{\bar{0}(2)} \supset \mathcal{T}_{\bar{0}(4)} \supset \dots, \tag{4}$$

where

$$\mathcal{T}_{\bar{0}(2p)} = \{v \in \mathcal{T}_{\bar{0}} \mid v(\mathcal{O}) \subset \mathcal{J}^{2p}\}.$$

As in the classical Lie theory, there exists a natural relationship between automorphisms and derivations of the sheaf $\mathcal{O}$. In particular, we have the exponential mapping

$$\exp : \mathcal{T}_{\bar{0}(2)} \to \mathcal{A}ut_{(2)}\,\mathcal{O}.$$

It is expressed by the usual exponential series which is actually a polynomial, since any $v \in \mathcal{T}_{\bar{0}(2)}$ satisfies $v^k = 0$ for any $k > \left[\frac{m}{2}\right]$. One proves that exp is bijective [14] and maps $\mathcal{T}_{\bar{0}(2p)}$ onto $\mathcal{A}ut_{(2p)}\,\mathcal{O}$, $p = 1, 2, \ldots$. Thus it is an isomorphism of sheaves of sets (but not of groups, in general). We denote $\log = \exp^{-1}$. For any $p \geq 1$, there is the following exact sequence of the sheaves of groups:

$$0 \to \mathcal{A}ut_{(2p+2)}\mathcal{O} \longrightarrow \mathcal{A}ut_{(2p)}\mathcal{O} \xrightarrow{\lambda_{2p}} (\mathcal{T}_{\mathrm{gr}})_{2p} \to 0, \tag{5}$$

where $\mathcal{T}_{\mathrm{gr}}$ is the $\mathbf{Z}$-graded tangent sheaf of the retract $(M, \mathcal{O}_{\mathrm{gr}})$ and λ_{2p} is the composition of the following mappings:

$$\lambda_{2p} : \mathcal{A}ut_{(2p)}\mathcal{O} \xrightarrow{\log} \mathcal{T}_{\bar{0}(2p)} \xrightarrow{\pi_p} \mathcal{T}_{\bar{0}(2p)}/\mathcal{T}_{\bar{0}(2p+2)} \xrightarrow{h_p} (\mathcal{T}_{\mathrm{gr}})_{2p},$$

π_p being the canonical projection and h_p a natural isomorphism. In fact, one checks easily that the mapping $\tilde{\lambda}_{2p} = \pi_p \circ \log$ is a homomorphism of sheaves of groups and that $\mathrm{Ker}\tilde{\lambda}_{2p} = \mathcal{A}ut_{(2p+2)}\mathcal{O}$.

Lemma 1 *If $H^1(M, (\mathcal{T}_{\mathrm{gr}})_{2p}) = 0$ for any $p \geq 2$, then $H^1(M, \mathcal{A}ut_{(2p)}\mathcal{O}) = \{e\}$ for any $p \geq 2$.*

Proof We will use the induction on p. Clearly, the claim is true for all sufficiently big p. We have to prove that if it is true for a certain $p \geq 3$, then it is true for $p - 1$ as well. The exact sequence (5) gives the cohomology exact sequence (see [7])

$$H^1(M, \mathcal{A}ut_{(2p)}\mathcal{O}) \longrightarrow H^1(M, \mathcal{A}ut_{(2p-2)}\mathcal{O}) \xrightarrow{\lambda^*_{2p-2}} H^1(M, (\mathcal{T}_{\mathrm{gr}})_{2p-2}).$$

Clearly, $H^1(M, \mathcal{A}ut_{(2p-2)}\mathcal{O}) = \{e\}$ follows from

$$H^1(M, (\mathcal{T}_{\mathrm{gr}})_{2p-2}) = 0, \qquad H^1(M, \mathcal{A}ut_{(2p)}\mathcal{O}) = \{e\}.$$

Theorem 2 *Suppose that $H^1(M, (\mathcal{T}_{\mathrm{gr}})_{2p}) = H^2(M, (\mathcal{T}_{\mathrm{gr}})_{2p}) = 0$ for any $p \geq 2$. Then the mapping $\lambda^*_2 : H^1(M, \mathcal{A}ut_{(2)}\mathcal{O}_{\mathrm{gr}}) \to H^1(M, (\mathcal{T}_{\mathrm{gr}})_2)$ is bijective.*

Proof The surjectivity is implied by Theorem 3 of [14], where the assumption about 2-cohomology is used. We now prove that the assumption about 1-cohomology implies the injectivity.

The sequence (5) for the sheaf $\mathcal{O} = \mathcal{O}_{\mathrm{gr}}$ gives the cohomology exact sequence

$$H^1(\mathcal{A}ut_{(4)}\mathcal{O}_{\mathrm{gr}}) \to H^1(M, \mathcal{A}ut_{(2)}\mathcal{O}_{\mathrm{gr}}) \xrightarrow{\lambda^*_2} H^1(M, (\mathcal{T}_{\mathrm{gr}})_2).$$

Suppose that $\gamma, \eta \in H^1(M, \mathcal{A}ut_{(2)}\mathcal{O}_{\mathrm{gr}})$ and that $\lambda_2^*(\gamma) = \lambda_2^*(\eta)$. Let γ be determined by the cocycle (g_{ij}) and η by the cocycle (h_{ij}) in an open cover $\mathbf{U} = (U_i)_{i \in I}$ of M. Then our assumption implies that $\lambda_2(g_{ij}) = \lambda_2(h_{ij}) + c_j - c_i$, where $c_i \in ((\mathcal{T}_{\mathrm{gr}})_2)_{U_i}$. We may assume that $c_i = \lambda_2(g_i)$, where $g_i \in (\mathcal{A}ut_{(2)}\mathcal{O}_{\mathrm{gr}})_{U_i}$. Then $\lambda_2(g_i g_{ij} g_j^{-1}) = \lambda_2(h_{ij})$. Thus, we may suppose from the beginning that $\lambda_2(g_{ij}) = \lambda_2(h_{ij})$. Consider the cochain $f \in C^1(\mathbf{U}, \mathcal{A}ut_{(2)}\mathcal{O}_{\mathrm{gr}})$, given by $f_{ij} = h_{ij} g_{ij}^{-1}$. Then $\lambda_2(f) = 0$.

Let $(M, \mathcal{O})$ be the supermanifold corresponding to the cohomology class of γ due to Theorem 1. Then $g_{ij} = h_i h_j^{-1}$, where $h_i : \mathcal{O}|U_i \to \mathcal{O}_{\mathrm{gr}}|U_i$, $i \in I$, are certain isomorphisms of sheaves of superalgebras inducing the identity mappings on $\mathcal{O}_{\mathrm{gr}}|U_i$. The equalities $h_{ij} = f_{ij} g_{ij} = f_{ij} h_i h_j^{-1}$ and $h_{ij} h_{jk} = h_{ik}$ imply that

$$f_{ij} h_i h_j^{-1} f_{jk} h_j h_k^{-1} = f_{ik} h_i h_k^{-1}$$

or

$$(h_i^{-1} f_{ij} h_i)(h_j^{-1} f_{jk} h_j) = h_i^{-1} f_{ik} h_i.$$

Clearly, $\lambda_2(h_i^{-1} f_{ij} h_i) = 0$, whence $(h_i^{-1} f_{ij} h_i) \in Z^1(\mathbf{U}, \mathcal{A}ut_{(4)}\mathcal{O})$. By Lemma 1, this latter cocycle is cohomologous to e, i.e. $h_i^{-1} f_{ij} h_i = u_i u_j^{-1}$, where $u_i \in (\mathcal{A}ut_{(4)}\mathcal{O})_{U_i}$. Thus, $f_{ij} = h_i u_i u_j^{-1} h_i^{-1}$. It follows that

$$\begin{aligned} h_{ij} &= f_{ij} g_{ij} = h_i u_i u_j^{-1} h_i^{-1} g_{ij} = h_i u_i u_j^{-1} h_j^{-1} \\ &= (h_i u_i h_i^{-1})(h_i h_j^{-1})(h_j u_j^{-1} h_j^{-1}) = v_i g_{ij} v_j^{-1}, \end{aligned}$$

where $v_i = h_i u_i h_i^{-1} \in (\mathcal{A}ut_{(4)}\mathcal{O}_{\mathrm{gr}})_{U_i}$. This implies that $\gamma = \eta$.

To calculate the quotient of $H^1(M, \mathcal{A}ut_{(2)}\mathcal{O}_{\mathrm{gr}})$ by $\mathrm{Aut}E$, the following assertion is useful. For any $c \in \mathbf{C}^\times$, denote by A_c the automorphism of E given by the multiplication by the scalar c.

Lemma 2 *We have*

$$\lambda_2^* \circ \mathrm{Int}\, A_c = c^2 \lambda_2^*, \qquad c \in \mathbf{C}^\times.$$

Remark The conditions of Theorem 2 are fulfilled, in particular, if $m = 2$ or 3. The corresponding special cases were proved in [10, Ch. 4], and [2], respectively. In the general case, the class $\lambda_2^*(\gamma)$ is closely related to the first obstruction to splitting considered in [10, Ch. 4]. If $\lambda_2^*(\gamma) = 0$, then $\gamma \in \mathrm{Im} H^1(M, \mathcal{A}ut_{(4)}\mathcal{O}_{\mathrm{gr}})$, and we can apply λ_4^*, and so on. The resulting obstruction theory is discussed in [1],[3],[13]. On the other hand, any non-split supermanifold can be regarded as a deformation of its retract, and λ_2^* can be interpreted as the corresponding Kodaira–Spencer mapping (for details see [3]).

3 Case of the cotangent bundle of an irreducible Hermitian symmetric space

We consider here a very special situation when the conditions of Theorem 2 are fulfilled and our problem can be completely solved.

Let us first discuss the case of the cotangent bundle $E = T(M)^*$ of a complex manifold M. The corresponding sheaf $\mathcal{E} = \Omega^1(M)$ is the sheaf of holomorphic 1-forms on M, and the sheaf $\mathcal{O} = \bigwedge \mathcal{E}$ coincides with the graded sheaf Ω of all holomorphic forms. The structure of the tangent sheaf $\mathcal{T} = \mathcal{D}er\, \Omega$ was studied in [4], where its relation with the sheaf of vector-valued forms was established. More precisely, let us denote by Θ the tangent sheaf and by $\Omega \otimes \Theta$ the graded sheaf of vector-valued forms on M. Using the contraction, one defines, for any $p \geq 0$, an injective homomorphism $i : \Omega^{p+1} \otimes \Theta \longrightarrow \mathcal{T}_p$. On the other hand, we have the mapping $l : \Omega^p \otimes \Theta \longrightarrow \mathcal{T}_p$ given by $l(\varphi) = [i(\varphi), d]$, where d is the exterior differentiation regarded as a section of $\mathcal{T}_1$. As a result, one gets

$$\mathcal{T}_p = i(\Omega^{p+1} \otimes \Theta) \oplus l(\Omega^p \otimes \Theta) \simeq (\Omega^{p+1} \otimes \Theta) \oplus (\Omega^p \otimes \Theta). \tag{6}$$

Now we suppose that M is a simply connected compact irreducible Hermitian symmetric space. Using the presentation $M = G/P$, where G is a simple complex Lie group and P its parabolic subgroup, one gets the following result (see [12]).

Theorem 3 *Let M be a simply connected compact irreducible Hermitian symmetric space,* $\dim M \geq 2$. *Then*
$H^1(M, \Omega^p \otimes \Theta) = 0$ *for* $p \geq 3$,
$H^2(M, \Omega^p \otimes \Theta) = 0$ *for* $p \geq 4$,
$H^1(M, \Omega^2 \otimes \Theta) = \mathbf{C}^2$ *if M is the complex Grassmannian* $\mathrm{Gr}_{n,k}$, $2 \leq k \leq n-2$, *and* $\mathbf{C}$ *otherwise.*

Theorem 3, together with (6), implies that the corresponding split supermanifold (M, Ω) satisfies the conditions of Theorem 2. By Theorem 1 and Lemma 2, we get the following classification theorem

Theorem 4 *Let M be a simply connected compact irreducible Hermitian symmetric space,* $\dim M \geq 2$.

If M is the complex Grassmannian $\mathrm{Gr}_{n,k}$, $2 \leq k \leq n-2$, *then the non-split supermanifolds with retract (M, Ω) are parametrized by* $\mathbf{CP}^1/\Sigma$ *where* $\Sigma = \mathbf{Z}_2$ *for $n = 2k$ and is trivial otherwise. For any other space M there exists precisely one non-split supermanifold with retract (M, Ω).*

A well known example of a supermanifold with retract (M, Ω) is the Π-symmetric super-Grassmannian $\Pi\mathrm{Gr}_{n|n,k|k}$, $1 \leq k \leq n-1$, defined in [10, Ch. 5]. It is non-split whenever $n \geq 3$. Theorem 4 gives other examples, for any simply connected compact irreducible Hermitian symmetric space M of dimension ≥ 2. In particular, we see that $\Pi\mathrm{Gr}_{n|n,k|k}$ is non-rigid whenever $2 \leq k \leq n-2$.

4 A family of non-split supermanifolds with retract (M, Ω)

Here we still are studying the case when $\mathcal{O}_{\text{gr}}$ is the sheaf of holomorphic forms Ω on M. Using closed (1,1)-forms on M, we will construct an abelian subsheaf of the sheaf of groups $\mathcal{A}ut_{(2)}\Omega$. The 1-cohomology of this subsheaf determines a family of supermanifolds with retract (M, Ω). This family is non-trivial whenever M is a compact Kähler manifold with $\dim M > 1$.

Let $\mathcal{Z}\Omega^1$ denote the subsheaf of Ω^1 consisting of closed forms. Consider the following sequence of sheaves and their homomorphisms:

$$\mathcal{Z}\Omega^1 \xrightarrow{\beta} \Omega^1 \xrightarrow{\nu} \mathcal{T}_2 \xrightarrow{\exp} \mathcal{A}ut_{(2)}\Omega,$$

where β is the identical inclusion and ν is given by

$$\nu(\psi) = \psi d.$$

The composition mapping $\mu : \mathcal{Z}\Omega^1 \to \mathcal{A}ut_{(2)}\Omega$ is expressed by

$$\mu(\psi) = \exp(\psi d) = \mathrm{id} + \psi d.$$

Clearly, this is a homomorphism of sheaves of groups. Therefore we have the cohomology mapping

$$\mu^* : H^1(M, \mathcal{Z}\Omega^1) \to H^1(M, \mathcal{A}ut_{(2)}\Omega)$$

taking 0 to the unit element.

Lemma 3 *We have $\lambda_2 \circ \mu = \nu \circ \beta$.*

Suppose that $\dim M > 1$ and that $\zeta, \zeta' \in H^1(M, \mathcal{Z}\Omega^1)$. If $\mu^*(\zeta) = \mu^*(\zeta')$, then $\beta^*(\zeta) = \beta^*(\zeta')$.

Let $\mathbf{U} = (U_i)$ be an open cover of M and let $\psi = (\psi_{ij})$ be a cocycle from $Z^1(\mathbf{U}, \mathcal{Z}\Omega^1)$. Then the above construction assigns to ψ the supermanifold given by the cocycle $g = (g_{ij}) \in Z^1(\mathbf{U}, \mathcal{A}ut_{(2)}\Omega)$, where

$$g_{ij} = \mathrm{id} + \psi_{ij} d.$$

Due to Theorem 1, we see from Lemma 3 that this supermanifold is non-split if and only if the cohomology class of ψ in $H^1(M, \Omega^1)$ is non-zero.

Now we pass to an important case, where a "closed cocycle" ψ appears. Let ω be a (1,1)-form on M satisfying $d\omega = 0$. Then, clearly, $\bar{\partial}\omega = 0$, and by the Dolbeault theorem ω determines a cohomology class $D(\omega) \in H^1(M, \Omega^1)$. It turns out that it can be given by a closed Čech cocycle. In fact, it is well known that there exists an open cover $\mathbf{U} = (U_i)$ of M such that $\omega = \bar{\partial}\psi_i$ in any U_i, where ψ_i is a $(1,0)$-form in U_i such that $\partial\psi_i = 0$. Then $D(\omega)$ is given by the cocycle $\psi = (\psi_{ij}) \in Z^1(\mathbf{U}, \mathcal{Z}\Omega^1)$, where $\psi_{ij} = \psi_j - \psi_i$ in $U_i \cap U_j \neq \emptyset$.

Denote by $\Phi^{p,q}$ the sheaf of smooth complex-valued (p,q)-forms on M and by $\mathcal{Z}\Phi^{1,1}$ the subsheaf of closed forms in $\Phi^{1,1}$. Then we have defined a linear mapping $\delta^* : \Gamma(M, \mathcal{Z}\Phi^{1,1}) \to H^1(M, \mathcal{Z}\Omega^1)$ such that $D = \beta^* \circ \delta^*$. Using δ^*, we get the mapping

$$\mu^* \circ \delta^* : \Gamma(M, \mathcal{Z}\Phi^{1,1}) \to H^1(M, \mathcal{A}ut_{(2)}\Omega).$$

Thus, any $(1,1)$-form ω on M such that $d\omega = 0$ determines a supermanifold with retract (M, Ω).

Note that any $\omega \in \Gamma(M, Z\Phi^{1,1})$ determines an element $[\omega]$ of the Dolbeault cohomology group $H^{1,1}(M, \mathbf{C})$. On the other hand, we may consider the *refined cohomology group*

$$\hat{H}^{1,1}(M, \mathbf{C}) = \Gamma(M, \mathcal{Z}\Phi^{1,1})/\bar{\partial}\partial\Gamma(M, \Phi^{0,0}).$$

We also have a linear mapping $\hat{\delta} : \hat{H}^{1,1}(M, \mathbf{C}) \to H^1(M, \mathcal{Z}\Omega^1)$ such that $\delta^*(\omega) = \hat{\delta}(\hat{\omega})$, where $\hat{\omega}$ is the class of the form ω in $\hat{H}^{1,1}(M, \mathbf{C})$. Using Lemma 3, we come to the following statement:

Theorem 5 *We have the commutative diagram*

$$\begin{array}{ccccc}
\Gamma(M, \mathcal{Z}\Phi^{1,1}) & \xrightarrow{\delta^*} & H^1(M, \mathcal{Z}\Omega^1) & \xrightarrow{\beta^*} & H^1(M, \Omega^1) \\
\| & & \uparrow \hat{\delta} & & \uparrow D \\
\Gamma(M, \mathcal{Z}\Phi^{1,1}) & \longrightarrow & \hat{H}^{1,1}(M, \mathbf{C}) & \longrightarrow & H^{1,1}(M, \mathbf{C}),
\end{array}$$

where the mappings in the top line are given by $\omega \mapsto \hat{\omega} \mapsto [\omega]$.

Suppose that $\dim M > 1$. *For any two* $\omega, \omega' \in \Gamma(M, \mathcal{Z}\Phi^{1,1})$, *the equation* $\mu^*\delta^*\omega = \mu^*\delta^*\omega'$ *implies* $[\omega] = [\omega']$. *In particular, the supermanifold corresponding to* ω *is non-split whenever* $[\omega] \neq 0$.

In particular, we see that any Kähler form ω on a compact manifold M determines a non-split supermanifold with retract (M, Ω). In the case when M is a compact Kähler manifold, the situation can be described in a more simple way.

Corollary If M is a compact Kähler manifold, then we have a linear mapping $\tilde{\delta} : H^{1,1}(M, \mathbf{C}) \to H^1(M, \mathcal{Z}\Omega^1)$ such that $\beta^* \circ \tilde{\delta} = D$. The mapping $\mu^* \circ \tilde{\delta} : H^{1,1}(M, \mathbf{C}) \to H^1(M, \mathcal{A}ut_{(2)}\Omega)$ is injective and takes 0 to e.

Applying the above construction, we will now associate with any holomorphic line bundle L over a complex manifold M a supermanifold with retract (M, Ω). The closed $(1,1)$-form ω will be here the curvature form of a metric on L.

Consider the sheaf $\mathcal{F}^\times$ of invertible elements of $\mathcal{F}$ and define a homomorphism of sheaves of groups $\mathbf{D} : \mathcal{F}^\times \longrightarrow \mathcal{Z}\Omega^1$ (logarithmic differentiation) by

$$\mathbf{D}f = \frac{1}{f}df = d\log f.$$

Hence we get the homomorphism of sheaves of groups

$$\mu \circ \mathbf{D} : \mathcal{F}^\times \longrightarrow \mathcal{A}ut_{(2)}\Omega,$$

whose explicit form is as follows:

$$\mu\mathbf{D}(f) = \mathrm{id} + \left(\frac{1}{f}df\right) d.$$

This homomorphism induces the homomorphism

$$\mu^* \circ \mathbf{D}^* : \mathrm{Pic}(M) = H^1(M, \mathcal{F}^\times) \longrightarrow H^1(M, \mathcal{A}ut_{(2)}\Omega)$$

that assigns to any line bundle $L \in \mathrm{Pic}(M)$ a supermanifold $(M, \mathcal{O})$ with retract (M, Ω), maybe split.

Let $L \in \mathrm{Pic}(M)$ be given by a cocycle $h = (h_{ij}) \in Z^1(\mathbf{U}, \mathcal{F}^\times)$ for an open cover $\mathbf{U} = (U_i)$ of M. Then $(M, \mathcal{O})$ is determined by the following cocycle $g = (g_{ij}) \in Z^1(M, \mathcal{A}ut_{(2)}\Omega)$:

$$g_{ij} = \mathrm{id} + \left(\frac{1}{h_{ij}}dh_{ij}\right) d. \tag{7}$$

Suppose a Hermitian metric on L be given. This is a 0-cochain (m_i) formed by smooth positive functions $m_i : U_i \longrightarrow \mathbf{R}$ such that

$$m_j = |g_{ij}|^2 m_i \quad \text{in} \quad U_i \cap U_j \neq \emptyset.$$

Then the curvature form $\omega \in \Gamma(M, \mathcal{Z}\Phi^{1,1})$ is defined by

$$\omega = \bar{\partial}\partial \log m_i \quad \text{in} \quad U_i.$$

The de Rham class of ω coincides with the class $\frac{i}{2\pi}c_1(L)$, where $c_1(L)$ is the Chern class of L [6, Sect. 1]. Now, the class $\hat{\omega} \in \hat{H}^{1,1}(M, \mathbf{C})$ does not depend on the choice of the metric. The element $\hat{c}_1(L) = -2\pi i\hat{\omega}$ is called the *refined Chern class* of L.

Theorem 6 *We have*

$$\begin{aligned} \delta^*\omega &= \mathbf{D}^*L, \\ \hat{\delta}(\hat{c}_1(L)) &= -2\pi i \mathbf{D}^*L, \end{aligned}$$

where ω is the curvature form of a metric in L. The supermanifold with retract (M, Ω) corresponding to the form ω is determined by the cocycle (7).

Example Consider the canonical line bundle $K_M = \bigwedge^n T(M)^*$ over a complex manifold M, $\dim M = n$. Let $\mathbf{U} = (U_i)$ be a coordinate cover of M and $x_1^{(i)}, \ldots, x_n^{(i)}$ holomorphic coordinates in U_i. Then K_M is trivial over any U_i and is determined in this cover by the 1-cocycle $(J_{ij}) \in Z^1(\mathbf{U}, \mathcal{F}^\times)$ given by

$$J_{ij} = \frac{D(x_1^{(i)}, \ldots, x_n^{(i)})}{D(x_1^{(j)}, \ldots, x_n^{(j)})}.$$

Consider a volume form on M given in U_i by

$$V = V_i dx_1^{(i)} \ldots dx_n^{(i)} d\overline{x}_1^{(i)} \ldots d\overline{x}_n^{(i)}.$$

Since $V_j = |J_{ij}|^2 V_i$ in $U_i \cap U_j$, we get, up to a constant factor, a Hermitian metric on K_M. The corresponding curvature form

$$\omega = \bar{\partial}\partial \log V_i \quad \text{in} \quad U_i$$

is just the canonical form defined in [9]. The supermanifold with retract (M, Ω) corresponding to K_M or to the canonical form will be called the *canonical supermanifold* corresponding to M.

5 Application to flag manifolds

In this section, we apply the construction of Sect. 4 to the case when M is a flag manifold of a connected simple complex Lie group G. We may identify M with the coset space G/P, where P is a parabolic subgroup of G. It is well known that M is Kähler. Moreover, the canonical form is a Kähler form on M ([9]). We also note that

$$H^{1,1}(M, \mathbf{C}) \simeq H^2(M, \mathbf{C}) \simeq \mathbf{C}^r,$$

where r is the dimension of the centre of the reductive part of P. We have $r = 1$ whenever P is a maximal subgroup. We may suppose that G coincides with the identity component of the group $\mathrm{Bih} M$ of all biholomorphisms of M. Then

$$\mathrm{Bih} M = G\Sigma,$$

where Σ is a finite group which can be easily described in terms of P (see, e.g., [11, §15, Theorem 3]).

Corollary of Theorem 5 shows that the construction of Sect. 4 gives rise to a non-void family of non-split supermanifolds having (M, Ω) as their retract.

Theorem 7 *Let $M = G/P$ is a flag manifold, where G is simple, and denote $r = \dim H^2(M, \mathbf{C})$. Then there exists a family of distinct non-split supermanifolds parametrized by $\mathbf{CP}^{r-1}/\Sigma$ and having (M, Ω) as their common retract.*

If P is maximal, then this family consists of a unique supermanifold, which is isomorphic to the canonical one.

This theorem applies, in particular, to simply connected compact irreducible Hermitian symmetric spaces M considered in Sect. 3. In this case, $r = 1$, and thus our construction gives only one non-split supermanifold with retract (M, Ω), namely the canonical one. By Theorem 4, the only non-split supermanifold with retract (M, Ω), where M is a simply connected compact irreducible Hermitian symmetric space different from $\mathrm{Gr}_{n,k}$, $2 \leq k \leq n-2$, is the canonical one. It turns out that in the case $M = \mathrm{Gr}_{n,k}$, $2 \leq k \leq n-2$, the canonical supermanifold is not isomorphic to the super-Grassmannian $\Pi\mathrm{Gr}_{n|n,k|k}$ (see [12]).

References

[1] F.A. Berezin. Introduction to Superanalysis. Reidel, Dordrecht 1987.

[2] V.A. Bunegina, A.L. Onishchik. Homogeneous supermanifolds associated with the complex projective line. *J. Math. Sciences,* 82(1996), 3503–3527.

[3] M. Eastwood, C. LeBrun. Thickening and supersymmetric extensions of complex manifolds. *Amer. J. Math.,* 108(1986), 1177–1192.

[4] A. Frölicher, A. Nijenhuis. Theory of vector-valued differential forms, P. 1. Derivations in the graded ring of differential forms. *Proc. Kon. Ned. Akad. Wet. Amsterdam,* 59(1956), 540–564.

[5] P. Green. On holomorphic graded manifolds. *Proc. Amer. Math. Soc.,* 85(1982), 587–590.

[6] P.R. Griffiths, J. Harris. Principles of Algebraic Geometry. J. Wiley & Sons, New York 1978.

[7] A. Grothendieck. Sur quelques points d'algèbre homologique. *Tôhoku Math. J.,* 9(1957), 119–221.

[8] S. Helgason. Differential Geometry, Lie Groups, and Symmetric Spaces. Academic Press, New York e.a. 1978.

[9] J.-L. Koszul. Sur la forme hermitienne canonique des espaces homogènes complexes. *Canad. J. Math.,* 7(1955), 562–576.

[10] Yu.I. Manin. Gauge Field Theory and Complex Geometry. Springer-Verlag, Berlin e.a. 1988.

[11] A.L. Onishchik. Topology of Transitive Transformation Groups. J.A. Barth Verlag, Leipzig e.a. 1994.

[12] A.L. Onishchik. Non-split supermanifolds associated with the cotangent bundle. *Prépubl. Université Poitiers, N 109.* Poitiers 1997.

[13] V.P. Palamodov. Invariants of analytic $\mathbf{Z}_2$-manifolds. *Funkc. Analiz i yego pril.*, 17(1983), 68–69 (in Russian).

[14] M.J. Rothstein. Deformations of complex supermanifolds. *Proc. Amer. Math. Soc.*, 95(1985), 255–260.

E-Mail: arkadiy@onishchik.msk.ru

COMPARAISON DE L'HOMOLOGIE DE HOCHSCHILD ET DE L'HOMOLOGIE DE POISSON POUR UNE DEFORMATION DES SURFACES DE KLEIN

J. ALEV, T. LAMBRE

J.A.: Université de Reims,
Département de Mathématiques, U.R.A. 1870 du C.N.R.S.,
B.P. 1039, 51687 Reims Cedex, France.
T.L.: Université de Paris Sud,
Département de Mathématiques, U.R.A. 1169 du C.N.R.S.,
91405 Orsay Cedex, France.

Abstract. Let P^G be the quotient variety of the affine plane by the action of a finite group $G \subset SL(2, \mathbb{C})$; then P^G inherits in a natural way a Poisson algebra structure. Let $A_1(\mathbb{C})$ be the first Weyl algebra $\mathbb{C}[p, q]$ with the relation $pq - qp = 1$, on which G acts by automorphisms in such a way that the invariant algebra $A_1(\mathbb{C})^G$ is a deformation of P^G. We prove that the trace group $HH_0(A_1(\mathbb{C})^G)$ is a deformation of the Poisson homology group $H_0^{Pois}(P^G)$. Moreover, these two groups are $\mathbb{C}$-vector spaces of finite dimension and $dim\,(HH_0(A_1(\mathbb{C})^G)) = dim\,(H_0^{Pois}(P^G)) = s(G) - 1$, where $s(G)$ denotes the number of irreducible representations of G.

1 Introduction

1.1

Le corps de base étant $\mathbb{C}$, soient S une algèbre commutative et A une algèbre filtrée, non nécessairement commutative, de gradué associé isomorphe à S. Le crochet de A, $[a, b] = ab - ba$, induit sur S un crochet de Poisson. Le but de ce travail est de comparer, en degré zéro, l'homologie de Hochschild de A et

Y. Khakimdjanov et al (eds.), Algebra and Operator Theory, 25-38.

l'homologie de Poisson de S dans le cas où S est l'anneau des fonctions régulières sur une singularité de Klein. Cette comparaison donne un isomorphisme dans les deux cas suivants :

(i) S est un anneau de polynômes et A une algèbre enveloppante ([7]) ;

(ii) S est l'anneau des fonctions régulières sur le fibré cotangent d'une variété algébrique affine lisse X et A est l'anneau des opérateurs différentiels sur X ([2], [14]).

Les singularités de Klein étant dans un sens relativement simples (point double rationnel), le résultat de ce travail soulève la question du degré de généralité d'un tel isomorphisme ([6],[12]).

1.2

Lorsqu'on représente les singularités de Klein comme les variétés quotients du plan affine par l'action d'un sous-groupe fini de $SL(2)$, ces surfaces héritent d'une structure de Poisson provenant de la structure symplectique du plan. Comme par ailleurs, un sous-groupe fini G de $SL(2)$ opère dans l'algèbre de Weyl $A_1(\mathbb{C})$ par automorphismes, l'algèbre des invariants $A_1(\mathbb{C})^G$ est une déformation de la surface de Klein correspondante. Nous montrons que la filtration de Bernstein de $A_1(\mathbb{C})$ induit naturellement une filtration sur le groupe $HH_0(A_1(\mathbb{C})^G)$ dont le gradué associé est isomorphe à l'homologie de Poisson en degré zéro de la surface de Klein. La preuve consiste à montrer que ces deux groupes d'homologie ont la même dimension finie en utilisant le fait que les algèbres considérées sont des algèbres d'invariants par des groupes finis d'automorphismes.

1.3

Nous tenons à remercier M. Farinati, T. Levasseur, A. Solotar et M. Vigué-Poirrier pour les fructueuses conversations que nous avons eues lors de l'élaboration de ce travail.

2 Rappels et Notations.

2.1

Soit G un sous-groupe fini *non trivial* de $SL(2,\mathbb{C})$; pour $n \geq 2$, notons $\omega_n = \exp(2i\pi/n)$ et considérons les éléments suivants de $SL(2,\mathbb{C})$:

$$\rho_n = \begin{pmatrix} \omega_n & 0 \\ 0 & \omega_n^{-1} \end{pmatrix}, \mu = \begin{pmatrix} 0 & i \\ i & 0 \end{pmatrix}, \phi_5 = \begin{pmatrix} -\omega_5^3 & 0 \\ 0 & -\omega_5^2 \end{pmatrix}, \nu = \begin{pmatrix} 0 & 1 \\ -1 & 0 \end{pmatrix}$$

$$\eta_8 = \frac{1}{\sqrt{2}} \begin{pmatrix} \omega_8^7 & \omega_8^7 \\ \omega_8^5 & \omega_8 \end{pmatrix}, \psi_5 = \frac{1}{\omega_5^2 - \omega_5^{-2}} \begin{pmatrix} \omega_5 + \omega_5^{-1} & 1 \\ 1 & -(\omega_5 + \omega_5^{-1}) \end{pmatrix}$$

D'après [11], G est conjugué à l'un des cinq types non isomorphes suivants :
Le groupe A_n d'ordre n engendré par ρ_n,
le groupe D_n d'ordre $4n$, engendré par ρ_{2n} et μ,
le groupe E_6 d'ordre 24, engendré par ρ_4, μ et η_8,
le groupe E_7 d'ordre 48, engendré par ρ_8, μ et η_8,
le groupe E_8 d'ordre 120, engendré par ϕ_5, ν et ψ_5.

En ce qui concerne les propriétés des algèbres d'invariants de G, on peut donc supposer que G est égal à l'un des groupes A_n, D_n, E_6, E_7, E_8 ; sauf mention expresse du contraire, dans toute la suite G va désigner l'un de ces groupes.

2.2

Le groupe G opère par automorphismes dans l'anneau de polynômes à deux variables $P = \mathbb{C}[X, Y]$ ainsi que dans l'algèbre de Weyl $A_1(\mathbb{C}) = \mathbb{C}[p, q]$, $pq - qp = 1$. Le résultat classique sur les surfaces de Klein admettant les P^G comme anneaux de fonctions régulières, est un énoncé de séparation complète ([10]) :

Théorème 1 *Soient G et G' deux groupes parmi A_n, D_n, E_6, E_7, E_8 ; alors :*

$$P^G \textit{ est isomorphe à } P^{G'} \textit{ si et seulement si } G = G'. \tag{1}$$

Dans [1], un résultat analogue est démontré pour les algèbres $A_1(\mathbb{C})^G$ et on a le théorème :

Théorème 2 *Soient G et G' deux groupes parmi A_n, D_n, E_6, E_7, E_8 ; alors :*

$$A_1(\mathbb{C})^G \textit{ est isomorphe à } A_1(\mathbb{C})^{G'} \textit{ si et seulement si } G = G'. \tag{2}$$

2.3

Dans [1], on calcule également les dimensions *finies* des espaces $HH_0(A_1(\mathbb{C})^G)$ pour $G = A_n, D_n$ par une méthode directe utilisant des générateurs explicites de l'algèbre $A_1(\mathbb{C})^G$; pour les cas de E_6, E_7, E_8 ce calcul devient très fastidieux vu la nature plus compliquée des générateurs de l'algèbre des invariants. Dans cette section, nous calculons le groupe $HH_0(A_1(\mathbb{C})^G)$ pour tous les groupes G en utilisant un résultat établi depuis dans [8].
L'algèbre $A_1(\mathbb{C})$ étant simple, l'algèbre des invariants $A_1(\mathbb{C})^G$ est équivalente au sens de Morita à l'algèbre produit croisé $A_1(\mathbb{C}) * G = \oplus_{g \in G} A_1(\mathbb{C})g$, de base G, à coefficients dans $A_1(\mathbb{C})$ et dans laquelle le produit est défini par la règle de redressement $ag.bh = ab^{g^{-1}}gh$ ([9],[1]). Le groupe $HH_0(A_1(\mathbb{C})^G)$ qui est un invariant de Morita sera calculé pour $A_1(\mathbb{C}) * G$.
Avec ces notations, le théorème de Lorenz ([8], page 74) nous donne :

Théorème 3 *Désignons par $T(G)$ l'ensemble des classes de conjugaison de G et par $[g]$ la classe de conjugaison de l'élément $g \in G$. Alors :*

$$HH_0(A_1(\mathbb{C}) * G) = \bigoplus_{[g] \in T(G)} H_0(Z_g/(g), HH_0(A_1(\mathbb{C}), A_1(\mathbb{C})g)). \tag{3}$$

Expliquons les différents termes de cette somme directe. $A_1(\mathbb{C})g$ est un $A_1(\mathbb{C})$-bimodule et $HH_0(A_1(\mathbb{C}), A_1(\mathbb{C})g)$ désigne l'homologie de Hochschild en degré zéro ; Z_g est le centralisateur de g dans G et on vérifie que le groupe $Z_g/(g)$ opère dans $HH_0(A_1(\mathbb{C}), A_1(\mathbb{C})g)$ par l'action induite par la conjugaison :

$$\forall x \in Z_g, x.(bg + [A, Ag]) = x^{-1}bgx + [A, Ag] = b^x g + [A, Ag]\,; \tag{4}$$

enfin, $H_0(Z_g/(g), HH_0(A_1(\mathbb{C}), A_1(\mathbb{C})g))$ désigne le 0-ème groupe d'homologie de $Z_g/(g)$ à valeurs dans $HH_0(A_1(\mathbb{C}), A_1(\mathbb{C})g)$. Pour plus de détails, le lecteur pourra consulter [8].

3 Calcul de l'homologie de Hochschild en degré zéro.

3.1 Crochet tordu par un automorphisme.

Soit A une $\mathbb{C}$-algèbre ; pour $a, b \in A$ et $h \in AutA$, introduisons le crochet de a et b tordu par h, défini par $[a, b]_h = ab - ba^h$. Les trois propriétés suivantes sont immédiates :

(i) si a est invariant par h, $[a, b]_h = [a, b]$;

(ii) $[ab, c]_h = [a, bc]_h + [b, ca^h]_h$;

(iii) pour $\sigma, h \in AutA$, $([a, b]_h)^\sigma = [a^\sigma, b^\sigma]_{\sigma^{-1}h\sigma}$.

Notons enfin par $[A, A]_h$ le sous-$\mathbb{C}$-espace de A engendré par les crochets tordus par h.

3.2

Le théorème suivant détermine la dimension de $HH_0(A_1(\mathbb{C})^G)$.

Théorème 4 *On suppose que G est l'un des groupes A_n, D_n, E_6, E_7, E_8. On a alors:*

(i)

$$\mathbb{C}g \oplus [A_1(\mathbb{C}), A_1(\mathbb{C})g] = A_1(\mathbb{C})g, \quad \forall g \in G, \; g \neq id, \tag{5}$$

et par conséquent $HH_0(A_1(\mathbb{C}), A_1(\mathbb{C})g)$ est le $Z_g/(g)$-module trivial pour tout $g \in G$, $g \neq id$.

(ii)

$$[A_1(\mathbb{C}), A_1(\mathbb{C})_{id}] = A_1(\mathbb{C}). \tag{6}$$

(iii) Soit $s(G)$ le cardinal de l'ensemble $T(G)$ des classes de conjugaison de G. Alors :

$$dim\, HH_0(A_1(\mathbb{C})^G) = s(G) - 1. \tag{7}$$

Preuve. Prouvons (i).
Remarquons d'abord que $[A_1(\mathbb{C}), A_1(\mathbb{C})g] = [A_1(\mathbb{C}), A_1(\mathbb{C})]_{g^{-1}}g$, de sorte qu'il suffit de montrer que

$$\mathbb{C} \oplus [A_1(\mathbb{C}), A_1(\mathbb{C})]_{g^{-1}} = A_1(\mathbb{C}), \quad \forall g \in G,\ g \neq id. \tag{8}$$

Ensuite, si g^{-1} est d'ordre $n > 1$ dans G, le sous-groupe (g) est conjugué à (ρ_n) dans $SL(2)$; il s'ensuit que g^{-1} est conjugué à une puissance $\rho = \rho_n^t$ de ρ_n, t premier avec n. En utilisant 3.1.(iii), on peut donc supposer $g^{-1} = \rho = \begin{pmatrix} \omega & 0 \\ 0 & \omega^{-1} \end{pmatrix}$, où $\omega = \omega_n^t$ est une racine primitive n-ème de l'unité.
La preuve se fait maintenant en trois étapes :
(1) Rappelons d'abord que $A_1(\mathbb{C})$ est $\mathbb{Z}$-graduée ; pour $i \in \mathbb{N} \setminus \{0\}$:

$$A_1(\mathbb{C})_0 = \mathbb{C}[pq],\ A_1(\mathbb{C})_i = \mathbb{C}[pq]p^i,\ A_1(\mathbb{C})_{-i} = q^i\mathbb{C}[pq]. \tag{9}$$

Puis, pq étant ρ-invariant, on a $[pq, f(pq)p^i]_\rho = [pq, f(pq)p^i] = -if(pq)p^i$; d'où l'inclusion :

$$\bigoplus_{i\in\mathbb{Z}\setminus\{0\}} A_1(\mathbb{C})_i \subseteq [A_1(\mathbb{C}), A_1(\mathbb{C})]_\rho. \tag{10}$$

(2) L'espace $[A_1(\mathbb{C}), A_1(\mathbb{C})]_\rho$ étant gradué, il suffit maintenant de montrer que $\mathbb{C}$ est un supplémentaire de $\sum_{r\geq 1}[\mathbb{C}[pq]p^r, q^r\mathbb{C}[pq]]_\rho$ dans $\mathbb{C}[pq]$. D'après le paragraphe précédent (i) et (ii), on a l'égalité :

$$\begin{aligned}[(pq)^k p^r, q^r(pq)^l]_\rho &= [(pq)^k, p^r q^r(pq)^l]_\rho + [p^r, q^r(pq)^l(pq)^k]_\rho \\ &= [p^r, q^r(pq)^l(pq)^k]_\rho,\end{aligned}$$

et il suffit de montrer que $\mathbb{C}$ est un supplémentaire de $\sum_{r\geq 1}[p^r, q^r\mathbb{C}[pq]]_\rho$ dans $\mathbb{C}[pq]$.
(3) Montrons par récurrence sur $r \geq 1$ que

$$[p^r, q^r(pq)^t]_\rho \in [p, q\mathbb{C}[pq]]_\rho. \tag{11}$$

En effet, on a

$$\begin{aligned}[p^r, q^r(pq)^t]_\rho &= [p^{r-1}, pq^r(pq)^t]_\rho + \omega^{r-1}[p, q^r(pq)^t p^{r-1}]_\rho \\ &= [p^{r-1}, q^{r-1}(pq + r - 1)(pq)^t]_\rho \\ &+\omega^{r-1}[p, q(pq - r + 1)^t q^{r-1}p^{r-1}]_\rho\end{aligned}$$

ce qui termine la récurrence considérée, puisque $q^{r-1}p^{r-1}$ est un polynôme en pq. Il suffit donc de montrer que $\mathbb{C}$ est un supplémentaire de $[p, q\mathbb{C}[pq]]_\rho$ dans $\mathbb{C}[pq]$.

Comme pour $t = 0, 1, 2, \ldots$, l'expression

$$\begin{aligned}[p, q(pq)^t]_\rho &= pq(pq)^t - \omega q(pq)^t p \\ &= pq(pq)^t - \omega qp(pq-1)^t \\ &= (pq)^{t+1} - \omega(pq-1)^{t+1}\end{aligned}$$

est un polynôme en pq de degré $t+1$, il est alors clair que $\mathbb{C}$ est un supplémentaire de $[p, q\mathbb{C}[pq]]_\rho$ dans $\mathbb{C}[pq]$.
(ii) C'est une propriété bien connue de l'algèbre de Weyl découlant par exemple du fait que la dérivation intérieure adp est surjective.
(iii) Le point (i) montre que $HH_0(A_1(\mathbb{C}), A_1(\mathbb{C})g)$ est le $Z_g/(g)$-module trivial pour $g \in G$, $g \neq id$; le point (ii) montre que l'identité n'a pas de contribution dans la formule de Lorenz, d'où le résultat.

4 Homologie de Poisson en degré zéro.

4.1 Structure de Poisson sur les surfaces de Klein.

Soit G l'un des groupes A_n, D_n, E_6, E_7, E_8. Le groupe G agit par automorphismes dans l'anneau de polynômes $P = \mathbb{C}[X, Y]$; cette action respecte la structure de Poisson de P provenant de la structure symplectique du plan et induit alors une structure de Poisson sur P^G. Remarquons que cette propriété est due au fait que G est inclus dans $SL(2)$ et que par exemple le crochet de Poisson $\{X+Y, XY\} = X - Y$ n'est pas une fonction symétrique. Par ailleurs, si $\mathcal{H}$ désigne l'hypersurface affine définie par le polynôme $f(X_1, X_2, \ldots, X_t)$, le nombre de Milnor de $\mathcal{H}$, $\mu(\mathcal{H})$, est la codimension de l'idéal engendré par les $\{f_i' \mid 1 \leq i \leq t\}$ dans $\mathbb{C}[X_1, X_2, \ldots, X_t]$. Rappelons enfin ([13]) qu'un polynôme $f(X_1, X_2, \ldots, X_t)$ est dit de poids homogène s'il existe une suite d'entiers $(n_1, n_2, \ldots, n_t)$ strictement positifs et un entier strictement positif d tels que

$$f(\lambda^{n_1} X_1, \lambda^{n_2} X_2, \ldots, \lambda^{n_t} X_t) = \lambda^d f(X_1, X_2, \ldots, X_t), \text{ pour tout } \lambda \in \mathbb{C}.$$

4.2

On peut maintenant énoncer la proposition suivante :

Proposition 1 *Soit H l'algèbre $\mathbb{C}[X_1, X_2, X_3]/(F)$, où F est un polynôme dans $\mathbb{C}[X_1, X_2, X_3]$ de poids homogène. Alors :*

(i) Il existe une structure d'algèbre de Poisson sur H, provenant d'une structure de Poisson canonique sur $\mathbb{C}[X_1, X_2, X_3]$;

(ii) Pour cette structure de Poisson, $H_0^{Pois}(H)$ s'identifie à l'algèbre $H' = \mathbb{C}[X_1, X_2, X_3]/(F_1', F_2', F_3')$.

(iii) Si F est à singularité isolée en l'origine, $\dim H_0^{Pois}(H) = \mu(\mathcal{H})$, où $\mathcal{H}$ désigne l'hypersurface définie par F.

Preuve. Prouvons (i).
Pour $P, Q \in \mathbb{C}[X_1, X_2, X_3]$, posons $\{P, Q\} = dP \wedge dQ \wedge dF/dX_1 \wedge dX_2 \wedge dX_3$, ce qui définit un crochet de Poisson dans $\mathbb{C}[X_1, X_2, X_3]$ pour lequel l'idéal (F) est un idéal de Poisson. Ainsi, H est muni de la structure de Poisson quotient.
(ii) On a d'abord les trois égalités

$$\{X_1, X_2\} = F_3', \ \{X_2, X_3\} = F_1', \ \{X_3, X_1\} = F_2'. \tag{12}$$

Ensuite, $\{,\}$ étant une dérivation en chacun des arguments, on a

$$X_1^{\alpha_1} X_2^{\alpha_2} F_3' = \frac{1}{\alpha_1 + 1}\{X_1^{\alpha_1+1} X_2^{\alpha_2}, X_2\} = \frac{1}{\alpha_2 + 1}\{X_1, X_1^{\alpha_1} X_2^{\alpha_2+1}\}, \tag{13}$$

et les formules analogues obtenues par permutation circulaire des indices. Pour montrer (ii), montrons l'égalité

$$\{\mathbb{C}[X_1, X_2, X_3], \mathbb{C}[X_1, X_2, X_3]\} + (F) = (F_1', F_2', F_3'). \tag{14}$$

Puisque F est de poids homogène, l'identité d'Euler implique l'existence de trois entiers n_1, n_2 et n_3 strictement positifs tels que

$$n_1 X_1 F_1' + n_2 X_2 F_2' + n_3 X_3 F_3' = (n_1 + n_2 + n_3) F, \tag{15}$$

ce qui implique l'inclusion facile

$$\{\mathbb{C}[X_1, X_2, X_3], \mathbb{C}[X_1, X_2, X_3]\} + (F) \subseteq (F_1', F_2', F_3'). \tag{16}$$

Il s'agit maintenant de montrer que

$$X_1^{\alpha_1} X_2^{\alpha_2} X_3^{\alpha_3} F_i' \in \{\mathbb{C}[X_1, X_2, X_3], \mathbb{C}[X_1, X_2, X_3]\} + (F), \tag{17}$$

pour tous $\alpha_1, \alpha_2, \alpha_3 \in \mathbb{N}$ et $i = 1, 2, 3$. Les formules (13) montrent que le seul cas à étudier est le "cas type" $X_1^{\alpha_1} X_2^{\alpha_2} X_3^{\alpha_3} F_1'$, avec $\alpha_1 \neq 0$. L'équation (15) implique d'abord

$$\begin{aligned} n_1 X_1^{\alpha_1} X_2^{\alpha_2} X_3^{\alpha_3} F_1' \quad & + n_2 X_2^{\alpha_1 - 1} X_2^{\alpha_2+1} X_3^{\alpha_3} F_2' + n_3 X_3^{\alpha_1 - 1} X_2^{\alpha_2} X_3^{\alpha_3+1} F_3' \\ & = X_1^{\alpha_1} X_2^{\alpha_2} X_3^{\alpha_3} (n_1 + n_2 + n_3) F. \end{aligned}$$

Puis, par dérivation on a

$$\{X_2, X_1^{\alpha_1} X_2^{\alpha_2} X_3^{\alpha_3+1}\} = (\alpha_3 + 1) X_1^{\alpha_1} X_2^{\alpha_2} X_3^{\alpha_3} F_1' - \alpha_1 X_1^{\alpha_1 - 1} X_2^{\alpha_2} X_3^{\alpha_3+1} F_3' \tag{18}$$

et

$$\{X_1^{\alpha_1} X_2^{\alpha_2+1} X_3^{\alpha_3}, X_3\} = (\alpha_2 + 1) X_1^{\alpha_1} X_2^{\alpha_2} X_3^{\alpha_3} F_1' - \alpha_1 X_1^{\alpha_1 - 1} X_2^{\alpha_2+1} X_3^{\alpha_3} F_2'. \tag{19}$$

Les équations (4.2), (18) et (19) constituent un système linéaire en les inconnues

$$X_1^{\alpha_1} X_2^{\alpha_2} X_3^{\alpha_3} F_1' \qquad X_1^{\alpha_1 - 1} X_2^{\alpha_2+1} X_3^{\alpha_3} F_2' \qquad X_1^{\alpha_1 - 1} X_2^{\alpha_2} X_3^{\alpha_3+1} F_3'$$

de déterminant $-\alpha_1(n_3(\alpha_3 + 1) + n_2(\alpha_2 + 1) + n_1 \alpha_1)$, entier strictement négatif. Il s'ensuit que $X_1^{\alpha_1} X_2^{\alpha_2} X_3^{\alpha_3} F_1' \in \{\mathbb{C}[X_1, X_2, X_3], \mathbb{C}[X_1, X_2, X_3]\} + (F)$.

(iii) F étant à singularité isolée en l'origine, $dim\, H'$ est finie et égale au nombre de Milnor $\mu(\mathcal{H})$ de $\mathcal{H}$.

Théorème 5 *Avec les notations du 4.1, on a :*

$$dim\, H_0^{Pois}(P^G) = s(G) - 1.$$

Preuve. Rappelons d'abord (cf. [11]) que P^G est engendré par trois générateurs homogènes que nous allons noter $g_1(X,Y), g_2(X,Y), g_3(X,Y)$ soumis à une seule relation $F_G(X_1, X_2, X_3)$ de sorte que l'on a :
- $F_G(g_1, g_2, g_3) = 0$;
- le polynôme F_G est irréductible dans $\mathbb{C}[X_1, X_2, X_3]$ et il admet une singularité isolée en l'origine ;
- le polynôme F_G est de poids homogène.

En considérant les dimensions de Krull des anneaux, il est facile de voir que l'application ϕ qui transforme X_1 en g_1, X_2 en g_2 et X_3 en g_3 est un isomorphisme d'algèbres. D'après le paragraphe 4.1 et la proposition 1, P^G et H sont des algèbres de Poisson : nous montrons dans l'annexe, par une étude au cas par cas, que quitte à normaliser g_1, g_2 et g_3, ϕ est un isomorphisme d'algèbres de Poisson. Si $\mathcal{H}$ désigne la surface définie par F_G, il vient alors :

$$dim\, H_0^{Pois}(P^G) = dim\, H_0^{Pois}(H) = \mu(\mathcal{H}),$$

d'après la proposition 1. Enfin, l'égalité $\mu(H) = s(G) - 1$ est immédiate par vérification au cas par cas, compte tenu de l'expression explicite de F_G donnée dans l'annexe.

5 HH_0 comme déformation de H_0^{Pois}.

5.1 $A_1(\mathbb{C})^G$ comme déformation de la surface de Klein P^G.

Il est bien connu que l'algèbre de Weyl $A_1(\mathbb{C})$ est munie de la filtration de Bernstein $\mathcal{B} = (A_1(\mathbb{C})_n)_{n\in\mathbb{N}}$ définie par $A_1(\mathbb{C})_n = Vect(p^i q^j \mid i + j \leq n)$. L'action de G étant linéaire, cette filtration est invariante et on en déduit une action sur le gradué associé qui est isomorphe à un anneau de polynômes en $X = gr_1 p$ et $Y = gr_1 q$; par ailleurs, $\mathcal{B}$ induit une filtration $\mathcal{B}^G$ sur $A_1(\mathbb{C})^G$ définie par $A_1(\mathbb{C})_n^G = A_1(\mathbb{C})_n \cap A_1(\mathbb{C})^G$ et par semi-simplicité de G, il vient :

$$gr_{\mathcal{B}^G}(A_1(\mathbb{C})^G) \equiv (gr_{\mathcal{B}} A_1(\mathbb{C}))^G. \tag{20}$$

Comme $gr_{\mathcal{B}} A_1(\mathbb{C})$ s'identifie à $P = \mathbb{C}[X,Y]$, l'algèbre $gr_{\mathcal{B}^G}(A_1(\mathbb{C})^G)$ s'identifie à l'anneau des fonctions régulières de la surface de Klein correspondant au groupe G. Il en résulte que $A_1(\mathbb{C})^G$ est une déformation de la surface de Klein correspondant au groupe G et induit alors sur P^G le crochet de Poisson donné par

$$\{a_n + A_1(\mathbb{C})_{n-1}^G, a_m + A_1(\mathbb{C})_{m-1}^G\} = gr_{n+m-2}([a_n, a_m]). \tag{21}$$

Ainsi, P^G se trouve muni de deux crochets de Poisson, l'un induit par la structure symplectique de P et l'autre induit par $A_1(\mathbb{C})^G$. Il est facile de vérifier que ces deux structures de Poisson coïncident.

5.2 Définitions.

Pour étudier les relations du paragraphe précédent, nous aurons besoin de quelques définitions générales.
Soient A une algèbre et $\mathcal{F} = (A_n)_{n\in\mathbf{N}}$ une $\mathbf{N}$-filtration sur A ; posons $S = gr_{\mathcal{F}} A$. Nous supposons que $A_0 = \mathbf{C}$ et convenons que $A_{-1} = (0)$.
Pour $r \geq 1$, nous dirons que $(A, \mathcal{F})$ est r-commutative si

$$[A_m, A_n] \subseteq A_{m+n-r}. \tag{22}$$

Il est clair que si $(A, \mathcal{F})$ est r-commutative, alors S est commutative ; dire que $(A, \mathcal{F})$ est r-commutative pour tout r équivaut à dire que A est commutative. On peut remarquer que l'algèbre de Weyl $A_1(\mathbf{C})$ est 2-commutative ([5]). Supposons dans la suite que $(A, \mathcal{F})$ est r-commutative. Pour $a_m \in A_m$ et $a_n \in A_n$, on obtient sur S le crochet de Poisson défini par :

$$\{gr_m a_m, gr_n a_n\} = gr_{m+n-r}[a_m, a_n]. \tag{23}$$

(gr_i désigne la projection canonique de A_i sur A_i/A_{i-1}.) Considérons maintenant $[A, A]$, filtrée par $(A_n \cap [A, A])_{n\in\mathbf{N}}$, puis la filtration quotient sur $HH_0(A) = A/[A, A]$ définie par $(A_n/A_n \cap [A, A])_{n\in\mathbf{N}}$; par ailleurs, on peut considérer $H_0^{Pois}(S) = S/\{S, S\}$ pour la structure de Poisson définie par (23). Afin de comparer $gr\, HH_0(A)$ et $H_0^{Pois}(S)$, nous aurons besoin de la propriété suivante exprimant une compatibilite graduée entre la filtration $\mathcal{F}$ sur A et la structure d'algèbre de Lie induite sur A. Nous dirons que l'algèbre filtrée r-commutative $(A, \mathcal{F})$ vérifie *la condition de Kassel* si :

$$\forall n \in N, \; A_n \cap [A, A] = \sum_{k+l=n+r} [A_k, A_l]. \tag{24}$$

Pour simplifier les notations, posons $C_m = \sum_{k+l=m} [A_k, A_l]$.

5.3 Le résultat principal.

Théorème 6 *Avec les notations ci-dessus, on suppose $HH_0(A)$ et $H_0^{Pois}(S)$ de dimension finie. Les trois conditions suivantes sont alors équivalentes:*
(i) $(A, \mathcal{F})$ vérifie la condition de Kassel ;
(ii) $gr\, HH_0(A) \equiv H_0^{Pois}(S)$;
(iii) $dim\, HH_0(A) = dim\, H_0^{Pois}(S)$.

Preuve. (i) implique (ii) : cette preuve est reprise dans [7] ; elle est reproduite ici pour la commodité du lecteur.
L'application gr_n induit l'application surjective $A_n \to S_n/\{S, S\}_n$; comme par ailleurs, C_{n+r} est inclus dans le noyau on obtient la suite exacte

$$\frac{A_n}{C_{n+r}} \to \frac{S_n}{\{S, S\}_n} \to 0 \tag{25}$$

qui induit la suite exacte

$$0 \longrightarrow \frac{A_{n-1}+C_{n+r}}{C_{n+r}} \longrightarrow \frac{A_n}{C_{n+r}} \longrightarrow \frac{S_n}{\{S,S\}_n} \longrightarrow 0. \qquad (26)$$

Comme $A_{n-1}+C_{n+r}/C_{n+r}$ s'identifie à $A_{n-1}/A_{n-1}\cap C_{n+r}$, la condition de Kassel donne finalement la suite exacte

$$0 \longrightarrow \frac{A_{n-1}}{A_{n-1}\cap[A,A]} \longrightarrow \frac{A_n}{A_n\cap[A,A]} \longrightarrow \frac{S_n}{\{S,S\}_n} \longrightarrow 0. \qquad (27)$$

(ii) implique (iii) : cette implication est claire.
(iii) implique (i) : $HH_0(A)$ étant de dimension finie, il existe un entier s tel que $HH_0(A)=A_s/A_s\cap[A,A]$. La suite exacte

$$0 \longrightarrow \frac{A_{s-1}+C_{s+r}}{C_{s+r}} \longrightarrow \frac{A_s}{A_s\cap C_{s+r}} \longrightarrow \frac{S_s}{\{S,S\}_s} \longrightarrow 0 \qquad (28)$$

implique

$$\begin{aligned} dim\,(HH_0(A)) &= dim\,(\tfrac{A_s}{A_s\cap[A,A]}) \leq dim\,(\tfrac{A_s}{A_s\cap C_{s+r}}) \\ &= dim\,(\tfrac{S_s}{\{S,S\}_s}) + dim\,(\tfrac{A_{s-1}+C_{s+r}}{C_{s+r}}) \\ &= dim\,(\tfrac{S_s}{\{S,S\}_s}) + dim\,(\tfrac{A_{s-1}}{A_{s-1}\cap C_{s+r}}) \\ &\leq dim\,(\tfrac{S_s}{\{S,S\}_s}) + dim\,(\tfrac{A_{s-1}}{A_{s-1}\cap C_{s-1+r}}) \\ &\leq dim\,(\tfrac{S_s}{\{S,S\}_s}) + dim\,(\tfrac{S_{s-1}}{\{S,S\}_{s-1}}) + dim\,(\tfrac{A_{s-2}}{A_{s-2}\cap C_{s-2+r}}) \\ &\leq \ldots \leq \textstyle\sum_{i=0}^{s} dim\,(\tfrac{S_i}{\{S,S\}_i}) \leq dim\,(H_0^{Pois}(S)). \end{aligned}$$

D'après l'hypothèse, toutes les inégalités sont des égalités et il vient d'abord

$$A_s\cap[A,A] = A_s\cap C_{s+r} = C_{s+r}, \qquad (29)$$

puis

$$\begin{aligned} A_{s-1}\cap[A,A] &= A_{s-1}\cap A_s\cap[A,A] = A_{s-1}\cap C_{s+r} \\ &= A_{s-1}\cap C_{s-1+r} = C_{s-1+r} \end{aligned}$$

et ainsi de suite. On a ainsi la condition de Kassel pour $n\leq s$.
Par ailleurs, $S_{s+1}/\{S,S\}_{s+1}=0$ de sorte que $A_{s+1}=A_s+C_{s+1+r}$. On a alors

$$A_{s+1}\cap[A,A] = A_s\cap[A,A]+C_{s+1+r} = C_{s+r}+C_{s+1+r} = C_{s+1+r} \qquad (30)$$

et on termine par une récurrence facile sur t pour traiter le cas de $A_{s+t}\cap[A,A]$.

Corollaire 1 *Soit G l'un des groupes A_n, D_n, E_6, E_7, E_8 agissant par automorphismes d'une part sur P et d'autre part sur $A_1(\mathbb{C})$. On a alors :*

$$gr\,(HH_0(A_1(\mathbb{C})^G)) \equiv H_0^{Pois}(P^G). \qquad (31)$$

Preuve. Le résultat découle des théorèmes 4, 5 et 6.

Annexe

Dans cette annexe, nous étudions au cas par cas l'algèbre P^G. Pour ce faire, nous allons utiliser les résultats et les notations donnés dans [11] et [4].

1) $G = A_n$: P^G est engendré par

$$g_1 = XY, g_2 = X^n \text{ et } g_3 = Y^n,$$

vérifiant la relation $F = X_1^n - X_2X_3$. On calcule facilement

$$\{g_1, g_2\} = -ng_2, \{g_2, g_3\} = n^2 g_1^{n-1} \text{ et } \{g_3, g_1\} = -ng_3.$$

En posant $h_1 = a_1g_1$, $h_2 = a_2g_2$, $h_3 = a_3g_3$ et

$$\{h_1, h_2\} = F_3'(h_1, h_2, h_3), \{h_2, h_3\} = F_1'(h_1, h_2, h_3), \{h_3, h_1\} = F_2'(h_1, h_2, h_3),$$

on obtient

$$a_1 = \frac{1}{n} \quad \text{et} \quad a_2a_3 = \frac{1}{n^n}.$$

2) $G = D_n$: P^G est engendré par

$$g_1 = X^2Y^2, g_2 = X^{2n} + (-1)^n Y^{2n} \text{ et } g_3 = X^{2n+1}Y - (-1)^n XY^{2n+1},$$

vérifiant la relation $F = X_1^{n+1} + X_1X_2^2 + X_3^2$. On calcule facilement

$$\{g_1, g_2\} = -4ng_3, \{g_2, g_3\} = 2ng_2^2 - (-1)^n 4n(2n+2)g_1^n \text{ et } \{g_3, g_1\} = 4ng_1g_2.$$

En posant $h_1 = a_1g_1$, $h_2 = a_2g_2$, $h_3 = a_3g_3$ et

$$\{h_1, h_2\} = F_3'(h_1, h_2, h_3), \{h_2, h_3\} = F_1'(h_1, h_2, h_3), \{h_3, h_1\} = F_2'(h_1, h_2, h_3),$$

on obtient

$$a_1 = \frac{-1}{4n^2}, \quad a_2^2 = \frac{-1}{2^{2(n+1)}n^{2n}} \quad \text{et} \quad a_3^2 = \frac{-1}{2^{2(n+2)}n^{2(n+1)}}.$$

3) $G = E_6$: P^G est engendré par

$$g_1 = XY^5 - X^5Y,$$

$$g_2 = X^8 + 14X^4Y^4 + Y^8$$

et

$$g_3 = X^{12} - 33X^8Y^4 - 33X^4Y^8 + Y^{12},$$

vérifiant la relation $F = X_1^4 + X_2^3 + X_3^2$. En utilisant le fait que le crochet de Poisson $\{,\}$ est gradué de degré -2 et en étudiant les degrés possibles des expressions considérées, on obtient l'existence de cinq scalaires α, β, γ, δ et λ tels que

$$\{g_1, g_2\} = \alpha g_3 + \beta g_1^2, \{g_2, g_3\} = \gamma g_1^3 + \delta g_1 g_3, \{g_3, g_1\} = \lambda g_2^2.$$

En écrivant les égalités $\{g_i, F(g_1, g_2, g_3)\} = 0$ pour $i = 1, 2, 3$, il vient

$$\beta = \delta = 0, \ \gamma = 2\alpha \ et \ 2\lambda = 3\alpha.$$

En calculant $\{g_1, g_2\}$, on trouve $\alpha = 8$. En posant $h_1 = a_1 g_1$, $h_2 = a_2 g_2$, $h_3 = a_3 g_3$ et

$$\{h_1, h_2\} = F_3'(h_1, h_2, h_3), \{h_2, h_3\} = F_1'(h_1, h_2, h_3), \{h_3, h_1\} = F_2'(h_1, h_2, h_3),$$

il vient

$$a_1 = 2^{-6}, a_2 = 2^{-8}, a_3 = 2^{-12}.$$

4) $G = E_7$: P^G est engendré par

$$g_1 = X^8 + 14X^4Y^4 + Y^8,$$

$$g_2 = X^{10}Y^2 - 2X^6Y^6 + X^2Y^{10}$$

et

$$g_3 = X^{17}Y - 34X^{13}Y^5 + 34X^5Y^{13} - XY^{17},$$

vérifiant la relation $F = X_1^3 X_2 + X_2^3 + X_3^2$. En utilisant le fait que le crochet de Poisson $\{,\}$ est gradué de degré -2 et en étudiant les degrés possibles des expressions considérées, on obtient l'existence de quatre scalaires α, β, γ et δ tels que

$$\{g_1, g_2\} = \alpha g_3, \{g_2, g_3\} = \beta g_1^2 g_2, \{g_3, g_1\} = \gamma g_1^3 + \delta g_2^2.$$

En écrivant les égalités $\{g_i, F(g_1, g_2, g_3)\} = 0$ pour $i = 1, 2, 3$, il vient

$$2\beta = 2\delta = 3\alpha \ et \ 2\gamma = \alpha.$$

En calculant $\{g_1, g_2\}$, on trouve $\alpha = 16$. En posant $h_1 = a_1 g_1$, $h_2 = a_2 g_2$, $h_3 = a_3 g_3$ et

$$\{h_1, h_2\} = F_3'(h_1, h_2, h_3), \{h_2, h_3\} = F_1'(h_1, h_2, h_3), \{h_3, h_1\} = F_2'(h_1, h_2, h_3),$$

il vient

$$a_1 = 2^{-12}, \ a_2 = 2^{-18} \ et \ a_3 = 2^{-27}.$$

5) $G = E_8$: P^G est engendré par

$$\begin{aligned} g_1 &= X^{11}Y + 11X^6Y^6 - XY^{11}, \\ g_2 &= X^{20} - 228X^{15}Y^5 + 494X^{10}Y^{10} + 228X^5Y^{15} + Y^{20} \\ g_3 &= X^{30} + 522X^{25}Y^5 - 10005X^{20}Y^{10} - 10005X^{10}Y^{20} - 522X^5Y^{25} + Y^{30}, \end{aligned}$$

vérifiant la relation $F = X_1^5 + X_2^3 + X_3^2$. En utilisant le fait que le crochet de Poisson $\{,\}$ est gradué de degré -2 et en étudiant les degrés possibles des expressions considérées, on obtient l'existence de trois scalaires α, β et γ tels que

$$\{g_1, g_2\} = \alpha g_3, \{g_2, g_3\} = \beta g_1^4, \{g_3, g_1\} = \gamma g_2^2.$$

En écrivant les égalités $\{g_i, F(g_1, g_2, g_3)\} = 0$ pour $i = 1, 2, 3$, il vient

$$3\alpha = 2\gamma \textit{ et } 2\beta = 5\alpha.$$

En calculant $\{g_1, g_2\}$, on trouve $\alpha = -20$. En posant $h_1 = a_1 g_1$, $h_2 = a_2 g_2$, $h_3 = a_3 g_3$ et

$$\{h_1, h_2\} = F_3'(h_1, h_2, h_3), \{h_2, h_3\} = F_1'(h_1, h_2, h_3), \{h_3, h_1\} = F_2'(h_1, h_2, h_3),$$

il vient enfin

$$a_1 = 10^{-6}, a_2 = 10^{-10} \textit{ et } a_3 = -10^{-15}.$$

References

[1] J. Alev, T.J. Hodges, J.-D. Velez, Fixed rings of the Weyl algebra $A_1(\mathbb{C})$, *J. Alg.*, **130**, 1990, 83-96.

[2] J.L. Brylinski, A differential complex for Poisson manifolds, *J. Diff. Geom.*, **28**, 1988, 93-114.

[3] H. Cartan & S. Eilenberg, *Homological algebra*, Princeton University Press, 1956.

[4] L. Chiang, H. Chu, M. Kang, Generation of Invariants, preprint.

[5] J. Dixmier, Sur les algèbres de Weyl, *Bull. Soc. Math. France*, **96**, 1968.

[6] A. Guichardet, Homologie de Hochschild des déformations d'algèbres de polynômes, *à paraître.*

[7] C. Kassel, L'homologie cyclique des algèbres enveloppantes, *Invent. Math.*, **91**, 1988, 221-251.

[8] M. Lorenz, Crossed Products: Characters, Cyclic Homology, and Grothendieck Groups, *Non commutative Rings, Math. Sciences Research Institute Publications*, **24**, Springer Verlag, 1992, 69-98.

[9] S. Montgomery, Fixed rings of finite automorphism groups of associative rings, *Lecture Notes in Math.*, **818**, Springer Verlag, 1980.

[10] P. Slodowy, Simple singularities and simple algebraic groups, *Lecture Notes in Math.*, **815**, Springer Verlag, 1980.

[11] T.A. Springer, Invariant theory, *Lecture Notes in Math.*, **585**, Springer Verlag, 1977.

[12] M. Van den Bergh, Non commutative homology of some three-dimensional quantum spaces, *K-theory*, **8**, 1994, 213-230.

[13] M. Vigué-Poirrier, Cyclic homology of algebraic hypersurfaces, *Journal of Pure and Applied Algebra*, **72**, 1991, 95-108.

[14] M. Wodzicki, Cyclic homology of differential operators, *Duke Math.*, **54**, 1987, 641-647.

Classification A.M.S. : 13A50, 13D10, 16E40.

E-mail: jacques.alev@univ-reims.fr

QUELQUES RESULTATS EN K-THEORIE REELLE

C.N.COSTINESCU

Département de Mathématiques,
Université Technique de Constructions,
Bucarest, ROUMANIE

Abstract. On étudie la suite spectrale d'Atiyah-Hirzebruch associée à KO* (**H**S(n,k)), où **H**S(n,k) désigne la variété quaternionique de Stiefel. Les isomorphismes de groupes abéliens : KO*(**H**S(n,k)) $\cong E_2^*$ (**H**S(n,k))$\cong$ $E_\infty^*(\mathbf{H}S(n,k))$ qui ont été établis,donnent la structure $\mathbf{Z}_8$- graduée de la K- théorie réelle.

Introduction

L'objet de ce travail est la variété quaternionique de Stiefel **H**S(n,k), où $0 \leq$ k $\leq$ n, constituée des applications linéaires isométriques de l'espace $\mathbf{H}^k$ dans $\mathbf{H}^n$ ou, ce qui revient au même, des k- repères orthonormés de $\mathbf{H}^n$; c'est-à-dire il s'agit de l'ensemble des $(n \times k)$-matrices (v_{ij}) telles que:

$$\sum_{s=1}^{n} v_{si}v_{sj} = \delta_{ij} \quad (1 \leq i \leq j \leq k)$$

Ces equations déterminent **H**S(n,k) comme un sous-espace de $\mathbf{H}^{nk} = \mathbf{R}^{4nk}$ et on obtient une variété sans bord de dimension $4nk - (2k^2 - k)$.

En 1971, A.ROUX a déterminé dans [11] les groupes de K- théorie complexe des variétés quaternioniques de Stiefel en termes de représentations des groupes symplectiques; l'outil principal utilisé a été une suite spectrale de L. HODGKIN.

Dans sa thèse [5], Raphael FREITAS s'intéresse à la K- théorie réelle des groupes symplectiques $Sp(n)$ et des variétés quaternioniques de Stiefel en utilisant notamment un travail de R.M. SEYMOUR [12] sur les groupes de Lie et leurs espaces homogènes dans le cadre plus général de la KR -théorie introduite

Y. Khakimdjanov et al (eds.), Algebra and Operator Theory, 39-48.

par M. ATIYAH dans [1]. Nous avons déterminé de plus la structure de $\mathbf{Z}_8$-graduée de $KO^*(Sp(n))$ dans [4]; le but de ce travail est de prouver que les résultats obtenus pour $Sp(n)$ valent encore pour les variétés quaternioniques de Stiefel.

Le plan est le suivant: une première partie contient les accessoires de K-théorie topologique, dans la deuxième on étudie la K-théorie complexe et ensuite la KO- théorie des variétés quaternioniques de Stiefel, la troisième partie établit les isomorphismes qui déterminent la structure $\mathbf{Z}_8$-graduée de $KO^*(\mathbf{H}S(n,k))$.

Partie 1

1. Dans toute la suite, le K-foncteur est définit sur la catégorie des CW-complexes finis à valeurs dans la catégorie des groupes abéliens; on rappelle que la K-théorie réelle est notée par KO, la K théorie complexe par KU(en notation abrégée K) et par KH on désigne la K-théorie quaternionique.

Soit X un CW-complexe fini et pointé dont le point base est x_0 ; la notation $\tilde{K}(X)$ désigne le noyau de l'homomorphisme:

$$K(X) \longrightarrow K(\{x_0\})$$

induit par l'inclusion $\{x_0\} \subset X$. On dit que $\tilde{K}$ est la K-théorie complexe réduite (respectivement $\tilde{K}$O la K-théorie réelle réduite) et on a la décomposition:

$$K(X) = \tilde{K}(X) \oplus K(\{x_0\}).$$

Les théories KO et KH sont jointes de la K- théorie complexe par les morphismes de BOTT (voir [3] et [10]):

$$K(X) \xrightarrow{r} KO(X) \xrightarrow{c} K(X)$$

$$K(X) \xrightarrow{q} KH(X) \xrightarrow{h} K(X)$$

qui vérifient les relations:

$$(r \circ c)(\alpha) = 2\alpha \quad \text{pour tout } \alpha \in KO(X)$$

$$(q \circ h)(\beta) = 2\beta \quad \text{si } \beta \in KH(X)$$

$$(c \circ r)(y) = (h \circ q)(y) = y + y* \quad \text{pour } y \in K(X)$$

où $* : K(X) \to K(X)$ désigne le morphisme induit par la conjugaison: à tout fibré vectoriel E sur X correspond le fibré conjugué $\bar{E}$.

Rappelons que si $X = S^{4n}$, le morphisme de complexification c (respectivement le morphisme de réelification r) est un automorphisme de $\mathbf{Z}$ pour n nombre pair (respectivement n impair) et la multiplication par 2 pour n nombre impair (respectivement pair).

La structure aditive de la K-théorie réduite de S^0 est décrite dans le tableau suivant:

i	0	1	2	3	4	5	6	7
$\tilde{K}U^{-i}$	**Z**	0	**Z**	0	**Z**	0	**Z**	0
$\tilde{K}O^{-i}$	**Z**	$\mathbf{Z}_2$	$\mathbf{Z}_2$	0	**Z**	0	0	0
$\tilde{K}H^{-i}$	**Z**	0	0	0	**Z**	$\mathbf{Z}_2$	$\mathbf{Z}_2$	0

puisqu'on a l'isomorphisme: $KO^i(X) = KH^{i+4}(X)$ et tenant compte du théorème de périodicité de BOTT (voir [3]).

En ce qui concerne la K- théorie complexe on a: $\tilde{K}^{-2}(S^0) = K(S^2) \cong \mathbf{Z}$ et le générateur est la classe stable du fibré en droites de Hopf sur S^2 : $[\eta] - 1$ qui vérifie la relation $([\eta]\text{-}1)^2 = 0$; l'isomorphisme de périodicité:

$$\beta : \tilde{K}^n(X) \longrightarrow \tilde{K}^{n+2}(X)$$

est défini par la multiplication par $[\eta] - 1$ (voir [2]).

L'anneau $\tilde{K}O^*(S^0)$ est une algèbre $\mathbf{Z}_8$ -graduée sur $\tilde{K}O^0(S^0) = KO^0(+) \cong \mathbf{Z}$ et il est engendré par 4 éléments: 1, e_1, e_4 et e_8 - avec $e_i \in \tilde{K}O^{-i}(S^0)$ - qui vérifient les relations:

$$2e_1 = e_1^3 = e_1e_4 = 0, \quad e_4^2 = 4e_8.$$

La notation $+$ désigne un espace réduit à un point.

La multiplication par e_8 induit l'isomorphisme de périodicité $\tilde{K}O^{-i}(X) \longrightarrow \tilde{K}O^{-i-8}(X)$ (voir [10]).

Pour tout CW-complexe fini X, il existe une suite exacte longue due a BOTT:

$$\cdots \longrightarrow KO^n(X) \xrightarrow{c} K^n(x) \xrightarrow{r\circ\beta^{-1}} KO^{n+2}(x) \xrightarrow{d} KO^{n+1} \longrightarrow \cdots$$

où c est le morphisme de complexification, r celui de réelification, β est l'isomorphisme de périodicité complexe et d une certaine opération de cohomologie (voir [3]).

En utilisant la relation $r \circ c = 2$ on démontre dans [4]:

Proposition 1.1.1 *Soit X un CW-complexe fini pour lequel $\tilde{K}^*(X)$ est sans torsion; alors le sous-groupe de torsion de $\tilde{K}O^*(X)$ noté par* $\mathrm{Tors}(\tilde{K}O^*(X))$ *est isomorphe à $(Z_2)^k$, avec $k \geq 0$.*

2. Soit X un CW-complexe fini; on munit l'anneau $\tilde{K}F(X)$ (où F est U ou O) de la filtration naturelle:

$$F_p(\tilde{K}F^n(X)) = \ker\ (\tilde{K}F^n(X) \xrightarrow{i!_{p-1}} \tilde{K}F^n(X_{p-1})$$

ou $i!_{p-1}$ désigne l'homomorphisme induit par l'inclusion de (p-1)-ième squelette X_{p-1} dans X. En utilisant cette filtration, à $\tilde{K}F^*(X)$ est associée une suite spectrale - dite d'*Atiyah-Hirzebruch* -dont le terme E_2 et de la forme:

$$E_2^{p,q}(X) = H^p(X; \tilde{K}F^q S^0)) \Rightarrow \tilde{K}F^{p+q}(X)$$

et le terme E_∞ est défini par:

$$E_\infty^{p,q}(X) = \frac{F_p(\tilde{K}F^{p+q}(X))}{F_{p+1}(\tilde{K}F^{p+q}(X))}.$$

Dans le cas de la K- théorie complexe, la différentielle d_3 de la suite spectrale d'Atiyah-Hirzebruch est donnée par le carré de Steenrod Sq^3; en KO -théorie, la différentielle d_2 est donnée par les formules suivantes:

$$d_2^{p,8k-1} = Sq^2 : E_2^{p,8k-1}(X) = H^p(X; \mathbf{Z}_2) \longrightarrow E_2^{p+2,8k-2}(X) = H^{p+2}(X; \mathbf{Z}_2)$$

$$d_2^{p,8k} = Sq^2 \circ \rho : E_2^{p,8k}(X) = H^p(X; \mathbf{Z}) \longrightarrow E_2^{p+2,8k-1}(X) = H^{p+2}(X; \mathbf{Z}_2)$$

où Sq^2 désigne le carré de Steenrod et $\rho : H^*(X; \mathbf{Z}) \to H^*(X; \mathbf{Z}_2)$ est l'homomorphisme de restriction induit par la projection canonique $\mathbf{Z} \to \mathbf{Z}_2$.

Si le CW-complexe fini X est sans torsion (c'est-à-dire $H^*(X; \mathbf{Z})$ est sans torsion) alors la suite spectrale d'Atiyah-Hirzebruch en K- théorie complexe est dégénérée - c.a.d. les différentielles d_s sont nulles pour tout $s \geq 2$ (voir [2]). Dans le cas de la KO- théorie, l'auteur a prouvé dans [4] le résultat suivant:

Proposition 1.2.1. *Soit X un CW-complexe fini sans torsion; alors dans la suite spectrale d'Atiyah-Hirzebruch associée a $KO^*(X)$ on a:*

a) $d_s^{p,q} = 0$ *pour tout p et tout* $s \geq 2$ *dès que* $q \equiv -2, -4$ (*modulo* 8)

b) $d_s^{p-q,q+s-1} = 0$ *pour tout p et tout* $s \geq 2$ *dès que* $q \equiv 0, 4$ (*modulo* 8)

3. P. HILTON donne dans [6] le résultat général suivant:

Théorème 1.3.1. *Soient X un CW-complexe fini et h une théorie arbitraire de cohomologie; alors $h \otimes \mathbf{Q}$ est aussi une théorie de cohomologie et pour tout q on a les isomorphismes*:

$$h^q(X) \otimes \mathbf{Q} \cong \bigoplus_{n+r=q} [H^r(X; Q) \otimes \check{h}^n]$$

où la notation $\check{h}^n$ désigne $h^n(S^0)$ - c'est a dire le n-ième groupe de coefficients de la théorie de cohomologie h (mentionnons que toutes les tensorisations sont sur **Z**).

En utilisant le théorème de Hilton 1.3.1 on obtient pour la K-théorie réelle (c.a.d. pour h = KO):

Proposition 1.3.2. *Si X est un CW -complexe fini, alors on a pour tout n:*

$$\tilde{K}O^{-n}(X) \otimes \mathbf{Q} \cong= \bigoplus_{p} [H^p(X; \mathbf{Z}) \otimes \mathbf{Q}]$$

avec $p + n \equiv 0,\ 4$ (*modulo* 8).

Demonstration (voir aussi [4]). Vu la proposition 1.2.1, l'homomorphisme de complexification $c : \tilde{K}O^n(S^0) \to \tilde{K}^n(S^0)$ est un isomorphisme pour $n \equiv 0$ (modulo8) et il est un monomorphisme pour $n \equiv 4$ (modulo 8).

Rappelons aussi que l'on a l'isomorphisme:

$$ch : K^*(X) \otimes \mathbf{Q} \to H^*(X; \mathbf{Q})$$

où *ch* désigne le caractère de Chern (voir [2]). Tenant compte du fait que:

$$\text{rang}_Z \tilde{K}O^{-n}(X) = \text{rang}_Z Gr(\tilde{K}O^{-n}(X))$$

le résultat cherché en résulte; la notation $\text{Gr}(\tilde{K}O^{-n}(X)$ désigne le groupe gradué associé:

$$\text{Gr}(\tilde{K}O^{-n}(X) = \bigoplus_{\text{p}} E_\infty^{p,-n-p}(X)$$

Remarque 1.3.3. On peut énoncer le théorème 1.3.1. sous la forme suivante:

$$E_\infty(X) \otimes \mathbf{Q} \cong E_2(X) \otimes \mathbf{Q}$$

où E_2 et E_∞ désignent les termes de la suite spectrale généralisée, associée à $h(X)$. Il en résulte que pour tout $\text{s} \geq 2$ les différentielles d_s prennent leurs valeurs dans le sous-groupe de torsion du terme E_s; on obtient pour tout $\text{s} \geq 2$ les epimorphismes canoniques:

$$H^*(\text{Tors}E_s) \to \text{Tors}(H^*E_s \cong E_{s+1})$$

Partie 2

Le but est la démonstration d'un résultat connu de LAZAROV (voir [9]): la **Z**- algèbre $\mathbf{Z}_8$ -graduée $KO^*(\mathbf{H}S(n,k))$ est aditivement isomorphe à une KO*(+) -algèbre extérieure sur k générateurs.

4. Nous allons utiliser un travail de R.M. SEYMOUR sur la KR -théorie des groupes de Lie; cette théorie a été introduite par Atiyah dans [1] et elle considère les fibrés complexes à l'involutions. Un tel fibré est un fibré vectoriel complexe (E, π, X) muni des involutions J_E et J_X sur E et respectivement sur X, telles que:

a) $\pi \circ J_E = J_X \circ \pi$

b) la restriction $(J_E)_X : E_x \to E_{J_X(X)}$ à la fibre E_x, au dessus du point $x \in X$, est semi- linéaire.

Proposition 2.4.1. *Si l'involution de la base X est l'identité alors on a l'isomorphisme* $KR^*(X) \cong KO^*(X)$.

Preuve. Vu l'hypothèse J_E est une involution semi- linéaire sur chaque fibre E_x; on en déduit que J_E est une involution de l'espace vectoriel réel E_x^R obtenu par restriction de scalaires de **C** à **R** Considérons: $F_x = \{u \in E_x^R \mid J_E(u) = u\}$ (c.a.d.: F_x est le sous-espace propre correspondant à la valeur propre +1 de J_E) et soit $F \xrightarrow{\pi} X$ le fibré vectoriel réel ainsi défini. Puisque (E, π, X) est le complexifié de (F, π, X) il résulte que cette association définit l'isomorphisme en question. La proposition 2.4.1. est démontrée. Dans ce qui suit nous nous placerons sous l'hypothèse de la dernière proposition; rappelons maintenant un théorème de structure (voir [12]) bien adapté à la K-théorie réelle.

Théorème 2.4.2 *Supposons que* Tors $(K^*(X)) = 0$ *et que la conjugaison décompose le groupe abélien* $(K^*(X)$ *sous la forme:*

$$(K^*(X) = M_+ \oplus T \oplus T^*$$

de sorte que soit la multiplication par +1 (*respectivement* -1) *sur* M_+(*respectivement sur* M_-) *et d'autre part échange* T et T^*; *en outre soient* $h_1, ..., h_n \in KO^*(X)$ *tels que les* $c(h_i)$ *forment une base pour* $K^*(+) \otimes (M_+ \oplus M_-)$ *en tant que* $K^*(+)$ *-module. Alors on a:*

$$KO^*(X) = E \oplus r(K^*(+) \otimes T)$$

où E *désigne le* $KO^*(+)$ *-module engendré par les éléments* $h_1, \ldots h_n$.

Corollaire 2.4.3. Sous les hypothèses du théorème 2.4.2., si $c(h_1), \ldots, c(h_n)$ forment une base pour $K^*(+) \otimes K^*(X)$, alors en tant que $KO^*(+)$ -module $KO^*(X)$ est engendré par $h_1, \ldots h_n$.

Définition 2.4.4 *Une représentation linéaire complexe d'un groupe de Lie* G *est dite de type quaternionique si elle est obtenue par restriction de scalaires (de* **H** *à* **C**) *dans une représentation quaternionique de* G.

Une représentation complexe de G *est dite de type réel si elle est la complexifiée d'une représentation réelle de* G.

Lemme 2.4.5. *Toute représentation de type quaternionique et de dimension finie d'un groupe de Lie compact* G, *est auto - conjuguée.*

Une démonstration du lemme 2.4.5. est donnée dans [5].

Considérons à présent l'anneau des représentations du groupe symplectique $Sp(n)$ et soit ρ_n la classe de la représentation où toute matrice $M \in Sp(n)$ opère dans $\mathbf{C}^{2n}$ comme la multiplication à gauche par M; désignons par $\alpha_j = \lambda^j \rho_n$ j -ième puissance extérieure, avec $j = 1, ..., n$.

Proposition 2.4.6. *La conjugaison sur* $RSp(n)$ *est l'identité.*

Démonstration. Vu le lemme 2.4.5. il suffit de montrer que ρ_n est une représentation complexe de type quaternionique; le résultat s'ensuit immédiatement du fait que les α_i avec $i = 1, ..., n$, engendrent l'anneau $RSp(n)$ (voir [8], page 193) et de plus la conjugaison est un homomorphisme d'anneaux.

Soient E un espace vectoriel de dimension complexe paire et $\rho_n : Sp(n) \longrightarrow \mathrm{Aut}_C(E)$; posons $\dim_C E = 2n$ et on considère $J : E \to E$

une anti-involution semi- linéaire définie par: $J(e_i) = e_{n+i}$ pour $1 \leq i \leq n$ et $J(e_i) = -e_{i-n}$ si $n \leq i \leq 2n$, où les e_i (avec $i = 1, ..., 2n$) forment la base de E.

En utilisant J on définit sur E une structure d'espace vectoriel quaternionique à gauche:

$$(\lambda + \mu j)z = \lambda z + \mu Jz \qquad \text{avec } \lambda + \mu j \in \mathbf{H} \text{ et } z \in E$$

Ensuite soit $H' : E \times E \longrightarrow \mathbf{C}$ la forme antihermitienne définie par:

$$H'(x, y) = H(J(x), y) \qquad x, y \in E$$

où H désigne la forme hermitienne usuelle sur l'espace de la représentation $E \cong \mathbf{C}^{2n}$ de ρ_n. Tenant compte de la définition de ρ_n il résulte que toute $M \in Sp(n) \subset \mathsf{U}(2n)$ est une isométrie de H', c'est-à-dire: $MJ = JM$; on en déduit que la multiplication par M définit un $\mathbf{H}$-automorphisme de E (muni de sa structure quaternionique). La démonstration de la proposition 2.4.6. est ainsi achevée.

5. Afin que rappeler le résultat de Roux sur la K -théorie complexe des variétés quaternioniques de Stiefel, notons par $A_{Sp(n)}$ l'idéal d'augmentation de l'application:

$$RSp(n) \to R\{1\} \cong \mathbf{Z}$$

induit par l'inclusion,de sorte que: $RSp(n) = A_{Sp(n)} \oplus \mathbf{Z}$ et P va designer le noyau de l'homomorphisme:

$$RSp(n) \to RSp(n-k)$$

induit par l'inclusion de $RSp(n-k)$ dans $RSp(n)$; alors les représentations virtuelles $\pi_i = \pi_i(\rho_n)$ de $A_{Sp(n)}$ avec $0 \leq i \leq n$ sont définies telles que $RSp(n)=\mathbf{Z}[\lambda^1(\rho_n), \cdots, \lambda^n(\rho_n)]$ s'identifie comme $\mathbf{Z}$- algèbre suplimentée à la $\mathbf{Z}$ algèbre $\mathbf{Z}[\pi_1, \cdots, \pi_n]$ et de plus les éléments $\pi_i(\rho_n)$ avec $n = k+1 \leq i \leq n$ appartiennent à P (voir [11]). Nous pouvons énoncer le:

Théorème 2.5.1. (ROUX) *En tant qu'algèbre* $\mathbf{Z}_2$*-graduée,* $K^*(\mathbf{H}S(n,k))$ *est isomorphe à l'algèbre extérieure* $\Lambda_Z[\pi_{n-k+1}, \cdots, \pi_n]$*, où chaque* π_i *est de degré* -1.

Roux a également montré que les classes de K-théorie respectivement associées aux π_i par l'isomorphisme de ci-dessus provient de l'image de β (voir 1.1). De plus chaque π_i est de la forme $\tilde{\pi}_i - \deg\tilde{\pi}_i$, où $\tilde{\pi}_i$ est une représentation irréductible et deg $\tilde{\pi}_i$ son degré; comme l'isomorphisme de périodicité complexe β est additive et $\beta(n) = 0$ pour tout $n \in \mathbf{Z}$ (voir [7], page 8) il s'ensuit que:

$$K^*(\mathbf{H}S(n,k)) \cong \Lambda_Z[\beta(\tilde{\pi}_{n-k+1}), \cdots, \beta(\tilde{\pi}_n)]$$

en tant que groupes abéliens. En utilisant maintenant la proposition 2.4.6. il résulte que les $\tilde{\pi}_i$ (avec $i = n-k+1, ..., n$) et par suite les $(\tilde{\pi}_i)$ sont auto- conjuguées.On en déduit que la conjugaison sur $K^*(\mathbf{H}S(n,k))$ est aussi l'identité.

Remarquons à cette fin que les représentations irréductibles $\tilde{\pi}_i$ sont ou bien de type réel ou bien de type quaternionique; la preuve se repose sur le fait qu'ils sont auto-conjuguées (voir [12], proposition 5.1.).

6. Désignons par t l'isomorphisme $KH^{i+4}(X) \cong KO^i(X)$ et par $h : KH(X) \to K(X)$ la complexification naturelle; nous avons la relation:

$$\mu_2^2 h(y) = (c \circ t)(y) \quad \text{pour tout } y \in KH(X)$$

(voir [12] lemme 5.2.), où μ_2 est un élément inversible de la $\mathbf{Z}$ - algèbre $K^*(+)$, plus précisément μ_2 remplit la condition $\mu_2^4 = 1$.

On identifie tout $x \in K^*(X)$ à l'élément $1 \otimes x$ de $K^*(+) \otimes K^*(X)$ et alors la relation de ci-dessus devient:

$$(6.1) \quad 1 \otimes h(y) = \mu_2^2(1 \otimes (c \circ t)(y)) \quad \text{pour tout } y \in KH^{-1}(X)$$

en utilisant que $\mu_2^4 = 1$.

Pour déterminer la structure additive du groupe $KO^*(\mathbf{H}S(n,k))$ nous allons montrer que les éléments $\beta(\tilde{\pi}_i)$, avec $n-k+1 \leq i \leq n$, sont des complexifications de classes réelles. Vu la remarque faite à la fin du paragraphe précèdent, soit $\tilde{\pi}_i$ est la complexification d'une représentation réelle de $Sp(n)$ et on a $\beta(\tilde{\pi}_i) = c(x_i)$ avec $x_i \in KO^{-1}(\mathbf{H}S(n,k))$, soit $\tilde{\pi}_i$ est de type quaternionique et on peut écrire: $\beta(\tilde{\pi}_i) = h(h_i)$ avec $y_i \in KH^{-1}(\mathbf{H}S(n,k))$ pour un certain $i = n-k+1, \cdots, n$.

Alors la formule (6.1.) prend la forme:

$$1 \otimes \beta(\tilde{\pi}_i) = \mu_2^2(1 \otimes (c \circ t)(y)).$$

En notant par π_i^r les$\tilde{\pi}_i$ qui sont de type réel et par π_i^h les $\tilde{\pi}_i$ qui sont de type quaternionique, nous obtenons:

$$\begin{aligned} & 1 \otimes \beta(\pi_{i_1}^r) \cdots \beta(\pi_{i_p}^r)\beta(\pi_{i_{p+1}}^h) \cdots \beta(\pi_{i_{p+q}}^h) = \\ (6.2) \quad & = (\mu_2^2)^{q-p}(1 \otimes c(x_{i_1}) \cdots c(x_{i_p})(c \circ t)(y_{i_{p+1}}) \cdots (c \circ t)(y_{i_q}) = \\ & = (\mu_2^2)^{q-p}(1 \otimes c(x_{i_1} \cdots x_{i_p} t(y_{i_{p+1}}) \cdots t(y_{i_q}))) \end{aligned}$$

avec $n-k+1 \leq i_1 < \cdots < i_p < \cdots < i_q \leq n$, $p < q$ et $1 \leq p < q \leq n$.

Vu le théorème 2.5.1. les monômes du membre gauche de la formule (6.2.) forment une base du $K^*(+)$ -module $K^*(+) \otimes K^*(\mathbf{H}S(n,k))$, base qui est constituée par des complexifiées de classes réelles. En conséquence on peut appliquer le corollaire 2.4.3. et nous avons donc démontré le:

Théorème 2.6.2. *Le* $\mathbf{Z}$- *algèbre* $\mathbf{Z}_8$ *-graduée* $KO^*(\mathbf{H}S(n,k))$ *est isomophe, en tant que groupe abélien, à l'algèbre extérieure sur k générateurs* $\Lambda_{KO^*(+)}(x_i, t(y_j))$.

Partie 3

7. En nous appuyant sur la structure $\mathbf{Z}_8$ -graduée de $KO^*(\mathbf{H}S(n,k))$, nous sommes maintenant en mesure d'établir le résultat principal de ce travail:

Théorème 3.7.1. *La suite spectrale d'Atiyah- Hirzebruch associée à $KO^*(\mathbf{H}S(n,k))$ est dégénérée et on a les isomorphismes de groupes abéliens:*

$$KO^q(\mathbf{H}S(n,k)) \cong E_2^q(\mathbf{H}S(n,k)) \cong E_\infty^q(\mathbf{H}S(n,k))$$

pour tout q=0,1,...,7.

Démonstration. Puisque $(\mathbf{H}S(n,k)$ est un CW-complexe fini sans torsion (sa cohomologie entière étant une algèbre extérieure engendrée par k éléments) nous pouvons appliquer les résultats énoncés dans 1.2. et 1.3. On obtient:

$$\text{Tors } E_2(\mathbf{H}S(n,k)) \cong [\text{Tors } KO^*(S^0)]^{2^k} = (\mathbf{Z}_2)^{2^{k+1}}$$

parce que Tors $KO^*(S^0) \cong (\mathbf{Z}_2)^2$ -voir le tableau de 1.1.

D'autre part, la remarque 1.3.3. nous donne:

$$\text{Tors } E_\infty(\mathbf{H}S(n,k)) \cong (\mathbf{Z}_2)^r \quad \text{avec } r \leq 2^{k+1}.$$

Vu le théorème de Roux 2.5.1. et la proposition 1.2.1., la torsion de la K -théorie réelle de $\mathbf{H}S(n,k)$ est 2 - torsion; enfin le théorème 2.6.2. nous permet d'écrire:

$$\text{Tors } KO^*(\mathbf{H}S(n,k)) \cong (\mathbf{Z}_2)^{2^{k+1}}.$$

En utilisant maintenant le monomorphisme canonique:

$$\text{Gr (Tors } KO^*(\mathbf{H}S(n,k))) \rightarrow \text{Tors Gr } (KO^*(\mathbf{H}S(n,k)))$$

nous en obtenons les inégalités:

$$(7.1) \qquad 2^{k+1} \leq r \leq 2^{k+1}$$

d'où il vient que Tors $E_2(\mathbf{H}S(n,k))$ =Tors $E_\infty(\mathbf{H}S(n,k))$; tenant compte de la torsion de $KO^*(S^0)$, la dernière égalité signifie que les différentielles $d_s^{p-s,q+s-1}$ et $d_s^{p,q}$ sont nulles pour tout p et tout $s \geq 2$ dès que $q \equiv -1,-2$ (modulo 8). En combinant ce résultat avec la proposition 1.2.2. on obtient l'isomorphisme:

$$E_2(\mathbf{H}S(n,k)) \cong E_\infty(\mathbf{H}S(n,k))$$

c'est -à-dire la première partie du théorème a été prouvée.

Les inégalités (7.1) nous donnent aussi que

$$\text{Tors} E_\infty(\mathbf{H}S(n,k)) = \text{Tors} KO^*(\mathbf{H}S(n,k))$$

et en utilisant de nouveau les monomorphismes canoniques:

$$\text{Gr (Tors } KO^q(\mathbf{H}S(n,k))) \rightarrow \text{Tors } E_\infty^q(\mathbf{H}S(n,k)))$$

pour $q = 0, 1, \ldots.., 7$, on en déduit que $KO^q(\mathbf{H}S(n,k))$, $E_2^q(\mathbf{H}S(n,k))$ et $E_\infty^q(\mathbf{H}S(n,k))$ ont la même torsion pour tout q. Or d'après la proposition 1.3.2. $KO^q(\mathbf{H}S(n,k))$, et $E_2^q(\mathbf{H}S(n,k))$ ont aussi les parties libres isomorphes pour $q = 0, 1, \cdots, 7$; enfin vu un problème d'extension de groupes on a:

$$\text{rang}_Z\ E_\infty^q(\mathbf{H}S(n,k)) = \text{rang}_Z\ KO^q(\mathbf{H}S(n,k)) \quad \text{pour tout } q.$$

Ainsi le théorème 3.7.1. a été démontré.

References

[1] ATIYAH M. F. K- theory and Reality, *Quart. J . Math. Oxford* (2), 17, (1966), 367-386.

[2] ATIYAH M. F. , HIRZEBRUCH F. Vector bundles and homogeneous spaces, *Proc. Symp. Pure Math.* 3, A M S Proc., (1961), 197-221.

[3] BOTT R. *Lectures on K (X)* Benjamin, New-York, 1969.

[4] COSTINESCU C .N. Aplicatii in KO -teorie, *St. Cerc. Mat.*, 39, 3, (1987), 234-239.

[5] FREITAS R. K théorie réelle des variétés de Stiefel sans torsion ; Thèse, U.S.T. de Lille 1, 1985.

[6] HILTON P. General cohomology theory and K -theory, *London Math. Soc.*, Lecture Notes 1, Cambridge (1971).

[7] HODGKIN L. K -theory of Lie groups, *Topology* 6 (1967), 1-36.

[8] HUSMOLLER L. *Fibre bundles*, Mc Graw-Hill, New - York, 1966.

[9] LAZAROV C. Secondary characteristic classes in K -theory, *Trans. Amer. Math. Soc.*, 136 (1968), 36-59.

[10] MAHAMMED N., PICCININI R. , SUTER U. *Some applications of topological K -theory*, North H olland Math. Studies 45, 1980.

[11] ROUX A. Application de la suite spectrale de Hodgkin au calcul de la K -théorie des variétés de Stiefel, *Bull. Soc. Math. France* 99 (1971), 345-368.

[12] SEYMOUR R.M. Real K -theory of Lie groups and homogeneous spaces *Quart. J. Math. Oxford* (2), 24 (1973).

SOME NILPOTENT LIE ALGEBRAS AND ITS APPLICATIONS

M.GOZE, Yu.KHAKIMDJANOV

M.G. et Yu.K. : Faculté des Sciences et Techniques,
Université de Haute Alsace,
à Mulhouse: 4, rue des Frères Lumière, 68093 Mulhouse cedex, France
à Colmar: 32, rue du Grillenbreit, 68000 Colmar, France
et I.R.M.A.(Unité de recherche associée au CNRS 001),
7 rue René Descartes, 67084 Strasbourg Cedex, France
Yu.K.: Institut of Mathematics, Uzbekistan Academy of Sciences
F.Hodjaev str. 29, Tashkent, 700143, Uzbekistan.

Abstract. We examine the problem of describing filiform Lie algebras over the field $\mathbb{C}$ admitting a semisimple derivation. As applications of obtained results we give a description of Lie algebras with a given filiform nilradical.

Introduction

The class of nilpotent Lie algebras is the union of two importent subclasses: subclass of characteristically nilpotent Lie algebras and subclass of nilpotent Lie algebras admitting a semisimple derivation. These subclasses have been studied in the papers [5], [6], [7], [8], [9]. We summarize some of these results and we give its applications. In what follows we suppose that all Lie algebras are finite-dimensional (with the dimension > 6 and defined over the field of complex numbers $\mathbb{C}$).

1 Filiform Lie algebras

In this section we resume certain results about filiform Lie algebras obtained in [7] and [8].

Y. Khakimdjanov et al (eds.), Algebra and Operator Theory, 49-64.

Let g be a nilpotent Lie algebra of dimension n . Let

$$\mathcal{C}^0 g \supset \mathcal{C}^1 g \supset \ldots \supset \mathcal{C}^{n-2} g \supset \mathcal{C}^{n-1} g = 0,$$

be the central descending sequence of g, where $\mathcal{C}^0 g = 0, \mathcal{C}^i g = [g, \mathcal{C}^{i-1} g]$, $1 \leq i \leq n-1$.

The Lie algebra g is called *filiform* if $\dim \mathcal{C}^k g = n-k-1$ for $1 \leq k \leq n-1$. We remark that the filiform Lie algebras have the maximal possible nilindex, that is $n-1$. These algebras are the "less" nilpotent.

Examples

1. Let L_n be the $(n+1)$-dimensional Lie algebra defined by

$$[X_0, X_i] = X_{i+1},\ i = 1, ..., n-1,$$

where $(X_0, X_1, \ldots, X_n)$ is a basis of L_n (the undefined brackets being zero excepted the brackets obtained by antisymmetry). This is, in a certain manner, the simplest filiform Lie algebra.

2. Let $n = 2k+1$ be an odd integer and let Q_n be the $(n+1)$-dimensional Lie algebra defined in the basis $(X_0, X_1, \ldots, X_n)$ by

$$[X_0, X_i] = X_{i+1}, \quad i = 1, \ldots, n-1;$$

$$[X_i, X_{n-i}] = (-1)^i X_n, \quad i = 1, \ldots, k.$$

This is a filiform Lie algebra. In the basis $(Z_0, Z_1, \ldots, Z_n)$, where $Z_0 = X_0 + X_1$, $Z_i = X_i$, $i =, \ldots, n$, this Lie algebra is defined by

$$[Z_0, Z_i] = Z_{i+1}, \quad i = 1, \ldots, n-2;$$

$$[Z_i, Z_{n-i}] = (-1)^i Z_n, \quad i = 1, \ldots, k.$$

Consider the algebraic Lie algebra Der$g$ of all derivations of g. Clearly, the central descending sequence of g is a flag that is invariant under all derivations. As we have supposed that $dim g \geq 6$, it follows that Der$g$ is solvable.

Definition 1 *A maximal abelian subalgebra of the derivations algebra Der$g$ constituted of the semisimple derivations is called maximal torus of derivations of g.*

All maximal tori of derivations of g are mutually conjugate in Der$g$ [14].

Definition 2 *The dimension of a maximal torus of derivation of g is called the rank of g.*

For a filiform Lie algebra g the rank of g is ≤ 2.

1.1 Filiform Lie algebra of rank 2

We have only two types of filiform Lie algebras of rank 2 : L_n and Q_n (this case is possible if $n = 2k + 1$ is odd).

Case $g = L_n$

Let $(X_0, X_1, \ldots, X_n)$ be a basis of L_n introduced in Example 1. We consider the endomorphisms d_1, d_2, t, h_2, h_3, $\ldots, h_{n-1}$ defined by

$$d_1(X_0) = 0 \ , \ d_1(X_i) = X_i \ , 1 \leq i \leq n.$$
$$d_2(X_0) = X_0 \ , \ d_2(X_i) = (i-1)X_i \ , \quad 1 \leq i \leq n$$
$$t(X_0) = X_1 \ , \ t(X_i) = 0 \ , 1 \leq i \leq n.$$
$$h_j(X_0) = 0 \ , \ h_j(X_i) = X_{i+j} \ , \quad 2 \leq j \leq n-1, \quad 1 \leq i \leq n-j.$$

Lemma 1 *The endomorphisms* $(d_1, d_2, t, h_2, h_3, \ldots, h_{n-1}, adX_0, adX_1, \ldots, adX_{n-1}))$ *form a basis of the derivations algebra* $DerL_n$ *and a maximal torus of derivations of* L_n *is spanned by* d_1 *and* d_2.

Case $g = Q_n, \quad n = 2k+1.$

Let $(Z_0, Z_1, \ldots, Z_n)$ be a basis of Q_n introduced in Example 2. We consider the endomorphisms d_1, d_2, t, h_3, h_5, $\ldots, h_{2k-1}$ defined by

$$d_1(Z_0) = Z_0 \ , \ d_1(Z_i) = (i-1)Z_i \ , \quad 1 \leq i \leq n-1 \ , \quad d_1(Z_n) = (n-2)Z_n.$$
$$d_2(Z_0) = 0, d_2(Z_i) = Z_i, 1 \leq i \leq n-1, d_2(Z_n) = 2Z_n$$
$$t(Z_0) = Z_n \ , \ t(X_i) = 0 \ , 1 \leq i \leq n.$$
$$h_{2j+1}(Z_0) = 0 \ , \ h_{2j+1}(Z_i) = Z_{i+2j+1} \ , \quad 1 \leq j \leq k-1, \quad 1 \leq i \leq n-j.$$

Lemma 2 *The endomorphisms* $(d_1, d_2, t, h_3, h_5, \ldots, h_{2k-1}, adZ_0, adZ_1, \ldots, adZ_{n-1})$ *form a basis of the derivations algebra* $DerQ_n$ *and a maximal torus of derivations of* L_n *is spanned by* d_1 *and* d_2.

1.2 Filiform Lie algebras of rank 1.

Let g be a filiform Lie algebra of dimension $n+1$ and of rank 1 . There is a basis $(Y_0, Y_1, \ldots, Y_n)$ of g such that g is one of the following Lie algebras:

(*i*) $g = \mathcal{A}^r_{n+1}(\alpha_1, \ldots, \alpha_t), \quad 1 \leq r \leq n-3, \quad t = \left[\frac{n-r-1}{2}\right],$
$[Y_0, Y_i] = Y_{i+1}, \quad 1 \leq i \leq n-1,$
$[Y_i, Y_j] = \left(\sum_{k=i}^{t} \alpha_k (-1)^{k-i} C^{k-i}_{j-k-1}\right) Y_{i+j+r}, \ 1 \leq i < j \leq n, \ i+j+r \leq n.$

(*ii*) $g = \mathcal{B}^r_{n+1}(\alpha_1, \ldots, \alpha_t) \ , \ n = 2m+1, \ 1 \leq r \leq n-4, \ t = \left[\frac{n-r-2}{2}\right],$
$[Y_0, Y_i] = Y_{i+1}, \quad 1 \leq i \leq n-2,$

$$[Y_i, Y_{n-i}] = (-1)^i Y_n, \quad 1 \le i \le m,$$
$$[Y_i, Y_j] = \left(\sum_{k=i}^{t} \alpha_k (-1)^{k-i} C_{j-k-1}^{k-i}\right) Y_{i+j+r},$$
$$1 \le i < j \le n-1, \quad i+j+r \le n-1.$$

$$(iii) \quad \boldsymbol{g} = \mathcal{C}_{n+1}(\alpha_1, \ldots, \alpha_t) \ , \quad n = 2m+1 \ , \quad t = m-1,$$
$$[Y_0, Y_i] = Y_{i+1}, \quad 1 \le i \le n-2,$$
$$[Y_i, Y_{n-i}] = (-1)^i Y_n, \quad 1 \le i \le m,$$
$$[Y_i, Y_{n-i-2k}] = (-1)^i \alpha_k Y_n, \ 1 \le k \le m-1, \ 1 \le i \le n-2k-1,$$

where C_q^s are the binomial coefficients (we suppose that $C_q^s = 0$ if $q < 0$ or $q < s$), $(\alpha_1, \ldots, \alpha_t)$ are the parameters satisfying polynomial relations emanated from Jacobi's identity and at least one parameter $\alpha_i \neq 0$. A maximal torus of derivations is spanned by d , where:

Case (i) :

$$d(Y_0) = Y_0 \ , \ d(Y_i) = (i+r) Y_i \ , \quad 1 \le i \le n.$$

Case (ii) :

$$d(Y_0) = Y_0, \ d(Y_i) = (i+r) Y_i \ , \ 1 \le i \le n-1, \ d(Y_n) = (n+2r) Y_n \ .$$

Case (iii) :

$$d(Y_0) = 0 \ , \ d(Y_i) = Y_i \ , \quad 1 \le i \le n-1 \ , \quad d(Y_n) = 2Y_n.$$

Remark 1 Let $r = 1$. Then, up to isomorphism, there are only four Lie algebras of rank 1 [8] .
Remark 2 The laws $\mathcal{C}_{n+1}(\alpha_1, \ldots, \alpha_t)$ satisfy the Jacobi's identity for all values of parameters $(\alpha_1, \ldots, \alpha_t)$.
Remark 3 Let $\boldsymbol{g}$ be a Lie algebra belonging to one of the families (i), (ii), (iii) and at least one of parameters α_i be different to zero. Then we can transformate one of these parameters to 1 using the automorphism ψ defined by $\psi(X_0) = aX_0$, $\psi(X_1) = bX_1$ (this is a unique type of automorphisms conserving the torus and the property of basis to be adapted). Modulo this transformation we have a classification up to isomorphism of filiform Lie algebras of rank 1.

1.3 Filiform Lie algebras of rank 0

Lie algebras of this family are characteristically nilpotent filiform Lie algebras (a Lie algebra is called *characteristically nilpotent* if all of its derivations are nilpotent) [7]. Let $\boldsymbol{g}$ be a filiform Lie algebra of dimension $n+1 \ge 6$ and of rank 0. We can associate to $\boldsymbol{g}$ a graded Lie algebra, denoted gr$\boldsymbol{g}$, obtained from the natural filtration of nilpotent Lie algebra $\boldsymbol{g}$ (filtration by the ideals of the descending central sequence). Then gr$\boldsymbol{g}$ is isomorphic to L_n or Q_n [16]. If gr$\boldsymbol{g}$ is isomorpic to L_n then we can suppose that $\boldsymbol{g}$ is defined by the relations:

$$[X_0, X_i] = X_{i+1}\ , \quad 1 \leq i \leq n-1,$$

$$[X_i, X_j] = \sum_{k=1}^{n-i-j} a_{ij}^k X_{i+j+k}, \quad 1 \leq i < j \leq n-2.$$

If gr$\boldsymbol{g}$ is isomorphic to Q_n ($n = 2m+1$) then we can suppose that $\boldsymbol{g}$ is defined by the relations (1):

$$[X_0, X_i] = X_{i+1}\ , \quad 1 \leq i \leq n-1, \quad [X_i, X_{n-i}] = (-1)^i X_n\ , \quad 1 \leq i \leq m,$$

$$[X_i, X_j] = \sum_{k=1}^{n-i-j} a_{ij}^k X_{i+j+k}, \quad 1 \leq i < j \leq n-2,$$

or by the relations (2):

$$[Z_0, Z_i] = Z_{i+1}, \quad i = 1, \ldots, n-2; \quad [Z_i, Z_{n-i}] = (-1)^i Z_n, \quad i = 1, \ldots, m.$$

$$[Z_i, Z_j] = \sum_{k=1}^{n-i-j} b_{ij}^k Z_{i+j+k}, \quad 1 \leq i < j \leq n-2.$$

Definition 3 *Let $\boldsymbol{g}$ be a filiform Lie algebra such that gr$\boldsymbol{g}$ is isomorphic to L_n but not isomorphic to $\boldsymbol{g}$. The sill algebra of $\boldsymbol{g}$ is defined by*

$$[X_0, X_i] = X_{i+1}\ , \quad 1 \leq i \leq n-1, \quad [X_i, X_j] = a_{ij}^r X_{i+j+r}\ ,$$

where r is the smallest index, $r \geq 1$ such that $a_{ij}^r \neq 0$ for some (i, j).

Definition 4 *Let $\boldsymbol{g}$ be a filiform Lie algebra such that gr$\boldsymbol{g}$ is isomorphic to Q_n but not isomorphic to $\boldsymbol{g}$ and suppose $\boldsymbol{g}$ is defined by the relations (2). If $b_{ij}^{n-i-j} = 0$ for all $1 \leq i < j \leq n-2$, then the filiform Lie algebra defined by*

$$[Z_0, Z_i] = Z_{i+1}\ , \quad 1 \leq i \leq n-2,$$

$$[Z_i, Z_{n-i}] = (-1)^i Z_n\ , \quad 1 \leq i \leq m,$$

$$[Z_i, Z_j] = b_{ij}^r Z_{i+j+r}\ , \ 1 \leq i < j \leq n-2,$$

where r is smallest index, $r \geq 1$ with $b_{ij}^r \neq 0$ is called sill algebra of first kind of $\boldsymbol{g}$. If $b_{ij}^{n-i-j} \neq 0$ for some i and j, $1 \leq i < j \leq n-2$, then the filiform Lie algebra defined by

$$[Z_0, Z_i] = Z_{i+1}\ , \quad 1 \leq i \leq n-2,$$

$$[Z_i, Z_{n-i}] = (-1)^i Z_n\ , \quad 1 \leq i \leq m,$$

$$[Z_i, Z_j] = b_{ij}^{n-i-j} Z_n\ , \quad 1 \leq i < j \leq n-2$$

is called sill algebra of second kind of $\boldsymbol{g}$.

We note that if g is a filiform Lie algebra non-isomorphic to Q_n but with gr$g \cong Q_n$, there is only one sill algebra of g (of first or second kind) nonisomorphic to Q_n .

Theorem 1 [7]. *Let g be a filiform Lie algebra of dimension $n+1 \geq 7$ non-isomorphic to L_n and Q_n. Then g is characteristically nilpotent if and only if g is not isomorphic to its sill algebra .*

2 Description of Lie algebras whose nilradical is filiform

Let g be a Lie algebra. Any maximal semisimple subalgebra $s \subset g$ is called a Levi subalgebra of g .The semidirect decomposition $g = s \oplus r$ holds, where r is the radical in g (the Levi decomposition), and all Levi subalgebras are mutually conjugate (Mal'tsev's theorem [10]). These theorems suggest to consider the problem of classification of Lie algebras with a fixed radical [15]. In this paper we study a similar problem: Description of Lie algebras with a given nilradical.

2.1 Description of the decomposable Lie algebras whose nilradical is filiform

A Lie algebra g with nilradical n is called decomposable if $g = s \oplus n$ (semidirect sum) where s is a reductive Lie algebra, whose action on n is completely reducible. Let $g = s \oplus n$ be a non-nilpotent decomposable Lie algebra whose nilradical n is filiform. For any $a \in s$ the mapping ad$a$ carries n to itself, i.e., ad$a\,|_n$ is a derivation of n. Thus, we have a Lie algebra homomorphism $\tau : s \longrightarrow$ Der$n$. The structure of g is uniquely defined by s, n and τ , which can be defined arbitrarily. If the homomorphism τ is given, then we denote by $g = s \oplus_\tau n$ the corresponding semidirect sum. The brackets of this semidirect sum are defined by the following formula:

$$[(x,y),(z,t)] = ([x,z],[y,t] + \tau(x)t - \tau(z)y), \; x,z \in s, \;\; y,t \in n.$$

We shall often identify s and n with the corresponding subalgebras in $s \oplus n$.

Definition 5 *The semidirect sum $g = s \oplus_\tau n$ is called faithful if the homomorphism τ is a monomorphism.*

As s is reductive, then any semidirect sum $g = s \oplus_\tau n$ can be represented in the form of the direct sum of a faithful semidirect sum and some ideal of s. In what follows we suppose that $g = s \oplus n$ is a faithful semidirect sum. As the dimension of a maximal torus of n is ≤ 2 , we have dim $s \leq 2$.

Case 1. $s = 0$. This class of Lie algebras was described in Theorem 1.

Case 2. $\dim \boldsymbol{s} = 1$ **or** 2. From the classification of filiform Lie algebras admitting a torus of derivations (see 1.1 and 1.2) and from Mal'cev theorem [10], we have

Theorem 2 *Let $\boldsymbol{g}$ be a faithful decomposable Lie algebra with a filiform nilradical of codimension ≥ 1. Then $\boldsymbol{g}$ is isomorphic to the algebra $\boldsymbol{g} = T \oplus_{id} \boldsymbol{n}$ where $\boldsymbol{n}$ is one of the following: L_n, Q_n, $\mathcal{A}^r_{n+1}(\alpha_1, \ldots, \alpha_t)$, $\mathcal{B}^r_{n+1}(\alpha_1, \ldots, \alpha_t)$, $\mathcal{C}_{n+1}(\alpha_1, \ldots, \alpha_t)$ and T is a subalgebra of the maximal torus of derivations of $\boldsymbol{n}$ described before. The Lie algebras $\boldsymbol{g}_1 = T_1 \oplus_{id} \boldsymbol{n}$ $\boldsymbol{g}_2 = T_2 \oplus_{id} \boldsymbol{n}$ are isomorphic if and only if $T_1 = T_2$.*

2.2 Description of non-decomposable Lie algebras with a filiform nilradical

Let $\boldsymbol{g}$ be a non-decomposable Lie algebra whose nilradical $\boldsymbol{n}$ is filiform and let $\boldsymbol{g} = \boldsymbol{p} \oplus \boldsymbol{m}$ be a Levi decomposition, where $\boldsymbol{m}$ is the radical of $\boldsymbol{g}$ and $\boldsymbol{p}$ is a Levi subalgebra of $\boldsymbol{g}$. We can also suppose that this is a faithful semidirect sum. In this case Lie algebra $\boldsymbol{g}$ is also called faithful. We can write $\boldsymbol{m} = \boldsymbol{s} \oplus \boldsymbol{n}$ (vectorial direct sum) where $\boldsymbol{n}$ is the nilradical of $\boldsymbol{g}$ and $\boldsymbol{s}$ is a complementary subspace.

Lemma 3 *Lie algebra $\boldsymbol{g}$ is solvable.*

Proof. Since $\boldsymbol{p}$ is semi-simple, we may suppose that $[\boldsymbol{p}, \boldsymbol{s}] \in \boldsymbol{s}$. As $[\boldsymbol{p}, \boldsymbol{s}] \subset \boldsymbol{n}$, we have $[\boldsymbol{p}, \boldsymbol{s}] = 0$. It follows that, for a faithful $\boldsymbol{g}$, the adjoint action of $\boldsymbol{p}$ in $\boldsymbol{n}$ is also faithful. Since Der$\boldsymbol{n}$ is solvable, $\boldsymbol{p} = 0$. Thus, $\boldsymbol{g}$ is solvable, whenever it is faithful. □

Lemma 4 $\dim \boldsymbol{s} \leq rank(\boldsymbol{n}) \leq 2$

Proof. We consider the Chevalley decomposition of the derivations algebra of $\boldsymbol{n}$:

$$Der\boldsymbol{n} = \boldsymbol{t} \oplus \boldsymbol{b},$$

where $\boldsymbol{t}$ is a maximal torus and $\boldsymbol{m}$ the ideal of all nilpotent elements. For any $s \in \boldsymbol{s}$, consider the decomposition $ads = t + b$, where $t \in \boldsymbol{t}$ and $b \in \boldsymbol{b}$. Then the linear mapping $\boldsymbol{s} \longrightarrow \boldsymbol{t}$ defined by $s \longrightarrow t$ is injective. In fact, if $\boldsymbol{t} = 0$, then ads is nilpotent on $\boldsymbol{n}$. But $[s, \boldsymbol{s}] \subset \boldsymbol{n}$. It follows that s is a nilpotent element of $\boldsymbol{g}$. Hence $s = 0$. Thus $\dim \boldsymbol{s} \leq \dim \boldsymbol{t}$. □

Corollary 1 *If exists a non-decomposable faithful Lie algebra $\boldsymbol{g}$ with* $\dim \boldsymbol{s} = 2$, *then $\boldsymbol{n} = L_n$ or Q_n.*

Lemma 5 *Let $\boldsymbol{g}$ be a faithful Lie algebra (decomposable or not) with a filiform nilradical $\boldsymbol{n}$ and let* $\dim \boldsymbol{g} - \dim \boldsymbol{n} = 2$. *Then we may choose a complementary subspace $\boldsymbol{s}$ to $\boldsymbol{n}$ in $\boldsymbol{g}$ such that $ad\boldsymbol{s}$ is a maximal torus of derivations of $\boldsymbol{g}$.*

Proof We study the two following possible cases:

$$(1) \quad \mathfrak{g} = \mathfrak{s} \oplus L_n, \quad (2) \quad \mathfrak{g} = \mathfrak{s} \oplus Q_n.$$

Case (1) Let (s_1, s_2) be a basis of $\mathfrak{s}$. Consider the Jordan decomposition $ads_1 = c_1 + n_1$, where t_1 is an element of the maximal torus of derivations T of L_n described in section 1, n_1 is a nilpotent derivation with $c_1 n_1 = n_1 c_1$. We consider also the Chevalley decomposition $ads_2 = c_2 + n_2$ with $c_2 \in T$, and n_2 is a nilpotent derivation of L_n. We may suppose that n_1 and n_2 are two exterior derivations of L_n whose belong to the space spanned by $S - \{d_1, d_2\}$, where S is the basis of $\mathrm{Der} L_n$ described in the section 1. Two cases are possible:

$$(i) \quad c_1 = \mu d_1, \ n_1 = \textstyle\sum_{i=2}^{n-1} \beta_i h_i \ , \ (\beta_2, \ldots, \beta_{n-1}) \in \mathbb{C}^{n-2} - \{0\}$$

$$c_2 = \alpha_1 d_1 + \alpha_2 d_2, \ n_2 = \nu t + \textstyle\sum_{i=2}^{n-1} \gamma_i h_i \ , \ (\gamma_2, \ldots, \gamma_{n-1}) \in \mathbb{C}^{n-2} - \{0\}$$

where $\mu \neq 0, \ \alpha_2 \neq 0, \quad t, \ h_i$ are the nilpotent derivations of L_n described in section 1.

$$(ii) \quad c_1 = \mu(d_1 + d_2), \ n_1 = \beta t,$$

$$c_2 = \alpha_1 d_1 + \alpha_2 d_2, \ n_2 = \nu t + \textstyle\sum_{i=2}^{n-1} \gamma_i h_i \ , \ (\gamma_2, \ldots, \gamma_{n-1}) \in \mathbb{C}^{n-2} - \{0\},$$

where $\mu \neq 0, \ \alpha_1 \neq \alpha_2$. The endomorphism $f = [ads_1, ads_2] = ad[s_1, s_2]$ is an interior derivation. In the case (i) we have

$$f(X_0) = \mu\nu X_1 + \sum_{i \geq 2} \tau_i X_i,$$

$$f(X_p) = \alpha_2 \sum_{2 \leq i \leq n-1} \beta_i X_{i+p}, \ p = 1, \ldots, n-2.$$

These relations are only possible, if $\nu = \beta_2 = \ldots = \beta_{n-1} = 0$. Thus, $f = 0$ and the endomorphisms $\mathrm{ad}s_1$, $\mathrm{ad}s_2$ belong in a same torus. In the case (ii) we have also $\nu = \gamma_2 = \ldots = \gamma_{n-1} = 0$ and $[ads_1, ads_2] = ad[s_1, s_2] = 0$.

Case (2) Analogously to case (1), by using the description of $\mathrm{Der} Q_n$ given in section 1, we have $\mathrm{ad}s_1\mathrm{ad}s_2 = \mathrm{ad}s_2\mathrm{ad}s_1$. □

Lemma 6 *Let $\mathfrak{g}$ be a faithful non-decomposable Lie algebra with a filiform nilradical $\mathfrak{n}$. Then* $\dim \mathfrak{g} - \dim \mathfrak{n} = 1$.

Proof Let $\boldsymbol{s}$ a complementary to $\boldsymbol{n}$ subspace in $\boldsymbol{g}$. We suppose $\dim \boldsymbol{s} \geq 2$. The Corollary 1 and the Lemma 5 show that $\boldsymbol{g}$ is isomorphic to one of the following Lie algebras:

$$\boldsymbol{g}_1 = (\mathbb{C}s_1 \oplus \mathbb{C}s_2) \oplus L_n \quad \boldsymbol{g}_2 = (\mathbb{C}c_1 \oplus \mathbb{C}c_2) \oplus Q_n$$

where $[s_i, x] = d_i(x)$, $x \in L_n$, $i = 1,2$, $[c_i, z] = d_i(z)$, $z \in Q_n$, $i = 1,2$, $[s_1, s_2] = \beta X_n$, $[c_1, c_2] = \gamma Z_n$, $(X_0, X_1, \ldots, X_n)$ is the basis of L_n, $(Z_0, Z_1, \ldots, Z_n)$ is the basis of Q_n, d_1 and d_2 are the elements of maximal torus T of derivation of $\boldsymbol{n}$, described in the section 1. But the mappings

$$\psi : T \oplus_{id} L_n \to \boldsymbol{g}_1 \quad \eta : T \oplus_{id} Q_n \to \boldsymbol{g}_2$$

defined by $\psi(d_1) = s_1$, $\psi(d_2) = s_2 - \beta X_n$, $\psi(x) = x$, $x \in L_n$, and by $\eta(d_1) = c_1 + \gamma Z_n$, $\eta(d_2) = c_2$, $\psi(z) = z$, $z \in Q_n$ are the isomorphisms. Thus, $\boldsymbol{g}$ is decomposable. □

Case $\dim \boldsymbol{s} = 1$ Let f be a nonzero element of $\boldsymbol{s}$ considered as derivation of

$\boldsymbol{n}$ and let $f = f_s + f_n$ be the Jordan's decomposition of f with f_s semisimple, f_n nilpotent, $[f_s, f_n] = 0$. We can suppose that $f_s \in T$, where T is the maximal torus of derivations of $\boldsymbol{n}$ described before. As $f_n \neq 0$, f_s admits an eigenvalue with a multiplicity greater than 2. This implies that $\boldsymbol{n} = L_n$ or Q_n or $\mathbb{C}_{n+1}(\alpha_1, \ldots, \alpha_t)$. Moreover, f_n should be an external derivation (that is $f_n \neq \mathrm{ad}X$ for any $X \in \boldsymbol{n}$). We deduce the following description of non-decomposable faithful Lie algebras with a filiform nilradical.

Theorem 3 *Let $\boldsymbol{g}$ be a non-decomposable faithful Lie algebra with a filiform nilradical $\boldsymbol{n}$. Then $\boldsymbol{g} = \{\mathbb{C} \cdot f\} \oplus \boldsymbol{n}$, where $f = f_s + f_n$, $[f, X] = f(X)$ for all $X \in \boldsymbol{n}$, and $\boldsymbol{n}$, f_s, f_n are one of the following:*
(i) $\boldsymbol{n} = L_n$, $f_s = d_1$, $f_n = \sum_{i=2}^{n-1} \beta_i h_i$, $(\beta_2, \ldots, \beta_{n-1}) \in \mathbb{C}^{n-2} - \{0\}$ *where h_i are the nilpotent derivations defined by* $h_i(X_0) = 0$, $h_i(X_j) = X_{i+j}$, $j = 1, \ldots, n-i$.
(ii) $\boldsymbol{n} = L_n$, $f_s = d_1 + d_2$ *and f_n is defined by* $f_n(X_0) = X_1$ *and* $f_n(X_i) = 0$ *for* $i \neq 0$.
(iii) $\boldsymbol{n} = Q_n$ $(n{=}2m{+}1)$, $f_s = d_2$, $f_n = \sum_{i=1}^{m-3} \beta_i h_{2i+1}$, $(\beta_1, \ldots, \beta_{m-1}) \in \mathbb{C}^{m-1} - \{0\}$, *where h_i is the nilpotent derivation of $\boldsymbol{n}$ defined by* $h_{2i+1}(Z_0) = 0$, $h_{2i+1}(Z_j) = Z_{2i+j+1}$, $j = 1, \ldots, n-2i-2$.
(iv) $\boldsymbol{n} = C_{n+1}(\alpha_1, \ldots, \alpha_t)$ $(n = 2m{+}1)$, $f_s = d$, $f_n = \sum_{i=1}^{m-1} \beta_i h_{2i+1}$, $(\beta_1, \ldots, \beta_{m-1}) \in \mathbb{C}^{m-1} - \{0\}$, *where* $h_{2i+1}(Y_0) = 0$, $h_{2i+1}(Y_j) = Z_{2i+j+1}$, $j = 1, \ldots, n-2i-2$.

We suppose here that d_1 and d_2 are the elements of maximal torus of derivation of $\boldsymbol{n}$, $(X_0, X_1, \ldots, X_n)$ is the basis of L_n, $(Z_0, Z_1, \ldots, Z_n)$ is the basis of Q_n, $(Y_0, Y_1, \ldots, Y_n)$ is the basis of $\mathbb{C}_{n+1}(\alpha_1, \ldots, \alpha_t)$, described in the section 1.

Remark 4. Analogously to Remark 3, we can transformate one of the parameters β_i to 1 using the automorphism ψ. Modulo this transformation theorem 3 give a classification up to isomorphism of non-decomposable faithful Lie algebras with a filiform nilradical.

3 Rigid Lie algebras with a filiform nilradical

3.1 Rigid Lie algebras

Let $\mathcal{L}_n$ be the variety of n-dimensional complex Lie algebras. This set is endowed of a structure of algebraic variety. Indeed, when a basis $\{e_1, ..., e_n\}$ of $\mathbb{C}^n$ is fixed, each n-dimensional complex Lie algebra can be identified with its structure constants C^i_{jk} defined by

$$\mu(e_i, e_j) = \sum_{k=1}^{n} C^k_{ij} e_k$$

where μ is the bracket of the given Lie algebra. The antisymetry of the law μ gives $C^k_{ij} = -C^k_{ji}$. The Jacobi conditions are expressed, in terms of structure constants, by the Jacobi polynomials :

$$\sum_{l=1}^{n} C^l_{ij} C^s_{lk} + C^l_{jk} C^s_{li} + C^l_{ki} C^s_{lj} = O$$

On the algebraic variety $\mathcal{L}_n$, we consider the action of the linear group $Gl(n)$ given by :

$$\begin{array}{ccc} Gl(n) \times \mathcal{L}_n & \longrightarrow & \mathcal{L}_n \\ (f, \mu) & \longrightarrow & f^{-1}(\mu(f, f)) \end{array}$$

where the last law is defined by

$$f^{-1}(\mu(f, f))(X, Y) = f^{-1}(\mu(f(X), f(Y)))$$

for all $X, Y \in \mathbb{C}^n$. The orbit of μ related to this action is noted by $\mathcal{O}(\mu)$. The topological notions are related with the Zariski topology.

Definition 6 *The law μ is rigid on $\mathcal{L}_n$ if this orbit $\mathcal{O}(\mu)$ is Zariski open on $\mathcal{L}_n$.*

Let us consider the ring $\mathbb{C}[C^i_{jk}]$ of polynomials of $(n^3 - n^2)/2$ formal variables C^i_{jk} ($C^i_{jk} = -C^i_{kj}$) and denote by J the ideal of $\mathbb{C}[C^i_{jk}]$ generated by the Jacobi polynomials. We can consider the scheme, also noted $\mathcal{L}_n$,

$$Spec \frac{\mathbb{C}[C^i_{jk}]}{J}.$$

The Zariski tangent of the scheme $\mathcal{L}_n$ at a point μ is equal to the space $Z^2(\mu, \mu)$ of the 2-cocycles of the Chevalley cohomology. As the tangent space to the point μ of the algebraic (smooth) variety $\mathcal{O}(\mu)$ corresponds to $B^2(\mu, \mu)$, then $H^2(\mu, \mu) = 0$ implies the rigidity of μ. If the scheme $\mathcal{L}_n$ is reduced at a point μ, then μ rigid implies $H^2(\mu, \mu) = 0$. But we can find many examples of rigid Lie algebras whose second cohomological group is not trivial (see [1] and [2]).

Theorem 4 *Every rigid Lie algebra is algebraic.*

This fundamental theorem is proved by Carles in this thesis. As consequence we can deduced, if the Lie algebra is rigid and solvable that its admits the following decomposition :

$$\boldsymbol{g} = \boldsymbol{t} \oplus \boldsymbol{n}$$

where $\boldsymbol{n}$ is the nilradical and $\boldsymbol{t}$ an external torus of deriavations, that is an abelian susalgebra whose corresponding adjoint operators are semisimple.

3.2 Classification of rigid Lie algebras

The previous theorem permits to build all the rigid Lie algebras up the dimension 7. Indeed, we well know the classification of nilpotent lie algebras up the dimension 6 and we can prove that the seven nilpotent Lie algebras are not rigid. For the dimension bigger or equal to 8, we find a problem. Probably, all the classifications of 7-dimensional nilpotent Lie algebras presents some mistakes. Classifications of rigid Lie algebras based on these lists are suspicious. In a paper presended in this book, we propose a direct method of construction of rigid Lie algebras. The more interesting result concerns the case where the nilradical is filiform. Indeed, in this case, we can use the previous theorem concerning the classification of filiform Lie algebras of nonnul rank. This permits to obtain the general classification of this family of rigid Lie algebras.

3.3 Classification of rigid Lie algebras whose nilradiacal is filiform.

We summarize the result developped in the following paper.

Theorem 5 *Every filiform Lie algebra* $\boldsymbol{n}$ *of rank* 2 *is the nilradical of a rigid Lie algebra. These algebras are isomorphic if and only if their nilradicals are isomorphic.*

Theorem 6 *A filiform Lie algebra* $\boldsymbol{n}$ *of rank* 1 *is the nilradical of a rigid Lie algebra if and only if* $\boldsymbol{n}$ *verifies one of the following cases :*
1st *case :* $\boldsymbol{n}$ *admits a diagonalizable derivation whose eigenvalues are* $1, k, \cdots, n+k-2$, $k \geq 2$, *when* $k = n-5$, $k = n-6$, *or* $k \leq (n-7)/3$ *if* $n-k$ *is odd and* $k \leq (n-6)/3$ *if* $n-k$ *is odd.*

2nd *case :* $\mathfrak{n}$ *is of odd dimension* n *and admits a diagonalizable derivation whose eigenvalues are* $1, k, k+1, \cdots, k+n-3, 2k+n-3$, *with* $k \in \mathbb{N}$ *odd verifying* $k = n-4$, *or* $k = n-6$, *or* ($k = 2$ *and* $n = 10$), *or* $k \leq \left[\frac{n-k}{6}\right] + \frac{n-k-8}{2}$.
3st *case :* $\mathfrak{n}$ *is of odd dimension* n *and admits a diagonalizable derivation whose eigenvalues are* $1, k, k+1, \cdots, k+n-3, 2k+n-3$, *with* $k \in \mathbb{N}$ *even,* $k \neq 1$ *and verifying* $k \leq \left[\frac{n-k-3}{6}\right] + \frac{n-k-9}{2}$, *or* $k = n-5$, *or* ($k = 3$ *and* $n = 14$) *.*

Remark 5
(i) The previous theorem gives the general classification of rigid Lie algebras admitting a filiform nilradical. Indeed, two Lie algebras corresponding to two dictinct values of the parameter k are not isomorphic.
(ii) A rigid Lie algebra whose nilradical is filifiform is necessary solvable, the Lie algebra of derivations of a filiform Lie algebra of dimension greater than 3 not having semisimple components.

4 Complex structure on a nilpotent Lie algebra

4.1 Definition

Definition 7 *Let* $\mathfrak{g}$ *be a real Lie algebra. A complex structure on* $\mathfrak{g}$ *is given by an endomorphism*

$$J : \mathfrak{g} \longrightarrow \mathfrak{g}$$

satisfying

$$\begin{cases} J^2 = -Id \\ [X, J(Y)] = J[X, Y] \end{cases}$$

for all $X, Y \in \mathfrak{g}$.

If exists a complex structure on the real Lie algebra $\mathfrak{g}$ we can endowe $\mathfrak{g}$ with a structure of complex Lie algebra, denoted by $\mathfrak{g}_{\mathbb{C}}$, of dimension

$$\dim \mathfrak{g}_{\mathbb{C}} = \frac{1}{2} \dim \mathfrak{g}$$

4.2 Complex structure on nilpotent Lie algebras of dimension less than 4

1. $\dim \mathfrak{g} = 2$
In this case, $\mathfrak{g}$ is abelian. Consider a basis $\{X_1, X_2\}$ of $\mathfrak{g}$. The endomorphism J defined by

$$\begin{cases} J(X_1) = X_2 \\ J(X_2) = -X_1 \end{cases}$$

is a complex structure on $\mathfrak{g}$.
2. $\dim \mathfrak{g} = 4$.

(i) If $\mathfrak{g}$ is abelian, there exists a complex structure.
(ii) If $\mathfrak{g}$ is filiform, it is defined by the foolwing brackets :

$$\begin{cases} [X_1, X_2] = X_3 \\ [X_1, X_3] = X_4 \end{cases}$$

Every endomorphisme J which satisfies $[X, J(Y)] = J[X, Y]$ is written in the basis $\{X_1, ..., X_4\}$:

$$\begin{pmatrix} a_1 & 0 & 0 & 0 \\ a_2 & b_2 & 0 & 0 \\ a_3 & b_3 & b_2 & 0 \\ a_4 & b_4 & b_3 & b_2 \end{pmatrix}$$

or

$$\begin{pmatrix} a_1 & b_1 & 0 & 0 \\ a_2 & b_2 & 0 & 0 \\ a_3 & b_3 & 0 & 0 \\ a_4 & b_4 & b_3 & 0 \end{pmatrix}$$

the parameters a_i and b_j being reals. Theses matrices cannot be verify $J^2 = -Id$. Then $\mathfrak{g}$ is not provided with a complex structure.
(iii) The Lie algebra $\mathfrak{g}$ is defined by

$$[X_1, X_2] = X_3$$

Every endomorphisme J which satisfies $[X, J(Y)] = J[X, Y]$ is written in the basis $\{X_1, ..., X_4\}$:

$$\begin{pmatrix} a_1 & b_1 & 0 & 0 \\ a_2 & b_2 & 0 & d_2 \\ a_3 & b_3 & b_2 & 0 \\ a_4 & b_4 & 0 & d_4 \end{pmatrix}$$

the parameters a_i, b_j and d_k being reals. The condition $J^2 = -Id$ implies :

$$\begin{aligned} a_1^2 + a_2 b_1 &= -1 \\ a_1 b_1 + b_1 b_2 &= 0 \\ b_1 d_2 &= 0 \\ d_2(b_2 + d_4) &= 0 \\ b_4 d_2 + d_4^2 &= -1 \end{aligned}$$

This gives $a_1^2 = -1$, or $d_4^2 = -1$. Then $\mathfrak{g}$ is not provided with a complex structure.

4.3 Complexe structure on a filiform real Lie algebra.

Theorem 7 *Every real filiform Lie algebra is not provided with a complex structure.*

Proof. If g is filiform, there is a basis $\{X_1, ..., X_n\}$ satisfying in particular :

$$\begin{cases} [X_1, X_2] = X_3 \\ [X_1, X_3] = X_4 \\ .... \\ [X_1, X_{n-1}] = X_n \end{cases}$$

This implies that every endomorphism J which verifies $[X, J(Y)] = J[X, Y]$ is written :

$$J = \begin{pmatrix} a_1 & b_1 & 0 & 0 & ... & 0 \\ a_2 & b_2 & 0 & 0 & ... & 0 \\ \vdots & \vdots & b_2 & 0 & ... & \vdots \\ \vdots & \vdots & b_3 & b_2 & .... & \vdots \\ \vdots & \vdots & \vdots & \vdots & \ddots & \vdots \\ a_n & b_n & b_{n-1} & b_{n-2} & ... & b_2 \end{pmatrix}$$

Then $J^2 = -Id$ implies $b_2^2 = -1$. This is impossible.

5 Locally flat Lie Groups and left-symmetric filiform Lie algebras

Let ∇ be a linear connection on the Lie group G. This connection is called *locally flat* if

$$C(X, Y) = \nabla_{[X,Y]} - [\nabla_X, \nabla_Y] = 0$$

$$T(X, Y) = \nabla_X Y - \nabla_Y X - [X, Y] = 0$$

for all elements $X$and Y in the space $\Gamma(TG)$ of sections of tangent bundle of G. We say that the group G is locally flat if there exists a locally flat connection on G. Let G be a locally flat Lie group. Then vector space $\Gamma(TG)$ is an algebra with product $XY = \nabla_X Y$ and we have the following property for this product:

$$(XY)Z - X(YZ) = (YX)Z - Y(XZ).$$

Any algebra with a product satisfying this property is called a *left-symmetric algebra* [17]. If A is a left-symmetric algebra, A is a Lie algebra under the product $[X, Y] = XY - YX$. This Lie algebra is said to be sub-adjacent to left-symmetric algebra A. The canonical Lie structure of $\Gamma(TG)$ coincides with the original Lie structure. G is locally flat if and only if $\Gamma(TG)$ has a left-symmetric product, which is compatible with its canonical Lie structure and satisfies the following conditions

$$(gX)Y = g(XY), \qquad X(fY) = X(f)Y + f(XY)$$

for all appropriate elements. Let G finite dimensional Lie group and let $\boldsymbol{g}$ be its Lie algebra. Existence of a left-symmetric product on the Lie algebra g compatible with its Lie structure is equivalent to existence of a left-invariant locally flat connexion on G.

Theorem 8 *Let $\boldsymbol{g}$ be a filiform Lie algebra admitting a semisimple derivations (see section 1 for its description) and let $\boldsymbol{g} \neq C_{n+1}(\alpha_1, \dots, \alpha_t)$. Then $\boldsymbol{g}$ provided with a left-symmetric product compatible with its Lie structure.*

In fact, let d be a semisimple derivation of $\boldsymbol{g}$. If $\boldsymbol{g}$ is a Lie algebra of rank 1 (see1.2), then all eigenvalues of d are different of zero. If $\boldsymbol{g}$ is a Lie algebra of rank 2, that is $\boldsymbol{g} = L_n$ or Q_n then always exist a derivation with eigenvalues different of zero and we can suppose that d is a such derivation. Finally, if $\boldsymbol{g}$ is a filiform Lie algebra satisfying the conditions of the Theorem then d is a non degenerate derivation. In this case always exists a left-symmetric product [13], [12].

Remark 6. In the paper [4], Benoist construct an 11-dimensional filiform Lie algebra no provided with a affine structure (for a filiform Lie algebra this is equivalent to non existence of a left-symmetric product compatible with its Lie structure). It is very interesting to determine all low-dimensional filiform Lie algebras (for example of dimension ≤ 11) no provided with such product.

References

[1] J.M. Ancochea-Bermudez, M. Goze, Nonrational rigid Lie algebras, Proceeding A.M.S., to appear.

[2] J.M. Ancochea-Bermudez, M. Goze, Le rang du système linéaire des racines d'une algèbre de Lie résoluble rigide. *Comm. Algebra*, 20:3(1992), 875-887.

[3] J.M. Ancochea-Bermudez. On the rigidity of solvable Lie algebras, in Deformation Theory of Algebras and Structures and Applications. ASI NATO, C.247, 403-445, Kluwer Academic Publishers.

[4] Y.Benoist, Une nilvariété non affine. *J. Differential Geometry*, 41(1995), 21-52.

[5] J.Dixmier, W.G.Lister, Derivations of nilpotent Lie algebras. *Proc.AMS*, 8(1957), 155-158.

[6] M.Goze, Yu.Khakimdjanov, *Nilpotent Lie algebras*. Kluwer Academic Publishers, MIA 361, 1996.

[7] M. Goze, Yu. Hakimjanov(Khakimdjanov) , Sur les algèbres de Lie nilpotentes admettant un tore de dérivations. *Manuscripta math.*, 84(1994), 115-224.

[8] Yu. Hakimjanov(Khakimdjanov), Variétés des lois d'algèbres de Lie nilpotentes. *Geom. Dedicata*, 40(1991), 269-295.

[9] Yu.Khakimdjanov, Characteristically nilpotent Lie algebras. *Math. USSR Sbornik*, 70(1991), 65-78.

[10] A. Mal'tsev, Semisimple subgroups of Lie groups. *Izv. Akad. Nauk SSSR, Matem.*, 9(1945), 329-352 (in Russian). English transl. in Transl. of AMS, 27(1950).

[11] A.Medina, Flat left-invariant connexions adapted to the automorphism structure of a Lie groups.*J. Differential Geometry*, 16(1981), 445-474.

[12] A.Medina, Autour des connexions plates invariantes à gauche sur les groupes de Lie. Thèse d'Etat, Montpellier, 1979.

[13] J.Milnor, On fundamental groups of complete affinely flat manifolds. *Advance in Math.*, 25(1977), 178-187.

[14] G.D. Mostow, Fully reductible subgroups of algebraic groups. *Amer. J. Math.*, 78(1956), 200-221.

[15] A. Onishchik, Yu. Khakimdjanov, Semidirect sums of Lie algebras. *Mat. Zametki*, 18:1(1975), 31-40 (russian). English transl. in Math. Notes of Acad. Sci. USSR,(1976), 600-604.

[16] M. Vergne, Cohomologie des algèbres de Lie nilpotentes. Application à l'étude de la variété des algèbres de Lie nilpotentes. *Bull. Soc. Math., France*, 98(1970), 81-116.

[17] E.B.Vinberg, Convex homogeneous cones. *Transl. of Moscow Math. Soc.*, 12, 340-403.

E-mail : M.Goze@univ-mulhouse.fr, Y.Hakimjanov@univ-mulhouse.fr

ALGEBRES DE LIE RIGIDES

M. GOZE, J.M. ANCOCHEA BERMUDEZ [2]

M.G.: Université de Haute Alsace,
Faculté des Sciences et Techniques
32, rue du Grillenbreit
F. 68000 Colmar. France

J.M.A.B.: Universidad Complutense
Departamento de Topologia y Geometria,
28040 Madrid. Espagne

1 Algèbres de Lie rigides

1.1 Le schéma $\mathcal{L}^n$

Soit $\mathcal{L}^n$ la variété des lois d'algèbres de Lie de dimension n sur un corps K algébriquement clos et de caractéristique nulle. La structure de variété algébrique de $\mathcal{L}^n$ est définie comme suit : soient $\mu \in \mathcal{L}^n$ et $(e_1, e_2, \cdots, e_n)$ une base fixée de K^n. Les constantes de structure de μ relatives à la base (e_i) vérifient :

$$(1)\left\{\begin{array}{l} C_{ij}^k = -C_{ji}^k\ , \quad 1 \leq i < j \leq n\ , \quad 1 \leq k \leq n \\ \sum_{l=1}^n C_{ij}^l C_{lk}^s + C_{jk}^l C_{li}^s + C_{ki}^l C_{jl}^s = 0\ , \quad 1 \leq i < j < k \leq n\ , \quad 1 \leq s \leq n. \end{array}\right.$$

et $\mathcal{L}^n$, paramétrisée par les C_{jk}^i vérifiant les conditions polynômiales (1), est une variété algébrique plongée dans K^{n^3}. Toutes les notions topologiques concernant $\mathcal{L}^n$ sont relatives à la topologie de Zariski.

[2] Travail effectué grace au soutien du DGICYT,PB 92-0220

Y. Khakimdjanov et al (eds.), Algebra and Operator Theory, 65-91.

Considérons les C_{jk}^i comme des variables formelles $(j < k)$, et soit J l'idéal de $\mathbb{C}[C_{jk}^i]$ engendré par les relations de Jacobi (avec $C_{jk}^i = -C_{kj}^i$ et $C_{jj}^i = 0$). On définit le schéma, noté également $\mathcal{L}^n$, par

$$\mathcal{L}^n = \frac{\mathbb{C}[C_{jk}^i]}{J}$$

L'espace tangent formel au schéma $\mathcal{L}^n$ en un point μ correspond à l'espace des 2-cocycles $Z^2(\mu, \mu)$ pour la cohomologie de l'algèbre de Lie $\boldsymbol{g} = (K^n, \mu)$ à valeurs dans le $\boldsymbol{g}$-module $\boldsymbol{g}$.

Le groupe linéaire $Gl(n, K)$ opère sur $\mathcal{L}^n$ par:

$$(f, \mu) \longrightarrow f^{-1}\mu(f, f)$$

correspondant au changement de bases pour une application bilinéaire. L'orbite $\mathcal{O}(\mu)$ de μ relative à cette action est constituée de toutes les lois isomorphes à μ.

1.2 Rigidité

Définition 1 *Une loi $\mu \in \mathcal{L}^n$ est rigide si l'orbite $\mathcal{O}(\mu)$ est ouverte dans $\mathcal{L}^n$.*

L'espace tangent à l'orbite $\mathcal{O}(\mu)$ au point μ s'identifie à l'espace des cobords $B^2(\mu, \mu)$. On en déduit que l'annulation du deuxième groupe de cohomologie $H^2(\mu, \mu)$ implique la rigidité de la loi μ [8]. L'existence de lois rigides vérifiant $dimH^2(\mu, \mu) \neq 0$ permet de montrer que le schéma $\mathcal{L}^n$ n'est pas réduit pour $n \geq 11$ ([2], [4]).

Dans tout ce qui suit, on identifiera une algèbre de Lie $\boldsymbol{g} = (K^n, \mu)$ et sa loi μ.

Les algèbres de Lie rigides constituent une famille remarquable d'algèbres de Lie à classer. En effet, il n'existe pour une dimension donnée qu'un nombre fini de classes d'isomorphie de telles algèbres. La raison en est simple : la variété algébrique $\mathcal{L}^n$ est une réunion finie de cmposantes algébriques irréductibles. Or l'adhérence d'une orbite d'une loi rigide est une de ces composantes. Il ne peut exister qu'un nombre fini de telles orbites et donc de classes d'isomorphie de lois rigides. Une telle classification est d'autant plus intéressante que la classification générale des algèbres de Lie est loin d'être établie. Les premières constructions des lois rigides reposaient sur le théorème de Nijenhuis-Richardson qui précise qu'une algèbre de Lie dont le deuxième groupe $H^2(\boldsymbol{g}, \boldsymbol{g})$ est nul est rigide. Si ce critère permet de construire toutes les lois jusqu'en dimension 7, l'existence de nombreux exemples de rigides à cohomologie non nulle réduit l'intêret de cette approche. Une deuxième méthode de construction fût basée sur le théorème d'algébricité de Carles: une algèbre de Lie rigide est décomposable sous la forme $\mathcal{S} \oplus \mathcal{T} \oplus \mathcal{N}$, et le cas résoluble se ramène à étendre le nilradical $\boldsymbol{n}$ par un tore de dérivations $\boldsymbol{t}$. La connaisance de toutes les algèbres de Lie nilpotentes de dimension 7 [9] permet d'établir la classification des rigides de dimension 8. Malheuresement cette méthode se heurte à l'inexistence de classification générale des algèbres de Lie nilpotentes de dimension au moins 8 (une telle classification est en fait impossible à établir).

Nous proposons ici une approche directe de la classification des rigides que l'on illustre en donnant cette classification lorsque le nilradical est filiforme. Rappelons que l'ensemble des lois filiformes est un ouvert dans la variété de toutes les lois nilpotentes.

2 Sur la structure des algèbres de Lie résolubles rigides

Une algèbre de Lie résoluble rigide se décompose sous la forme

$$\boldsymbol{g} = \boldsymbol{t} \oplus \boldsymbol{n}$$

où $\boldsymbol{n}$ est le nilradical de $\boldsymbol{g}$ et $\boldsymbol{t}$ une sous algèbre abélienne dont les éléments sont $ad_{\boldsymbol{g}}$-semi simples [5]. Dans [2] on présente une théorie des racines quelque peu semblables à celle des algèbres simples (qui sont rigides).

Considérons une algèbre de Lie résoluble, non nilpotente et décomposable :

$$\boldsymbol{g} = \boldsymbol{t} \oplus \boldsymbol{n}$$

$\boldsymbol{n}$ étant le nilradical et $\boldsymbol{t}$ comme ci dessus.

Définition 2 *Un vecteur $X \in \boldsymbol{t}$ est régulier si*

$$dim(Ker\ adX) = min\{dim(Ker\ adY), \forall Y \in \boldsymbol{t}\}$$

L'opérateur adX étant diagonalisable et $\boldsymbol{n}$ étant adX-invariant, il existe une base $(X_1, X_2, \cdots, X_{p+q}, \cdots, X_n = X)$ de vecteurs propres de adX telle que $(X_1, \cdots, X_{p+q})$ soit une base de $\boldsymbol{n}$, $(X_{p+q+1}, \cdots, X_n)$ une base de $\boldsymbol{t}$ et $(X_{p+1}, \cdots, X_n)$ une base de $Ker\ adX$.

Définition 3 *Soit X un vecteur régulier de l'algèbre décomposable $\boldsymbol{g}$. Le système linéaire de racines associé à $(X_1, \cdots, X_{n-1})$ est le système linéaire à $(n-1)$ variables x_i dont les équations sont*

$$x_i + x_j = x_k$$

si la composante du vecteur $\mu(X_i, X_j)$ sur X_k est non nulle.

On notera $S(X)$ ce système linéaire.

Théorème 1 *(Théorème du rang [2]) Si l'algèbre résoluble décomposable $\boldsymbol{g} = \boldsymbol{t} \oplus \boldsymbol{n}$ est rigide, alors pour tout vecteur régulier X on a :*

$$rang(S(X)) = dim\boldsymbol{n} - 1.$$

Si $\boldsymbol{g}$ est rigide, le rang du système linéaire ne dépend ni du choix du vecteur régulier X, ni du choix de la base $\{X_i\}$.

Corollaire 1 *Si l'algèbre résoluble décomposable $\boldsymbol{g} = \boldsymbol{t} \oplus \boldsymbol{n}$ est rigide, alors $\boldsymbol{t}$ est une sous algèbre abélienne maximale dont les éléments sont $ad\boldsymbol{g}$-semi-simples ($\boldsymbol{t}$ est un tore externe de dérivations dans la terminologie de Malcev).*

Remarque
Le système $S(X)$ étant à coefficients entiers, les solutions peuvent être choisies rationnelles. Comme les valeurs propres de adX pour $X \in \boldsymbol{t}$ sont solutions de $S(X)$, on peut toujours les supposer entières :

$$[X_j, X_l] = n_j^l X_l \quad , \quad j = p+q+1, \cdots, n \ ; \quad l = 1, \cdots n \ ; \quad n_j^l \in \mathbb{Z}.$$

Mais ceci n'implique nullement la rationalité des algèbres rigides.
Exemple. Soit $\boldsymbol{g}$ l'algèbre de Lie résoluble de dimension 16 définie dans la base $(X_1, X_3, X_4, \cdots, X_{16}, X)$ par

$$[X, X_i] = iX_i \qquad i = 1, 3, 4, \cdots, 16$$

$$[X_1, X_i] = X_{i+1} \qquad i = 3, \cdots, 15$$

$$[X_3, X_4] = \lambda_1 X_7, \quad [X_4, X_5] = \lambda_2 X_9, \quad [X_5, X_6] = \lambda_3 X_{11},$$

$$[X_6, X_7] = \lambda_4 X_{13}, \quad [X_7, X_8] = \lambda_5 X_{15}, \quad [X_i, X_j] = a_{i,j} X_{i+j}$$

les $a_{i,j}$ étant entièrement déterminés par les λ_i , $i = 1, \cdots, 5$ à partir des relations

$$a_{i,j} = a_{i,j+1} + a_{i+1,j}, \quad a_{i,i+1} = \lambda_{i-2} \quad i = 3, \cdots, 7.$$

L'algèbre de Lie correspondant à

$$\lambda_1 = \frac{2+\sqrt{11}}{3}, \quad \lambda_2 = 1, \quad \lambda_3 = \frac{10-2\sqrt{11}}{7},$$

$$\lambda_4 = \frac{62-18\sqrt{11}}{35}, \quad \lambda_5 = \frac{136-44\sqrt{11}}{35}$$

est rigide et non rationnelle ([3]).

3 Classification des algèbres de Lie rigides résolubles de dimension inférieure ou égale à 7

3.1 Comment prouver la rigidité

Le théorème du rang nous donne une condition nécessaire de rigidité et permet de construire une famille finie d'algèbres de Lie décomposables contenant entièrement la famille des algèbres de Lie rigides. La vérification de la rigidité peut se faire, en petite dimension, par le calcul du deuxième groupe de cohomologie. Pour les dimensions inférieures ou égales à 7 ceci suffit pour établir la classification. Pour les dimensions plus grandes, l'existence de rigides à cohomologie non nulle, conduit à une approche directe de la rigidité. C'est ce que nous proposons ici dans un exemple "générique". Soit μ_0 la loi d'algèbre de Lie dans L^9 définie par :

$$\begin{cases} \mu_0(X, X_i) = iX_i, & for \quad i = 1,2,4,5,6,7,8,9 \\ \mu_0(X_1, X_i) = X_{i+1}, & for \quad i = 4,5,6,7,8 \\ \mu_0(X_2, X_i) = X_{i+2}, & for \quad i = 4,5,6,7 \\ \mu_0(X_4, X_5) = X_9, & \end{cases}$$

les autres crochets étant nuls, et où $(X, X_1, X_2, X_4, X_5, X_6, X_7, X_8, X_9)$ est une base de $\mathbb{C}^9$ (Attention aux indices).

Proposition 1 *μ_0 est rigide.*

Notons, tout d'abord, que μ_0 est résoluble, décomposable : $\boldsymbol{t}$ est engendré par X et $\boldsymbol{n}$ par les vecteurs X_i et le théorème du rang est vérifié. Afin de prouver la rigidité, il suffit de montrer que toute loi dont les constantes de structures sont dans un voisinage suffisamment petit de μ_0 dans $\mathcal{L}_\backslash$ est dans l'orbite de μ_0. Considérons une telle loi μ. L'opérateur

$$ad_{\mu_0}X : \mathbb{C}^9 \longrightarrow \mathbb{C}^9$$

défini par

$$ad_{\mu_0}X(Y) = \mu_0(X, Y)$$

a pour valeurs propres $\{0,1,2,4,5,6,7,8,9\}$. Si μ est dans un voisinage suffisamment petit de μ_0, ces valeurs propres

$$\{0, \lambda_1, \lambda_2, \lambda_4, \lambda_5, \lambda_6, \lambda_7, \lambda_8, \lambda_9\}$$

sont telles que λ_i est proche de i (plus petit qu'un ε donné). Elles sont donc deux à deux distinctes et l'op''erateur $ad_\mu X$ est diagonalisable. Considérons une base $\{X, U_1, U_2, U_4, U_5, U_6, U_7, U_8, U_9\}$ of $\mathbb{C}^9$ telle que tha $\mu(X, U_i) = \lambda_i U_i$ for $i = 1,2,4,5,6,7,8,9$. Comme la grassmannienne est une variété compacte, il existe, par chaque valeur propre λ_i de $ad_\mu X$, un vecteur propre, U_i' dont les composantes relatives à la base $\{X, X_i\}$ sont $(\varepsilon_1, .., \varepsilon_{i-1}, 1 + \varepsilon_i, \varepsilon_{i+1}, .., \varepsilon_9)$, chaque ε_i étant en module plus petit que ε. Or les conditions de Jacobi montrent que $\mu(U_i', U_j')$ est un vecteur propre associé à la valeur propre $\lambda_i + \lambda_j$. Les relations

$$\mu_0(X_1, X_i) = X_{i+1} \quad , \quad i = 4,5,6,7,8,$$

$$\mu_0(X_2, X_i) = X_{i+2} \quad , \quad i = 4,5,6,7$$

$$\mu_0(X_4, X_5) = X_9$$

montre que

$$\begin{cases} \mu(X,U_i') = \lambda_i U_i' & , i = 1,2,4,5,6,7,8,9 \\ \mu(U_1',U_i') = a_{1i}U_{i+1}' & , a_{1i} \neq 0 \quad , i = 4,5,6,7,8 \\ \mu(U_2',U_i') = a_{2i}U_{i+2}' & , a_{2i} \neq 0 \quad , i = 4,5,6,7 \\ \mu(U_4',U_5') = a_{45}U_9' & , a_{45} \neq 0, \end{cases}$$

les autres crochets étant nuls. Ceci implique

$$\lambda_i + \lambda_j = \lambda_{i+j}$$

pour $(i,j) = (1,4),(1,5),(1,6)(1,7),(1,8),(2,4),\ (2,5),\ (2,6),(2,7),(4,5)$. On obtient donc le système linéaire suivant :

$$(S)\begin{cases} \lambda_1+\lambda_4=\lambda_5 \\ \lambda_1+\lambda_8=\lambda_9 \\ \lambda_2+\lambda_7=\lambda_9 \end{cases} \begin{cases} \lambda_1+\lambda_5=\lambda_6 \\ \lambda_2+\lambda_4=\lambda_6 \\ \lambda_4+\lambda_5=\lambda_9 \end{cases} \begin{cases} \lambda_1+\lambda_6=\lambda_7 \\ \lambda_2+\lambda_5=\lambda_7 \end{cases} \begin{cases} \lambda_1+\lambda_7=\lambda_8 \\ \lambda_2+\lambda_6=\lambda_8 \end{cases}$$

dont les solutions sont $\lambda_i = i\lambda_1$ pour $i = 2,4,5,6,7,8,9$. Les valeurs propres non nulles $ad_\mu X$ sont $i\lambda_1$ pour $i = 1,2,4,5,6,7,8,9$. Soit $Y = X/\lambda_1$. L'opérator $ad_\mu Y$ a pour valeurs propres $1,2,4,5,6,7,8,9$, et pour vecteurs propres U_i'. D'où:

$$\begin{cases} \mu(Y,U_i') = iU_i' & , i = 1,2,4,5,6,7,8,9 \\ \mu(U_1',U_i') = a_{1i}U_{i+1}' & , i = 4,5,6,7,8 \quad a_{1i} \neq 0 \\ \mu(U_2',U_i') = a_{2i}U_{i+2}' & , i = 4,5,6,7 \quad a_{2i} \neq 0 \\ \mu(U_4',U_5') = a_{45}U_9' & , a_{45} \neq 0 \end{cases}$$

Le changement de bases $Y_i = b_i U_i'$ avec $b_1 b_i = b_{i+1}/a_{1i}$ pour $i = 4,5,6,7,8$, $b_2 b_4 = b_6/a_{24}$ et $b_4 b_5 = b_9/a_{45}$ permet de supposer que $a_{1i} = a_{24} = a_{45} = 1$ for $i = 4,5,6,7,8$. Alors, les conditions de Jacobi impliquent :

$$\mu(Y_2,Y_i) = Y_{i+2} \quad i = 5,6,7.$$

Dans la base $(Y,Y_1,Y_2,Y_4,Y_5,Y_6,Y_7,Y_8,Y_9)$ la loi μ vérifie :

$$\begin{cases} \mu(Y,Y_i) = iY_i & , i = 1,2,4,5,6,7,8,9 \\ \mu(Y_1,Y_i) = Y_{i+1} & , i = 4,5,6,7,8 \\ \mu(Y_2,Y_i) = Y_{i+2} & , i = 4,5,6,7 \\ \mu(Y_4,Y_5) = Y_9 \end{cases}$$

ce qui montre que μ est isomorphe à μ_0. Ainsi μ_0 est rigide.

3.2 Algèbres résolubles rigides de dimension au plus 7

dimension 2

Il n'existe, à isomorphisme près, qu'une seule loi rigide

$$\mu_1^2(X,X_1) = X_1$$

dimension 3

Il n'existe aucune loi rigide résoluble de dimension 3.

dimension 4

Toute loi d'algèbre de Lie rigide sur $\mathbf{C}^4$ est isomorphe à l'une des deux lois :

$$\begin{cases} \mu_4^1(X, X_1) = X_1, \\ \mu_4^1(X', X_2) = X_2 \end{cases}$$

les autres crochets étant nuls.

dimension 5

Toute loi d'algèbre de Lie rigide sur $\mathbf{C}^5$ est isomorphe à :

$$\begin{cases} \mu_5^1(X, Y_i) = iY_i & , i = 1, 2, 3, \\ \mu_5^1(X_0, Y_i) = Y_i & , i = 2, 3, \\ \mu_5^1(Y_1, Y_2) = Y_3. \end{cases}$$

les autres crochets étant nuls.

dimension 6

Toute loi d'algèbre de Lie résoluble rigide dans $\mathbf{C}^6$ est isomorphe à l'une des lois suivantes :

$$\begin{cases} \mu_6^1(X, Y_i) = iY_i & , i = 1, 2, 3, 4, 5, \\ \mu_6^1(Y_1, Y_i) = Y_{i+1} & , i = 2, 3, 4, \\ \mu_6^1(Y_2, Y_3) = Y_5 \end{cases}$$

$$\begin{cases} \mu_6^2(X, Y_i) = iY_i & , i = 1, 2, 3, 4, \\ \mu_6^2(X_0, Y_i) = Y_i & , i = 2, 3, 4, \\ \mu_6^2(Y_1, Y_i) = Y_{i+1} & , i = 2, 3. \end{cases}$$

$$\begin{cases} \mu_6^3(X, Y_i) = iY_i & , i = 1, 2, 3, \\ \mu_6^3(X_0, Y_2) = Y_2, \\ \mu_6^3(X_0', Y_3) = Y_3. \end{cases}$$

dimension 7

Toute loi d'algèbre de Lie résoluble rigide dans $\mathbf{C}^7$ est isomorphe à l'une des lois suivantes :

$$\begin{cases} \mu_7^1(X, Y_i) = iY_i, & i = 1, 2, 3, 4, 5, 6, \\ \mu_7^1(Y, Y_i) = Y_{i+1}, & i = 2, 3, 4, 5, \\ \mu_7^1(Y_2, Y_i) = Y_{i+2}, & i = 3, 4. \end{cases}$$

$$\begin{cases} \mu_7^2(X, Y_i) = iY_i, & i = 1, 2, 3, 4, 5, 7, \\ \mu_7^2(Y_1, Y_i) = Y_{i+1}, & i = 2, 3, 4, \\ \mu_7^2(Y_2, Y_i) = Y_{i+2}, & i = 3, 5, \\ \mu_7^2(Y_3, Y_4) = -Y_7. \end{cases}$$

$$\begin{cases} \mu_7^3(X, Y_i) = iY_i, & i = 1, 3, 4, 5, 6, 7, \\ \mu_7^3(Y_1, Y_i) = Y_{i+1}, & i = 3, 4, 5, 6, \\ \mu_7^3(Y_4, Y_4) = Y_7. \end{cases}$$

$$\begin{cases} \mu_7^4(X, Y_i) = iY_i, & i = 1,2,3,4,5, \\ \mu_7^4(X, Y_3') = 3Y_3', \\ \mu_7^4(Y_1, Y_i) = Y_{i+1}, & i = 2,3,4, \\ \mu_7^4(Y_2, Y_3) = \mu_4(Y_2, Y_3') = Y_5. \end{cases}$$

$$\begin{cases} \mu_7^5(X, Y_i) = iY_i, & i = 1,2,3,4,5, \\ \mu_7^5(X_0, Y_i) = Y_i, & i = 2,3,4,5, \\ \mu_7^5(Y_1, Y_i) = Y_{i+1}, & i = 2,3,4. \end{cases}$$

$$\begin{cases} \mu_7^6(X, Y_i) = iY_i, & i = 1,2,3,4,5, \\ \mu_7^6(X_0, Y_i) = Y_i, & i = 2,3,4, \\ \mu_7^6(X_0, Y_5) = 2Y_5, \\ \mu_7^6(Y_1, Y_i) = Y_{i+1}, & i = 2,3, \\ \mu_7^6(Y_2, Y_3) = Y_5. \end{cases}$$

$$\begin{cases} \mu_7^7(X, Y_i) = iY_i, & i = 1,2,3,4,5, \\ \mu_7^7(X_0, Y_i) = Y_i, & i = 3,4,5, \\ \mu_7^7(Y_1, Y_i) = Y_{i+1}, & i = 3,4, \\ \mu_7^7(Y_2, Y_3) = Y_5. \end{cases}$$

$$\begin{cases} \mu_7^8(X, Y_i) = iY_i, & i = 1,2,3,4, \\ \mu_7^8(X_0, Y_i) = Y_i, & i = 2,3, \\ \mu_7^8(X_0', Y_4) = Y_4, \\ \mu_7^8(Y_1, Y_2) = Y_3. \end{cases}$$

De plus, ces lois sont deux à deux non isomorphes.

Remarque Toutes les lois rigides résolubles de dimension inférieure ou égale à 7 vérifient $dimH^2(\mu, \mu) = 0$.

4 Classification des algèbres de Lie résolubles de dimension 8

Dans sa thèse, Ancochea a proposé une classification de ces algèbres basèe sur le théorème du rang. Une erreur dans l'examen des systèmes lin'eaires associés à une algèbre rigide fait que cette liste est incomplète. Ceci a été remarqué par Carles [4]. Ce dernier, utilisant la classification récente des nilpotentes de dimension 7, a pu établir une liste de rigides de dimension 8. Comme la liste des nilpotentes comporte également des erreurs, la classification de Carles peut poser problèmes. Néanmoins, la comparaison des résultats d'Ancochea et de Carles permet de donner une liste fiable.

Théorème 2 *Toute loi d'algèbre de Lie résoluble rigide dans $\mathbb{C}^8$ est isomorphe à l'une des lois suivantes :*

$$\begin{cases} \mu_8^1(X,Y_i) = iY_i, \quad i \in \{1,2,3,5,6,7,8\}, \quad \mu_8^1(Y_1,Y_i) = Y_{i+1}, \ i \in \{2,6,7\}, \\ \mu_8^1(Y_2,Y_3) = Y_5, \quad \mu_8^1(Y_2,Y_5) = Y_7, \\ \mu_8^1(Y_2,Y_6) = Y_8, \quad \mu_8^1(Y_3,Y_3) = Y_8. \end{cases}$$

$$\begin{cases} \mu_8^2(X,Y_i) = iY_i, \ i \in \{1,3,4,5,6,7,8\}, \ \mu_8^2(Y_1,Y_i) = Y_{i+1}, \ i \in \{3,4,5,6,7\}, \\ \mu_8^2(Y_3,Y_i) = Y_{i+3}, \quad i = 4,5. \end{cases}$$

$$\begin{cases} \mu_8^3(X,Y_i) = iY_i, \ i \in \{1,3,4,5,6,7,9\}, \quad \mu_8^3(Y_1,Y_i) = Y_{i+1}, \ i \in \{3,4,5,6\}, \\ \mu_8^3(Y_3,Y_i) = Y_{i+3}, \quad i = 4,6, \quad \mu_8^3(Y_4,Y_5) = -Y_9. \end{cases}$$

$$\begin{cases} \mu_8^4(X,Y_i) = iY_i, \ i \in \{1,4,5,6,7,8,9\}, \ \mu_8^4(Y_1,Y_i) = Y_{i+1}, \ i \in \{4,5,6,7,8\}, \\ \mu_8^4(Y_4,Y_5) = Y_9. \end{cases}$$

$$\begin{cases} \mu_8^5(X,Y_i) = iY_i, \ i \in \{2,3,4,5,6,7,8\}, \quad \mu_8^5(Y_2,Y_i) = Y_{i+2}, \ i \in \{3,4,5,6\}, \\ \mu_8^5(Y_3,Y_i) = Y_{i+3}, \quad i = 4,5. \end{cases}$$

$$\begin{cases} \mu_8^6(X,Y_i) = iY_i, \quad i \in \{2,3,4,6,7,8,10\}, \quad \mu_8^6(Y_2,Y_i) = Y_{2+i}, \quad i = 4,6,8, \\ \mu_8^6(Y_3,Y_i) = Y_{i+3}, \quad i = 4,7, \quad \mu_8^6(Y_4,Y_6) = Y_{10}. \end{cases}$$

$$\begin{cases} \mu_8^7(X,Y_i) = iY_i, \quad i \in \{2,3,5,6,7,8,9\}, \quad \mu_8^7(Y_2,Y_i) = Y_{i+2}, \ i = 3,5,6,7, \\ \mu_8^7(Y_3,Y_i) = Y_{i+3}, \quad i = 5,6. \end{cases}$$

$$\begin{cases} \mu_8^8(X,Y_i) = iY_i, \ i \in \{2,3,5,7,8,9,11\}, \quad \mu_8^8(Y_2,Y_i) = Y_{i+2}, i \in \{3,5,7,9\}, \\ \mu_8^8(Y_3,Y_i) = Y_{i+3}, \quad i = 5,8. \end{cases}$$

$$\begin{cases} \mu_8^9(X,Y_i) = iY_i, \quad i \in \{1,2,3,4,5,6\}, \quad \mu_8^9(X,Y_3') = 3Y_3', \\ \mu_8^9(Y_1,Y_i) = Y_{i+1}, \quad i = 2,3,4,5, \quad \mu_8^9(Y_2,Y_i) = Y_{i+2}, i = 3,4, \\ \mu_8^9(Y_2,Y_3') = Y_5, \quad \mu_8^9(Y_3,Y_3') = Y_6. \end{cases}$$

$$\begin{cases} \mu_8^{10}(X,Y_i) = iY_i, \quad i = 1,2,3,4,5,6, \quad \mu_8^{10}(X,Y_4') = 4Y_4', \\ \mu_8^{10}(Y_1,Y_i) = Y_{i+1}, \quad i = 2,3,4,5, \quad \mu_8^{10}(Y_2,Y_i) = Y_{i+2}, \quad i = 3,4, \\ \mu_8^{10}(Y_2,Y_4') = Y_6. \end{cases}$$

$$\begin{cases} \mu_8^{11}(X,Y_i) = iY_i, \quad i = 1,2,3,4,5,6, \quad \mu_8^{11}(X,Y_5') = 5Y_5', \\ \mu_8^{11}(Y_1,Y_i) = Y_{i+1}, \quad i = 2,3,4,5, \quad \mu_8^{11}(Y_1,Y_5') = Y_6, \\ \mu_8^{11}(Y_2,Y_3) = Y_5', \quad \mu_8^{11}(Y_2,Y_4) = Y_6. \end{cases}$$

$$\begin{cases} \mu_8^{12}(X,Y_i) = iY_i, \quad i=1,2,3,4,5,7, \quad \mu_8^{12}(X,Y_3') = 3Y_3', \\ \mu_8^{12}(Y_1,Y_2) = Y_3', \quad \mu_8^{12}\left(Y_1,Y_3'\right) = Y_4, \quad \mu_8^{12}(Y_1,Y_4) = Y_5, \\ \mu_8^{12}(Y_2,Y_3) = Y_5, \quad \mu_8^{12}\left(Y_2',Y_3'\right) = Y_5, \mu_8^{12}(Y_2,Y_5) = \mu_8^{12}\left(Y_3',Y_4\right) = Y_7. \end{cases}$$

$$\begin{cases} \mu_8^{13}(X,Y_i) = iY_i, \quad i=1,2,3,4,5,7, \quad \mu_8^{13}(X,Y_5') = 5Y_5', \\ \mu_8^{13}(Y_1,Y_i) = Y_{i+1}, \quad i=2,3,4, \quad \mu_8^{13}(Y_2,Y_3) = Y_5', \\ \mu_8^{13}(Y_2,Y_5) = \mu_8^{13}(Y_2,Y_5') = Y_7, \quad \mu_8^{13}(Y_3,Y_4) = -Y_7. \end{cases}$$

$$\begin{cases} \mu_8^{14}(X,Y_i) = iY_i, \quad i=1,3,4,5,6,7, \quad \mu_8^{14}(X,Y_4') = 4Y_4', \\ \mu_8^{14}(Y_i,Y_i) = Y_{i+1}, \quad i=3,4,5,6, \quad \mu_8^{14}(Y_3,Y_4) = \mu_8^{14}(Y_3,Y_4') = Y_7. \end{cases}$$

$$\begin{cases} \mu_8^{15}(X,Y_i) = iY_i, \quad i=1,2,3,4,5, \quad \mu_8^{15}(X,Y_3') = 3Y_3', \\ \mu_8^{15}(X,Y_4') = 4Y_4', \quad \mu_8^{15}(Y_1,Y_i) = Y_{i+1}, \quad i=2,3,4, \\ \mu_8^{15}(Y_1,Y_3') = Y_4', \quad \mu_8^{15}(Y_1,Y_4') = \mu_8^{15}(Y_2,Y_3) = Y_5. \end{cases}$$

$$\begin{cases} \mu_8^{16}(X,Y_i) = iY_i, \quad i=1,2,3,4,5, \quad \mu_8^{16}(X,Y_1') = Y_1', \\ \mu_8^{16}(X,Y_3') = 3Y_3', \quad \mu_8^{16}(Y_1,Y_i) = Y_{i+1}, \quad i=2,4, \\ \mu_8^{16}(Y_1,Y_i') = Y_{i+1}', \quad i=1,3, \quad \mu_8^{16}(Y_1',Y_i) = Y_{i+1}, \quad i=3,4, \\ \mu_8^{16}(Y_1',Y_2) = Y_3', \quad \mu_8^{16}(Y_2,Y_3) = Y_5, \quad \mu_8^{16}(Y_2,Y_3') = -Y_5. \end{cases}$$

$$\begin{cases} \mu_8^{17}(X,Y_i) = iY_i, \quad i=1,2,3,4,5, \quad \mu_8^{17}(X,Y_i') = iY_i', \quad i=3,5, \\ \mu_8^{17}(Y_1,Y_i) = Y_{i+1}, \quad i=2,3,4, \quad \mu_8^{17}(Y_1,Y_3') = Y_4, \\ \mu_8^{17}(Y_2,Y_3) = Y_5', \quad \mu_8^{17}(Y_2,Y_3') = Y_5 + 2Y_5'. \end{cases}$$

$$\begin{cases} \mu_8^{18}(X,Y_i) = iY_i, \quad \mu_8^{18}(X,Y_i') = iY_i', \quad i=1,2,3, \mu_8^{18}(X,Y_1'') = Y_1'' \\ \mu_8^{18}\left(Y_1,Y_1'\right) = Y_2, \quad \mu_8^{18}(Y_1,Y_1'') = Y_2', \quad \mu_8^{18}\left(Y_1'',Y_2'\right) = Y_3', \\ \mu_8^{18}(Y_1,Y_2') = Y_3', \quad \mu_8^{18}(Y_1',Y_2) = Y_3', \quad \mu_8^{18}(Y_1',Y_2') = \mu_8^{18}\left(Y_1'',Y_2\right) = Y_3. \end{cases}$$

$$\begin{cases} \mu_8^{19}(X,Y_i) = iY_i, \ i=0,1,2,3,4,5,6, \quad \mu_8^{19}(Y_0,Y_i) = Y_i, \quad i=2,3,4,5,6, \\ \mu_8^{19}(Y_1,Y_i) = Y_{i+1}, \quad i=2,3,4,5. \end{cases}$$

$$\begin{cases} \mu_8^{20}(X,Y_i) = iY_i, \quad i=0,1,2,3,4,5,6, \quad \mu_8^{20}(Y_0,Y_i) = Y, \quad i=3,4,5,6, \\ \mu_8^{20}(Y_1,Y_i) = Y_{i+1}, \quad i=3,4,5, \quad \mu_8^{20}(Y_2,Y_i) = Y_{i+2}, \quad i=3,4. \end{cases}$$

$$\begin{cases} \mu_8^{21}(X,Y_i) = iY_i, \quad i=0,1,2,3,4,5,6, \quad \mu_8^{21}(Y_0,Y_i) = Y_i, \quad i=4,5,6, \\ \mu_8^{21}(Y_1,Y_i) = Y_{i+1}, \quad i=2,4,5, \quad \mu_8^{21}(Y_2,Y_4) = Y_6. \end{cases}$$

$$\begin{cases} \mu_8^{22}(X,Y_i) = iY_i, \quad i=0,1,2,3,4,5,6, \quad \mu_8^{22}(Y_0,Y_i) = Y_i, \quad i=2,3,4, \\ \mu_8^{22}(Y_0,Y_i) = 2Y_i, \quad i=5,6, \quad \mu_8^{22}(Y_1,Y_i) = Y_{i+1}, \quad i=2,3,5, \\ \mu_8^{22}(Y_2,Y_i) = Y_{i+2}, i=3,4 \end{cases}$$

$$\begin{cases} \mu_8^{23}(X,Y_i) = iY_i, \quad i=0,1,2,3,4,5,6, \quad \mu_8^{23}(Y_0,Y_6) = Y_6, \\ \mu_8^{23}(Y_1,Y_i) = Y_{i+1}, \quad i=2,3,4, \quad \mu_8^{23}(Y_2,Y_3) = Y_5. \end{cases}$$

$$\begin{cases} \mu_8^{24}(X,Y_i)=iY_i, & i=0,1,2,3,4,5,7, \quad \mu_8^{24}(Y_0,Y_i)=Y_i, \quad i=3,4,5, \\ \mu_8^{24}(Y_0,Y_7)=2Y_7, & \mu_8^{24}(Y_1,Y_i)=Y_{i+1}, \quad i=3,4, \\ \mu_8^{24}(Y_2,Y_3)=Y_5, & \mu_8^{24}(Y_3,Y_4)=Y_7. \end{cases}$$

$$\begin{cases} \mu_8^{25}(X,Y_i)=iY_i, & i=0,1,2,3,4,5,7, \quad \mu_8^{25}(Y_0,Y_i)=Y_i, \quad i=2,3,4, \\ \mu_8^{25}(Y_0,Y_5)=2Y_5, & \mu_8^{25}(Y_0,Y_7)=3Y_7, \quad \mu_8^{25}(Y_1,Y_i)=Y_{i+1}, \quad i=2,3, \\ \mu_8^{25}(Y_2,Y_i)=Y_{i+2}, & i=3,5. \end{cases}$$

$$\begin{cases} \mu_8^{26}(X,Y_i)=iY_i, & i=0,1,2,3,4,5,7, \quad \mu_8^{26}(Y_0,Y_i)=Y_i, \quad i=2,3,4,5, \\ \mu_8^{26}(Y_0,Y_7)=2Y_7, & \mu_8^{26}(Y_1,Y_i)=Y_{i+1}, \quad i=2,3,4, \\ \mu_8^{26}(Y_2,Y_5)=Y_7, & \mu_8^{26}(Y_3,Y_4)=-Y_7. \end{cases}$$

$$\begin{cases} \mu_8^{27}(X,Y_i)=iY_i, & i=0,1,2,3,5,6,7, \quad \mu_8^{27}(Y_0,Y_i)=Y_i, \quad i=2,3, \\ \mu_8^{27}(Y_0,Y_5)=2Y_5, & \mu_8^{27}(Y_0,Y_i)=3Y_i, \quad i=6,7, \\ \mu_8^{27}(Y_1,Y_i)=Y_{i+1}, & i=2,6, \quad \mu_8^{27}(Y_2,Y_i)=Y_{i+2}, \quad i=3,5. \end{cases}$$

$$\begin{cases} \mu_8^{28}(X,Y_i)=iY_i, & i=0,1,2,3,4,5,7, \quad \mu_8^{28}(Y_0,Y_2)=Y_2, \\ \mu_8^{28}(Y_0,Y_i)=2Y_i, & i=3,4, \quad \mu_8^{28}(Y_0,Y_5)=3Y_5, \quad \mu_8^{28}(Y_0,Y_7)=4Y_7, \\ \mu_8^{28}(Y_1,Y_3)=Y_4, & \mu_8^{28}(Y_2,Y_i)=Y_{i+2}, \quad i=3,5, \mu_8^{28}(Y_3,Y_4)=-Y_7 \end{cases}$$

$$\begin{cases} \mu_8^{29}(X,Y_i)=iY_i, & i=0,1,2,3,4,5, \quad \mu_8^{29}(X,Y_3')=3Y_3', \\ \mu_8^{29}(Y_0,Y_i)=Y_i, & i=2,3,4, \quad \mu_8^{29}(Y_0,Y_3')=Y_3', \\ \mu_8^{29}(Y_0,Y_5)=2Y_5, & \mu_8^{29}(Y_1,Y_i)=Y_{i+1}, \quad i=2,3 \\ \mu_8^{29}(Y_1,Y_3')=Y_4, & \mu_8^{29}(Y_2,Y_3)=Y_5. \end{cases}$$

$$\begin{cases} \mu_8^{30}(X,Y_i)=iY_i, \; i=0,1,2,3,4,5, \quad \mu_8^{30}(X,Y_0')=0, \quad \mu_8^{30}(Y_0,Y_5)=Y_5, \\ \mu_8^{30}(Y_0',Y_i)=Y_i, \quad i=2,3,4, \quad \mu_8^{30}(Y_1,Y_i)=Y_{i+1}, \quad i=2,3. \end{cases}$$

$$\begin{cases} \mu_8^{31}(X,Y_i)=iY_i, & i=0,1,2,3,4,5, \quad \mu_8^{31}(X,Y_0')=0, \\ \mu_8^{31}(Y_0,Y_i)=Y_i, & i=2,3, \quad \mu_8^{31}(Y_0',Y_i)=Y_i, \quad i=4,5, \\ \mu_8^{31}(Y_1,Y_i)=Y_{i+1}, & i=2,4. \end{cases}$$

$$\begin{cases} \mu_8^{32}(X,Y_i)=iY_i, & i=0,1,2,3,4,5, \quad \mu_8^{32}(X,Y_0')=0, \\ \mu_8^{32}(Y_0,Y_i)=Y_i, & i=2,4,5, \quad \mu_8^{32}(Y_0',Y_i)=Y_i, \quad i=3,4,5, \\ \mu_8^{32}(Y_1,Y_4)=Y_5, & \mu_8^{32}(Y_2,Y_3)=Y_5. \end{cases}$$

$$\begin{cases} \mu_8^{33}(X,Y_i)=iY_i, & i=0,1,2,3,4, \quad \mu_8^{33}(X,Y_0')=0, \\ \mu_8^{33}(X,Y"_0)=0, & \mu_8^{33}(Y_0,Y_2)=Y_2, \\ \mu_8^{33}(Y_0',Y_3)=Y_3, & \mu_8^{33}(Y"_0,Y_4)=Y_4. \end{cases}$$

De plus, ces lois sont deux à deux non isomorphes.

Ces algèbres ont aussi un deuxième groupe de cohomologie trivial. Le calcul de la dimension a été effectué sur Mathematica, en utilisant le programme élaboré à l'université de Séville par G. Valeiras.

5 Suite caractéristique d'une algèbre de Lie nilpotente

5.1 Définition

La suite caractéristique $c(\boldsymbol{n})$ du nilradical $\boldsymbol{n}$ est définie ainsi: soit $Y \in \boldsymbol{n} - \mathcal{D}(\boldsymbol{n})$ un vecteur de $\boldsymbol{n}$ n'appartenant pas à l'algèbre dérivée et considérons

$$c(Y) = (h_1, h_2, \cdots,)$$

la suite ordonnée des blocs de Jordan de l'opérateur nilpotent adY. Alors

$$c(\boldsymbol{n}) = Sup\{c(Y), Y \in \boldsymbol{n} - \mathcal{D}(\boldsymbol{n})\}$$

l'ordre étant l'ordre lexicographique sur l'ensemble des suites finies, est un invariant de $\boldsymbol{n}$ appelé *suite caractéristique.* Cet invariant de $\boldsymbol{n}$ est adapté aux déformations: si $\boldsymbol{n}'$ est une déformation de $\boldsymbol{n}$, alors $c(\boldsymbol{n}') \geq c(\boldsymbol{n})$.

5.2 Détermination des algèbres de lie rigides à partir de la suite caractéristique du nilradical

5.2.1 c=(1,1,..,1).

Dans ce cas $\boldsymbol{n}$ est abélienne. Si $\boldsymbol{g} = \boldsymbol{t} \oplus \boldsymbol{n}$ est rigide alors $rangS(X) = dim\boldsymbol{n} - 1$. Si $\boldsymbol{n}$ est abélienne, les seules équations de $S(X)$ sont relatives à l'action de $\boldsymbol{t}$ sur $\boldsymbol{n}$. D'où $dim\boldsymbol{t} = dim\boldsymbol{n}$ et $\boldsymbol{t}$ correspond à la sous algèbre des dérivations diagonales de l'algèbre abelienne $\boldsymbol{n}$. Ainsi $dim\boldsymbol{g} = 2p$, $\boldsymbol{g} = \boldsymbol{t} \oplus \boldsymbol{n}$ et si $\mathbf{r_2}$ est l'algèbre de Lie de dimension 2 définie par $[X, Y] = Y$, on a :

$$\boldsymbol{g} = \mathbf{r_2} \oplus \mathbf{r_2} \oplus \cdots \oplus \mathbf{r_2}$$

la somme étant directe. Cette algèbre vérifie $H^2(\boldsymbol{g}, \boldsymbol{g}) = 0$. Elle est bien rigide.

5.2.2 c = (2,1,...,1)

Ici $\boldsymbol{n}$ est une somme directe d'une algèbre de Heisenberg et d'une algèbre abélienne. Le cas abélien étant étudié ci dessus, on peut supposer que $\boldsymbol{n} = H_{2p+1}$ est l'algèbre de Heisenberg de dimension $2p + 1$. Ainsi $rgS(X) = 2p$. Compte tenue de la structure de H_{2p+1} :

$$[Y_{2i+1}, Y_{2i+2}] = Y_{2p+1} \quad i = 0, \cdots, p-1$$

le système linéaire associé

$$\begin{cases} a_1 + a_2 & = a_{2p+1} \\ a_3 + a_4 & = a_{2p+1} \\ \vdots & \vdots \\ a_{2p-1} + a_{2p} & = a_{2p+1} \end{cases}$$

est de rang p. D'où, sachant que $rgS(X) = 2p$, $dim t = p+1$. L'algèbre décomposable correspondante

$$\boldsymbol{g} = \boldsymbol{t}_{p+1} \oplus H_{2p+1}$$

définie dans la base correspondante

$$(X_0, X_1, \cdots, X_p, Y_1, Y_2, \cdots, Y_{2p+1})$$

par

$$\begin{cases} [X_1, X_j] = 0 \\ [Y_{2i-1}, Y_{2i}] = Y_{2p} \quad i = 1, \cdots, p \\ [X_0, Y_{2i-1}] = iY_{2i-1} \quad i = 1, \cdots, p \\ [X_0, Y_{2i}] = (2p+1-i)Y_{2i} \quad i = 1, \cdots, p \\ [X_0, Y_{2p+1}] = (2p+1)Y_{2p+1} \\ [X_i, Y_{2i-1}] = Y_{2i-1} \quad i = 1, \cdots, p \\ [X_i, Y_{2i}] = -Y_{2i} \quad i = 1, \cdots, p \end{cases}$$

est rigide. Notons que si $\boldsymbol{n} = H_{2p+1} \oplus \mathbf{a}_r$ où $\mathbf{a}_r$ est l'algèbre abelienne de dimension r, alors

$$\boldsymbol{g} = \boldsymbol{t} \oplus H_{2p+1} \oplus \mathbf{a}_r$$

est rigide dès que $dim t = r + p + 1$ et dans ce cas $\boldsymbol{g}$ est de la forme

$$\boldsymbol{g} = \boldsymbol{t}_{p+1} \oplus H_{2p+1} \oplus \mathbf{r}_2 \oplus \mathbf{r}_2 \cdots \oplus \mathbf{r}_2.$$

5.2.3 Le nilradical est métabelien

On a $c(\boldsymbol{n}) = (2, 2, \cdots, 2, 1, \cdots, 1)$. Ces algèbres sont stucturées par les algèbres de Lie métabéliennes dont la suite caractéristique est du type $(2, 2, \cdots, 2, 1)$. Supposons donc que $\boldsymbol{n}$ admette une telle suite caractéristique. Il existe une base $(Y_1, \cdots, Y_p, Z_1, \cdots, Z_p, Y)$ de $\boldsymbol{n}$ telle que

$$\begin{cases} [Y, Y_i] = Z_i \quad 1 \leq i \leq p \\ [Y_i, Y_j] = \sum a_{ij}^k Z_k \quad 1 \leq i < j \leq p \end{cases}$$

(les a_{ij}^k étant des paramètres libres). Si $a_{ij}^k = 0$, alors $rg(S(X)) = p$ et $dim t = p + 1$. L'algèbre $\boldsymbol{g}$ s'écrit dans ce cas:

$$\begin{cases} [X, Y_i] = iY_i \ ; \ [X, Z_i] = (1+p+1)Z_i \ ; \ [X, Y] = (p+1)Y \\ [X_i, Y] = 0 \quad i = 1, \cdots, p \\ [X_i, Y_j] = 0 \ ; \ [X_i, Z_j] = 0 \quad i = 1, \cdots, p \quad j = 1, \cdots, p \quad i \neq j \\ [X_i, Y_i] = Y_i \ ; \ [X_i, Z_i] = Z_i \quad i = 1, \cdots, p \\ [X_{p+1}, Y_i] = iY_i \ ; [X_{p+1}, Z_i] = (i+p+1)Z_i \quad i = 1, \cdots, p \\ [X_{p+1}, Y] = (p+1)Y \\ [Y, Y_i] = Z_i \quad i = 1, \cdots, p \end{cases}$$

et elle est rigide.

Le cas $a_{ij}^k \neq 0$ se traite de manière analogue. Par exemple, supposons $dim\boldsymbol{n} = 5$. Si l'un des $a_{ij}^k \neq 0$, alors $\boldsymbol{n}$ est isomorphe à :

$$[Y, Y_i] = Z_i \ , i = 1, 2 \ \ ; \ \ [Y_1, Y_2] = Z_1.$$

Le tore maximal est de dimension 3. Une base de $\boldsymbol{n}$ réduisant simultanément les opérateurs de $\boldsymbol{t}$ peut être donnée par :

$$Y' = Y + Y_2 \ \ , \ Y_i' = Y_i \ i = 1, 2 \ \ , \ Z_i' = Z_i \ , \ i = 1, 2, 3.$$

L'algèbre décomposable correspondante $\boldsymbol{g} = \boldsymbol{t} \oplus \boldsymbol{n}$ définie par

$$\begin{cases} [X, Y_1'] = 2Y_2' \ ; \ [X, Y_2'] = 3Y_2' \ ; \ [X, Z_1] = 5Z_1 \quad [X, Z_2] = 4Z_2 \\ [X, Y'] = Y' \\ [X_2, Y_1'] = Y_1' \ ; [X_2, Y_2'] = 2Y_2' \ ; \ [X_2, Z_1] = 3Z_1 \ ; [X_2, Z_2] = 2Z_2 \\ [X_3, Y_2] = Y_2 \ ; [X_3, Z_1] = Z_1 \ ; \ [X_3, Z_2] = Z_2 \\ [Y', Y_2'] = Z_2 \ ; \ [Y', Y_1'] = 0 \ ; \ [Y_1', Y_2'] = Z_1 \end{cases}$$

est rigide.

La dimension du tore maximal dépend du rang du tenseur (a_{ij}^k). Par exemple, le nilradical de l'algèbre métabélienne suivante est de dimension 7 et de rang 2:

$$\begin{cases} [X, Y_i] = (i+1)Y_i \ ; \ [X, Z_i] = (i+2)Z_i \ \ i = 1, 2, 3 \\ [X, Y] = Y \\ [X_2, Z_i] = iZ_i \ \ i = 1, 2, 3 \\ [X_2, Y_i] = Y_i \ \ i = 1, 2, 3 \\ [Y, Y_i] = Z_i \ \ i = 1, 2, 3 \\ [Y_1, Y_2] = Z_3 \end{cases}$$

Cette algèbre de Lie est aussi rigide.

Dans le paragraphe suivant, nous allons décrire toutes les algèbres de Lie résolubles dont le nilradical $\boldsymbol{n}$ est filiforme, c'est-à-dire vérifiant

$$c(\boldsymbol{n}) = (n-1, 1) \ \ et \ \ dim\boldsymbol{n} = n.$$

6 Conditions nécessaires et suffisantes pour qu'une algèbre de Lie filiforme soit le nilradical d'une algèbre de Lie rigide

Soit $\boldsymbol{n}$ une algèbre de Lie nilpotente. Les tores maximaux de dérivations externes sont deux à deux conjugués. Leur dimension commune est par définition *le rang de* $\boldsymbol{n}$. L'algèbre $\boldsymbol{n}$ est dite *filiforme* si sa suite caractéristique est égale à $(n-1, 1)$ où $n = dim\boldsymbol{n}$. Pour une telle algèbre le rang est au plus égal à 2.

Théorème 3 *Toute algèbre de Lie filiforme* $\mathfrak{n}$ *de rang* 2 *est le nilradical d'une algèbre de Lie rigide. Deux telles algbres de Lie sont isomorphes si et seulement si leurs nilradicaux sont isomorphes.*

Théorème 4 *Une algèbre de Lie filiforme* $\mathfrak{n}$ *de rang* 1 *est le nilradical d'une algèbre de Lie rigide si et seulement si* $\mathfrak{n}$ *vérifie l'un des trois cas suivants:*
1er *cas:* $\mathfrak{n}$ *admet une dérivation diagonalisable dont les valeurs propres sont* $1, k, \cdots, n+k-2$, $k \geq 2$, *dès que* $k = n-5$, $k = n-6$, *ou* $k \leq (n-7)/3$ *si* $n-k$ *est impair et* $k \leq (n-6)/3$ *si* $n-k$ *est pair.*
2eme *cas:* $\mathfrak{n}$ *est de dimension paire* n *et admet une dérivation diagonalisable dont les valeurs propres sont* $1, k, k+1, \cdots, k+n-3, 2k+n-3$, *avec* $k \in \mathbb{N}$ *pair vérifiant soit* $k = n-4$, *soit* $k = n-6$, *soit* $k = 2$ *et* $n = 10$, *soit* $k \leq \left[\frac{n-k}{6}\right] + \frac{n-k-8}{2}$.
3eme *cas:* $\mathfrak{n}$ *est de dimension paire* n *et admet une dérivation diagonalisable dont les valeurs propres sont* $1, k, k+1, \cdots, k+n-3, 2k+n-3$, *avec* $k \in \mathbb{N}$ *impair,* $k \neq 1$ *et vérifiant soit* $k \leq \left[\frac{n-k-3}{6}\right] + \frac{n-k-9}{2}$, *soit* $k = n-5$, *soit* $k = 3$ *et* $n = 14$.

Une algèbre de Lie rigide dont le niradical est filiforme est nécessairement résoluble. En effet, l'algèbre de dérivations d'une algèbre de Lie filiforme n'admet pas de composantes semi-simples. Nous pouvons nous restreindre, dans la démonstration, au seul cas résoluble.
Remarque : Le théorème précédent fournit la classification générale des rigides dont le nilradical est filiforme après avoir remarqué que deux valeurs distinctes du paramêtres k donnent deux algèbres rigides non isomorphes.
Le reste de ce travail est consacré à la démonstration de ce théorème. Nous allons dans un premier temps rappeler la liste des algèbres filiformes de rang non nul.

7 Algèbres de Lie filiformes de rang non nul

Théorème 5 *[7] Soit* $\mathfrak{n}$ *une algèbre de Lie filiforme de dimension* n *et admettant une dérivation diagonalisable* f *non nulle. Il existe alors une base* $\{Y_1, Y_2, \cdots, Y_n\}$ *de* $\mathfrak{n}$ *dont les crochets vérifient l'un des cas suivants:*

$$i) \mathfrak{g} = L_n$$

$$[Y_1, Y_i] = Y_{i+1} \quad 2 \leq i \leq n-1$$
$$[Y_i, Y_j] = 0 \quad 1 \leq i, j \leq n$$

$$ii) \mathfrak{g} = A_n^k(\lambda_1, \cdots, \lambda_t) \ , \quad 2 \leq k \ , \quad t = [\frac{n-k+1}{2}]$$
$$[Y_1, Y_i] = Y_{i+1} \quad 2 \leq i \leq n-1$$

$$[Y_i, Y_{i+1}] = \lambda_{i-1} Y_{2i+k-1} \ , \quad 2 \leq i \leq [\frac{n-k+1}{2}]$$

$$[Y_i, Y_j] = a_{ij} Y_{i+j+k-2} \ , \quad 2 \leq i < j \quad et \quad i+j+k-2 \leq n.$$

$$iii) \boldsymbol{g} = Q_n \ , \quad n = 2m$$

$$[Y_1, Y_i] = Y_{i+1} \ , \quad 2 \leq i \leq n-2$$

$$[Y_i, Y_{n-i+1}] = (-1)^{i+1} Y_n \ , \quad 2 \leq i \leq n-1$$

$$iv) \boldsymbol{g} = B_n^k(\lambda_1, \cdots, \lambda_t) \ , \ n = 2m \ , \ 2 \leq k \leq n-3 \ , \ t = [\frac{n-k}{2}]$$

$$[Y_1, Y_i] = Y_{i+1} \quad 2 \leq i \leq n-2$$

$$[Y_i, Y_{n-i+1}] = (-1)^{i+1} Y_n \quad 2 \leq i \leq n-1$$

$$[Y_i, Y_{i+1}] = \lambda_{i-1} Y_{2i+k-1} \ , \quad i = 2, \cdots, [\frac{n-k}{2}]$$

$$[Y_i, Y_j] = a_{ij} Y_{i+j+k-2} \ , \quad 2 \leq i, j \leq n-2 \ et \ i+j+k-2 \leq n-1 \ , \ j \neq i+1$$

$$v) \boldsymbol{g} = C_n(\lambda_1, \cdots, \lambda_t) \ , \ n = 2m+2 \ , \ t = m-1$$

$$[Y_1, Y_i] = Y_{i+1} \quad 2 \leq i \leq n-2$$

$$[Y_i, Y_{n-i+1}] = (-1)^{i-1} Y_n \quad 2 \leq i \leq m+1$$

$$[Y_i, Y_{n-i-2k+1}] = (-1)^{i+1} \lambda_k Y_n \ , \ 2 \leq i \leq n-2-2k \ et \ 1 \leq k \leq m-1.$$

où $(\lambda_1, \cdots, \lambda_t)$ *sont des paramètres vérifiant les relations polynomiales découlant des identités de Jacobi pour les éléments de base* $\{Y_1, \cdots, Y_n\}$.

Il est aisé de vérifier que le rang d'une algèbre de Lie filiforme est au plus égal à 2. Plus précisément, il est égal à 2 si $\boldsymbol{n}$ est isomorphe à L_n ou Q_{2m} et 1 dans les autres cas. L'étude des algèbres de Lie rigides dont le nilradical est filiforme se réduit donc à l'étude de la rigidité des algèbres résolubles décomposables $\boldsymbol{g} = \boldsymbol{t} \oplus \boldsymbol{n}$ où $\boldsymbol{n}$ est une des algèbres définies ci dessus et $\boldsymbol{t}$ le tore maximal de $\boldsymbol{n}$.

8 Algèbres de Lie rigides dont le nilradical filiforme est de rang 2

8.1

Proposition 2 *Toute algèbre de Lie rigide non scindée dont le nilradical est isomorphe à l'algèbre filiforme L_n est de dimension $n+2$ et est isomorphe à l'algèbre de Lie définie par :*

$$\begin{cases} [X_1, Y_i] = iY_i & 1 \leq i \leq n \\ [X_2, Y_i] = Y_i & 2 \leq i \leq n \\ [Y_1, Y_i] = Y_{i+1} & 2 \leq i \leq n-1 \end{cases}$$

De plus ces algèbres vérifient $H^2(\mathfrak{g}, \mathfrak{g}) = 0$.

Ces algèbres ont été étudiées dans [2].

8.2

Proposition 3 *Toute algèbre de Lie rigide non scindée dont le nilradical est isomorphe à l'algèbre de Lie Q_n $(n = 2m)$ est résoluble de dimension $n+2$ et isomorphe à l'algèbre de Lie définie par:*

$$\begin{cases} [X_1, Y_1] = Y_1 \\ [X_1, Y_i] = (i-2)Y_i \ , \ 3 \leq i \leq n-1 \\ [X_1, Y_n] = (n-3)Y_n \\ [X_2, Y_n] = 2Y_n \\ [X_2, Y_i] = Y_i \ , \ i = 2, \cdots, n-1 \\ [Y_1, Y_i] = Y_{i+1} \ , \ 2 \leq i \leq n-2 \\ [Y_i, Y_{n+1-i}] = (-1)^i Y_n \ , \ i = 1, \cdots, m \end{cases}$$

En effet, si $\boldsymbol{n} = Q_{2m}$, le tore $\boldsymbol{t}$ est engendré par les dérivations

$$f_1(Y_1) = Y_1 \ , \ f_1(Y_i) = (i-2)Y_i \ , \ i = 2, \cdots, 2m-1 \ , \ f_1(Y_{2m}) = (2m-3)Y_{2m}$$

$$f_2(Y_1) = 0 \ , \ f_1(Y_i) = Y_i \ , \ i = 2, \cdots, 2m-1, f_2(Y_{2m}) = 2Y_{2m}.$$

Comme Q_{2m} est la seule algèbre de Lie, à isomorphisme près, admettant un tel tore de dérivations, l'algèbre de Lie ci dessus est rigide.

9 Algèbres de Lie rigides dont le nilradical filiforme est de rang 1

9.1

Soit $(Y_1, \cdots, Y_n)$ une base de $\boldsymbol{n}$ définie dans le théorème 1. Comme le rang de A_n^k est 1, on a $dim \boldsymbol{t} = 1$. Une dérivation diagonale de $\boldsymbol{n}$ est donnée par:

$$f(Y_1) = Y_1 \ , \ f(Y_2) = kY_2 \ , \ \cdots \ , \ f(Y_{n-1}) = (n-3+k)Y_{n-1},$$
$$f(Y_n) = (n-2+k)Y_n.$$

Les classifications des algèbres de Lie rigides du type $\boldsymbol{g} = \boldsymbol{t} \oplus A_n^k$ passe par la résolution des équations polynômiales que doivent vérifier les paramètres $\lambda_1, \cdots, \lambda_t$. On définit le poids ρ de l'équation de Jacobi $\sum [X_i, [X_j, X_l]] = 0$ par $\rho = i + j + l$. Il est clair que $6 \leq \rho \leq 3n - 3$.

Lemme 1 *i) Le nombre d'équations de Jacobi de poids ρ est :*

$$N(\rho) = 3\rho'^2 - 3\rho' + 1 \quad si \quad \rho = 6\rho'$$
$$N(\rho) = 3\rho'^2 - 2\rho' \quad si \quad \rho = 6\rho' + 1$$
$$N(\rho) = 3\rho'^2 - \rho' \quad si \quad \rho = 6\rho' + 2$$
$$N(\rho) = 3\rho'^2 \quad si \quad \rho = 6\rho' + 3$$
$$N(\rho) = 3\rho'^2 + \rho' \quad si \quad \rho = 6\rho' + 4$$
$$N(\rho) = 3\rho'^2 + 2\rho' \quad si \quad \rho = 6\rho' + 5$$

Comme conséquence, on en déduit que le nombre d'équations de Jacobi concernant les paramètres $\lambda_1, \cdots, \lambda_t$ et définissant l'algèbre de Lie $\boldsymbol{g}$ (ou A_n^k) est égal à $\sum_{\rho=6}^{3n-3} N(\rho)$. Ce système d'équations polynômiales a été entièrement résolu dans [2] et [3] dans les cas $k = 1$ et $k = 2$. On se propose de déterminer l'ensemble algébrique paramétré par $\lambda_1, \cdots, \lambda_t$ et défini par les N équations algébriques de degré 1 ou 2 et homogènes.

Choisissons une base adaptée. Les équations de poids $\rho = 1 + i + j$ sont les seules qui soient linéaires. Elles s'écrivent :

$$a_{ij} = a_{ij+1} + a_{i+1j}$$

ce qui implique

$$a_{ij} = \sum_{s=0}^{j+h} C_{j-i-h-1}^{s} \lambda_{i-1+s}.$$

On est donc conduit à n'étudier que les équations de poids $\rho = i + j + l$ avec $i \geq 2, j > i$. La première équation non triviale est celle de poids $\rho = 9 = 2+3+4$. Cette équation concerne le vecteur de poids $3k + 3$.

Proposition 4 *Si $k > n-3$, l'algèbre de Lie $\mathfrak{g}$ n'est pas rigide.*
Si $k = n-3$, il n'existe à isomorphisme près qu'une seule algèbre de type $\mathfrak{t}\oplus A_n^{n-3}$ qui est définie par:

$$[X,Y_1] = Y_1 \ , \ [X,Y_2] = kY_2 \ , \ \cdots \ , \ [X,Y_{n-1}] = (n-3+k)Y_{n-1}$$

$$[X,Y_n] = (n-2+k)Y_n.$$

$$[Y_1,Y_i] = Y_{i+1} \quad 2 \leq i \leq n-2$$

$$[Y_2,Y_3] = Y_{n-1}.$$

Si $k = n-4$, il n'existe à isomorphisme près qu'une seule algèbre de type $\mathfrak{t}\oplus A_n^{n-4}$ qui est définie par:

$$[X,Y_1] = Y_1 \ , \ [X,Y_2] = kY_2 \ , \ \cdots \ , \ [X,Y_{n-1}] = (n-3+k)Y_{n-1}$$

$$[X,Y_n] = (n-2+k)Y_n.$$

$$[Y_1,Y_i] = Y_{i+1} \quad 2 \leq i \leq n-2$$

$$[Y_2,Y_3] = Y_{n-1} \ , \quad [Y_2,Y_4] = Y_n.$$

En effet si $2k+1 > n-k-2$, $[Y_i,Y_j] = 0$ et le tore est de dimension 2.

Supposons $k \leq n-5$

Lemme 2 *Si $n < 2k+5$, alors $\mathfrak{g}$ n'est pas rigide.*

En effet, l'équation de Jacobi de poids $\rho = 9$ concerne le vecteur de poids $k+k+1+k+2 = 3k+3$. Cette équation est triviale si $3k+3 > n+k-2$ qui est le poids maximum.

On suppose donc

$$\frac{n-5}{2} < k < n-4$$

et on va chercher les équations de Jacobi indépendantes. Nous avons vu que la première équation non triviale est celle de poids $\rho = 9$. Elle s'écrit:

$$-a_{2,s}a_{3,4} + a_{3,s-1}a_{2,4} - a_{4,s-2}a_{2,3} = 0 \quad s = k+5$$

Lemme 3 *L'équation E_{10} correspondant au poids $\rho = 10 = 2+3+5$ est liée à l'équation E_9.*

En effet E_{10} s'écrit :

$$-a_{2,s+1}a_{3,5} + a_{3,s}a_{2,5} - a_{5,s-2}a_{2,3} = 0$$

Mais $a_{ij} = a_{ij+1} + a_{i+1j}$. D'où :

$$E_{10} = -(a_{2,s} - a_{3,s})a_{3,4} + (a_{4,s-2} - a_{4,s-1}a_{2,3} - (a_{3,s-1} - a_{4,s-1})(a_{2,4} - a_{2,3})$$

$$= E_9 + -a_{4,s_1}a_{3,4} - a_{4,s-1}(a_{2,4} - a_{2,3}) + a_{4,s-1}a_{3,4}$$

et donc

$$E_{10} = E_9.$$

Conséquences

Dans l'idéal de Jacobi, on récupère une relation de dépendance entre les équations pour $\rho \geq 10$ et $k = 2$. Ainsi *le schéma $\mathcal{L}^n$ n'est pas réduit au point $\boldsymbol{n} \oplus \boldsymbol{t}$ dès que $n \geq 11$.*

Remarques.

1. L'équation E_9 est polynômiale de degré 2 en les variables $\lambda_1, \cdots, \lambda_{t_0}$ où $t_0 = \frac{4+k}{2}$ si k est pair, ou $\frac{3+k}{2}$ si k est impair.

2. Il existe deux équations de poids 11 : $\rho = 2 + 3 + 6$ et $\rho = 2 + 4 + 5$. La première éqution est évidemment indépendante de E_9. On notera E_{11}^{236} et E_{11}^{245} ces équations.

Lemme 4

$$E_{11}^{245} = -E_{11}^{236} + E_{10}^{235}.$$

En effet $E_{11}^{245} = -a_{2,s+2}a_{4,5} + a_{4,s}a_{2,5} - a_{5,s-1}a_{2,4}$
$= -a_{2,s+2}(a_{3,5} - a_{3,6}) + (a_{3,s} - a_{3,s-1})a_{2,5} - (a_{5,s-2} - a_{6,s-2})a_{2,4}$
$= a_{2,s+2}a_{3,6} - a_{3,s+1}a_{2,5} - a_{6,s-2}a_{2,4} - a_{2,s+2}a_{3,5} + a_{3,s}a_{2,5} - a_{5,s-2}a_{2,4}$
$= a_{2,s+2}a_{3,6} - a_{3,s+1}(a_{2,6} + a_{3,5}) + a_{6,s-2}a_{2,3} - a_{2,s+2}a_{3,5} + a_{3,s}a_{2,5} - a_{5,s-2}a_{2,4}$
$= -E_{11}^{236} - a_{3,s+1}a_{3,5} - a_{2,s+2}a_{3,5} + a_{3,s}a_{2,5} - a_{5,s-2}a_{2,4}$
$= -E_{11}^{236} - a_{2,s+1}a_{3,5} + a_{3,s}a_{2,5} - a_{5,s-2}a_{2,4}$
$= -E_{11}^{236} + E_{10}$

Afin de bien comprendre les relations de dépendance, on décrit les relations correspondant au poids $\rho = 12$.

Lemme 5

$$E_{12}^{246} = -E_{12}^{237} + E_{11}^{236}.$$

$$E_{12}^{345} = E_{12}^{237} - 2E_{11}^{236} + E_9^{234}.$$

Le système $\{E_9, E_{11}^{236}, E_{12}^{237}\}$ est un système réduit d'équations à $t_0 + 1$ ou $t_0 + 2$ paramètres suivant la parité de $n + k$.

Les deux premières relations se traitent comme précédemment. Quant au système réduit, il s'écrit dans le cas $k = 2$:

$$\begin{cases} -2\lambda_1\lambda_3 + 3\lambda_2^2 - \lambda_2\lambda_3 = 0 \\ 2\lambda_1\lambda_4 + \lambda_2(-4\lambda_3 - \lambda_4) + \lambda_3(6\lambda_3 - \lambda_4) = 0 \\ -4\lambda_3^2 + 3\lambda_3\lambda_4 + 3\lambda_2\lambda_4 = 0 \end{cases}$$

Ce système a été résolu dans [2].

Ceci étant notons, pour un poids ρ fixé, S_ρ le système réduit des équations de Jacobi de poids inférieur ou égal à ρ et déterminons le système réduit $S_{\rho+1}$. Posons $\rho' = \rho - 5$ et supposons que

$$\rho' < n - 2k - 1.$$

Lemme 6

$$E_{\rho+1}^{ijk} = -E_{\rho+1}^{ij-1k} + E_{\rho}^{ijk-1} \quad (i,j,k) > (2,3,\rho'+1).$$

La démonstration est analogue à celle du lemme 2 ou du lemme 3.

Proposition 5 *Si $\rho < n - 2k + 4$ alors S_ρ est formé des équations:*

$$S_\rho = \{E_9^{234}, E_{11}^{236}, \cdots, E_\rho^{23\rho'}\}.$$

Ainsi, pour ρ donné, le système réduit est formé de $\rho - 9$ équations. Ceci permet d'affirmer :

Proposition 6 *Soit $\mathfrak{g} = \mathfrak{t} \oplus \mathcal{A}_n^k$ une algèbre décomposable où $\mathfrak{t}$ est un tore externe de dérivations. Alors :*

1) Si $2k+5 \leq n \leq 3k+7$ si $n-k$ est impair, ou si $2k+5 \leq n \leq 3k+6$ si $n-k$ est pair, alors $\mathfrak{g}$ n'est pas rigide.

2) Si $n \geq 3k+7$ si $n-k$ est impair, ou si $n \geq 3k+6$ si $n-k$ est pair, alors $\mathfrak{g}$ est rigide dès qu'il existe i avec $\lambda_i \neq 0$.

Supposons $n \geq 3k+7$ (ou $3k+6$). Il existe au moins deux solutions non triviales du système de Jacobi dont les algèbres de Lie correspondantes sont rigides et non isomorphes. Il suffit de prendre $(\lambda_1 = 1, \lambda_i = 0, i \neq 0)$ et $(\lambda_i = 0, i \neq t, \lambda_t = 1)$. En conclusion, on peut affirmer que:

Proposition 7 *Soit $\mathfrak{g}$ une algèbre de Lie isomorphe à $\mathfrak{t} \oplus \mathcal{A}_n^k$ avec $n \geq 3k+7$ si $n-k$ est impair, ou si $n \geq 3k+6$ si $n-k$ est pair, dont les paramètres $(\lambda_1, \cdots, \lambda_t)$ sont non tous nuls. Alors $\mathfrak{g}$ est rigide et deux telles algèbres dont les paramètres $(\lambda_1, \cdots, \lambda_t)$ et $(\lambda_1', \cdots, \lambda_t')$ correspondent à des points distincts de l'espace projectif $\mathcal{P}^n(\mathbb{C})$ ne sont pas isomorphes.*

9.2

Considérons l'algèbre de Lie B_n^k définie par :

$$[Y_1, Y_i] = Y_{i+1} \quad 2 \leq i \leq n-2$$

$$[Y_i, Y_{n-i+1}] = (-1)^{i+1} Y_n \quad 2 \leq i \leq n-1$$

$$[Y_i, Y_j] = a_{ij} Y_{k+i+j-2}$$

$$2 \leq i, j \leq n-2 \ et \ i+j+k-2 \leq n-1$$

Elle est définie pour n pair. Posons $n = 2p+4$. Il existe une dérivation f de cette algèbre de Lie $\boldsymbol{n}$, engendrant le tore externe maximal, dont les valeurs propres sont :

$$1, k, k+1, k+2, \cdots, k+2p+1, 2k+2p+1$$

la base ci dessus étant la base des vecteurs propres correspondants.
Il est clair que si $k \geq 2p$, l'algèbre B_n^k est isomorphe à Q_n. Ce cas a déjà été étudié. Supposons donc $k < 2p$.
Premier cas : k est pair
Posons $k = 2p - 2l$ ce qui implique $n = k + 2l + 4$; les valeurs propres de f sont

$$1, k, k+1, k+2, \cdots, 2k, \cdots, 2k+2l+1, 3k+2l+1.$$

Les constantes de structure a_{ij} sont toutes des combinaisons linéaires des constantes génératrices $\lambda_i = a_{i+1,i+2}$ pour $i = 1, \cdots, l+1$.
Commençons à examiner les trois cas particuliers $l = 0, l = 1, l = 2$.

$l = 0$.
L'algèbre de Lie $\boldsymbol{g}$ est définie par:

$$[X, Y_1] = Y_1; \ [X, Y_i] = (k+i-2)Y_i, \ i = 2, \cdots, n-1; \ [X, Y_n] = (3k+1)Y_n$$

$$[Y_1, Y_i] = Y_{i+1}, \ i = 2, \cdots, n-2; \ [Y_i, Y_{n-i+1}] = (-1)^i Y_n, \ i = 2, \cdots, n-2;$$

$$[Y_2, Y_3] = Y_{n-1}$$

avec $k = n-4$ et n pair. *Cette algèbre est rigide.*

$l = 1$.
Il n'existe aussi qu'une seule algèbre (à isomorphisme près) vérifiant ces hypothèses. elle est donnée par :

$$[X, Y_1] = Y_1; \ [X, Y_i] = (k+i-2)Y_i, \ i = 2, \cdots, n-1; \ [X, Y_n] = (3k+3)Y_n$$

$$[Y_1, Y_i] = Y_{i+1}, \ i = 2, \cdots, n-2; \ [Y_i, Y_{n-i+1}] = (-1)^i Y_n, \ i = 2, \cdots, n-2;$$

$$[Y_2, Y_3] = Y_{n-3};\ [Y_2, Y_4] = Y_{n-2};\ [Y_2, Y_5] = 9Y_{n-1};\ [Y_3, Y_4] = 2Y_n - 1$$

avec $k = n - 6$ et n pair. *Cette algèbre est rigide.*

$l = 2$.
Ce cas diffère des précédents.

Lemme 7 *Si $k = n-8$, alors les algèbres de Lie $\mathfrak{g}$ correspondantes sont rigides si et seulement si $k = 2$ (et $n = 10$).*

Il existe dans ce cas deux algèbres non isomorphes (de dimension 11) :
$\mathfrak{g}_{11}^1$

$$[X, Y_1] = Y_1;\ [X, Y_i] = iY_i,\ i = 2, \cdots, 9;\ [X, Y_{10}] = 11Y_{10}$$

$$[Y_1, Y_i] = Y_{i+1},\ i = 2, \cdots, 8;\ [Y_i, Y_{11-i}] = (-1)^i Y_{10},\ i = 2, \cdots, 5;$$

$$[Y_2, Y_i] = Y_{i+2},\ i = 3, 4;\ [Y_2, Y_i] = -Y_{i+2},\ i = 6, 7;$$

$$[Y_3, Y_i] = Y_{i+3},\ i = 4, 5;\ [Y_4, Y_5] = Y_9.$$

$\mathfrak{g}_{11}^2$

$$[X, Y_1] = Y_1;\ [X, Y_i] = iY_i,\ i = 2, \cdots, 9;\ [X, Y_{10}] = 11Y_{10}$$

$$[Y_1, Y_i] = Y_{i+1},\ i = 2, \cdots, 8;\ [Y_i, Y_{11-i}] = (-1)^i Y_{10},\ i = 2, \cdots, 5;$$

$$[Y_2, Y_i] = Y_{i+2},\ i = 3, 4;\ [Y_2, Y_5] = 2Y_7;\ [Y_2, Y_6] = 3Y_8;\ [Y_2, Y_7] = 7Y_9;$$

$$[Y_3, Y_i] = -Y_{i+3},\ i = 4, 5;\ [Y_3, Y_6] = -4Y_9;\ [Y_4, Y_5] = 3Y_9.$$

Retournons au cas général en supposant $l \geq 3$.

Proposition 8 *Si $2l < k + 2$, l'algèbre décomposable $\mathfrak{g}$ de nilradical B_n^k n'est pas rigide.*

Ceci est une conséquence du paragraphe précédent. On considère les paramètres $\lambda_1 = a_{23}, \lambda_2 = a_{34}, \cdots, \lambda_{l+1} = a_{k+l+2,k+l+3}$. Ces paramètres engendrent tous les a_{ij}. Ils sont en nombres $l + 1$. D'après le paragraphe précédent le nombre d'équations de Jacobi indépendantes est $2l - k - 2$ si ce nombre est positif, sinon il est égal à 0.
Supposons donc $2l \geq k + 2$.

Proposition 9 *Soit $\mathfrak{g} = \mathfrak{t} \oplus B_n^k$ une algèbre décomposable de dimension $n+1 = k + 2l + 5$ avec $\dim \mathfrak{t} = 1$.*
1) Si $k > \left[\frac{l+2}{3}\right] + l - 2$, alors $\mathfrak{g}$ n'est pas rigide.
2) Si $k \leq \left[\frac{l+2}{3}\right] + l - 2$, alors $\mathfrak{g}$ est rigide.

En effet les coefficient a_{ij} sont définis à partir des paramêtres λ_i par

$$a_{r+2,s+2} = \sum_{i=1}^{[\frac{s-r+1}{2}]} (-1)^{i+1} S^i_{r-s-2i+2} \lambda_{s+i}, \quad s \geq r$$

où la suite S^j_i est déterminée par les relations

$$S^i_p = S^i_{p-1} + S^{i-1}_p \quad i = 1, \cdots, l+1, \quad S^1_p = S^p_1 = 1 \ \forall p \in \mathbb{N}.$$

Comme $B^k_n/\mathbb{C}Y_n$ est isomorphe à A^k_{n-1}, les coefficients a_{ij} vérifient les mêmes relations de Jacobi correspondant au cas A^k_{n-1}. Ces coefficients vérifient de plus les équations linéaires suivantes, qui ne sont rien d'autre que les équations de Jacobi relatives aux vecteurs Y_2, Y_{2+h}, Y_{3+2l-h}, pour $2h < 2l+1$ et $h \geq 1$, $Y_{2+s}, Y_{2+s'}, Y_{3+2l-5s'}$ pour $0 \leq s \leq s'$ et $s + 2s' \leq 2l+1$.

Lemme 8 *Le rang du système linéaire relatif aux λ_i est égal à $\left[\frac{l+2}{3}\right]$.*

En effet, les équations de Jacobi relatives aux triplets $(Y_2, Y_{2+h}, Y_{3+2l-h})$ sont:

$$a_{h+2,2l+3-h} - (-1)^h a_{2,2l+3-h} + (-1)^{h-1} a_{2,h+2} = 0,$$

On remplace les a_{ij} par leurs valeurs fontion des λ_i. On obtient:

$$\sum_{i=1}^{l-h+1} (-1)^{i+1} S^i_{2(l-h-i+1)+1} \lambda_{h+i} - (-1)^h \sum_{i=1}^{[\frac{l+2-h}{2}]} (-1)^{i+1} S^i_{2(l-i+1)+1-h} \lambda_i$$

$$+(-1)^{h-1} \sum_{i=1}^{[\frac{h+1}{2}]} (-1)^{i+1} S^i_{h-2(i-1)} \lambda_i = 0.$$

Les équations relatives aux triplets $(Y_{2+s}, Y_{2+s'}, Y_{2+2l-s-s'+1}$ s'écrivent:

$$(-1)^{s+1} a_{2+s',2l+3-s-s'} - (-1)^{s'+1} a_{2+s,2l+3-s-s'} + (-1)^{4+2l-s-s'} a_{2+s,2+s'} = 0,$$

soit, en fonction des λ_i :

$$(-1)^{s+1} \sum_{i=1}^{2l-s-2s'+2} (-1)^{i+1} S^i_{2(l-s-s'-2i+3)+1} \lambda_{s'+i}$$

$$- (-1)^{s'} \sum_{i=1}^{\left[\frac{2l+2-2s-s'}{2}\right]} (-1)^{i+1} S^i_{2l-2i+3-2s-s'} \lambda_{s+i}$$

$$+(-1)^{4+2l-s-s'} \sum_{i=1}^{\left[\frac{s'-s+1}{2}\right]} (-1)^{i+1} S^i_{s'-s-2i+2)} \lambda_{s+i} = 0.$$

En particulier, les équations relatives aux triplets $(Y_{2+2s}, Y_{3+2s}, Y_{2+2l-4s})$ avec $6s \leq 2l - 1$ sont de la forme:

$$-2\lambda_{2s+1} + (2l - 6s - 3)\lambda_{2s+2} + \cdots = 0,$$

ainsi le système est de rang supérieur ou égal à $\left[\frac{l-1}{3}\right] + 1$. Si nous notons par $E_{s,s'}$ l'équation $(Y_{2+s}, Y_{2+s'}, Y_{3+2l-s-s'})$, nous obtenons les relations suivantes:

$$E_{s,s'} = -E_{s,s'-1} + E_{s-1,s'}$$

ce qui montre que toutes les équations de Jacobi se déduisent des équations $E_{s,s+1}$. Comme de plus les équations $E_{2s+1,2s+2}$ coincident avec les équations $E_{2s+2,2s+3}$, le système de Jacobi se réduit au système des équations $E_{2s,2s+1}$ qui sont indépendantes, comme nous l'avons vu plus haut. d'où le lemme.
Remarque. Si $k < \left[\frac{l+2}{3}\right] + l - 2$ les équations de Jacobi ne semblent pas avoir de solutions (ceci est vrai pour $l = 2, 3$ et 4). Ainsi, sous ces hypothèses, non seulement il n'existerait pas d'algèbres de Lie rigides dont le nilradical est isomorphe à B_n^k, mais il n'existerait pas d'algèbres de Lie décomposables de ce type.

Corollaire 2 *Soit $\mathbf{g} = \mathbf{t} \oplus B_n^k$ (k et n pairs) une algèbre de Lie décomposable de dimension $n + 1$. Elle est rigide si et seulement si $k \leq \left[\frac{l+2}{3}\right] + l - 2$ et deux solutions $(\lambda_1, \cdots, \lambda_{l+1})$ distinctes dans $\mathcal{P}^{l+1}$ correspondent à des algèbres non isomorphes.*

Deuxième cas : k est impair

Il est clair que $k + 1 < 2p$. Posons $k = 2p - 2l - 1$ ce qui donne $n = k + 2l + 5$. Les valeurs propres de la dérivation diagonale f sont

$$1, k, k + 1, \cdots, 2k + 2l + 2, 3k + 2l + 2.$$

Le nombre de constantes de structure indéterminées est égal à $l + 1$. Les cas correspondants à $l = 0, 1, 2$ et 3 étant d'un type particulier, nous les étudions dans un premier temps.
$l = 0$ $(k = 2p - 1)$. Il n' existe qu'une seule loi rigide à isomorphisme près :

$$[X, Y_1] = Y_1, [X, Y_i] = (k - i - 2)Y_i \; i = 2, \cdots, n - 1, \; [X, Y_n] = (2k - n - 3)Y_n$$

$$[Y_1, Y_i] = Y_{i+1} \; i = 2, \cdots, n - 2 \;\; [Y_2, Y_3] = Y_{n-2}, \; [Y_2, Y_4] = Y_{n-1}$$

$$[Y_i, Y_{n-i+1}] = (-1)^i Y_n.$$

$l = 1$ $(k = 2p - 3)$. Il n'existe pas de loi rigide.
$l = 2$ $(k = 2p - 5)$. Il n'existe que deux lois rigides si $k = 3$. Sinon, les algèbres correspondantes ne sont pas rigides.
$l = 3$ $(k = 2p - 7)$. Il existe 4 lois rigides dès que $k = 3$ $(dim\mathbf{g} = 15)$. Ces lois définissent des algèbres rigides non rationnelles et non réelles.
Supposons à présent $l \geq 4$. On a alors le théorème suivant:

Proposition 10 *Soit $g = t \oplus B_n^k$ une algèbre de Lie décomposable de dimension $n+1 = k+2l+6$ avec $dim t = 1$.*

1) Si $k > \left[\frac{l+1}{3}\right] + l - 2$, alors g n'est pas rigide.

2) Si $k \leq \left[\frac{l+1}{3}\right] + l - 2$, alors g est rigide.

Notons $E'_{s,s'}$ l'équation $(Y_{2+s}, Y_{2+s'}, Y_{4+2l-s-s'})$. La démonstration du théorème se calque alors sur celle du théorème 8 en remarquant que l'équation $E'_{s,s+1}$ est égale à l'équation $E_{s+1,s+2}$ du lemme 8.

Remarques.

1. Les équations $E'_{s,s+1}$ ne concernent que les paramètres $\lambda_2, \cdots, \lambda_l$. Alors les algèbres de Lie correspondant aux cas $\lambda_1 = 1, \lambda_i = 0, i \neq 1$ et $\lambda_{l+1} = 1, \lambda_i = 0, i \neq l+1$ et vérifiant $k \leq \left[\frac{l+1}{3}\right] + l - 2$ sont rigides.

2. Si k est pair, il peut ne pas exister des algèbres de Lie vérifiant les hypothèses de rigidité du théorème 8 (exemple si $k = 2$, il n'existe pas de telles algèbres dès que $n \geq 14$).

9.3

L'algèbre de Lie nilpotente $n = C_n(\lambda_1, \cdots, \lambda_t)$, $n = 2m + 2$ $t = m - 1$ est définie par:

$$[Y_1, Y_i] = Y_{i+1} \quad 2 \leq i \leq n-2$$

$$[Y_i, Y_{n-i+1}] = (-1)^{i-1} Y_n \quad 2 \leq i \leq m+1$$

$$[Y_i, Y_{n-i-2k+1}] = (-1)^{i+1} \lambda_k Y_n \ , \ 2 \leq i \leq n-2-2k \ et \ 1 \leq k \leq m-1.$$

Elle est de rang 1, une dérivation diagonale est donné par :

$$f(Y_1) = 0, \quad f(Y_i) = Y_i \quad i = 2, \cdots, n-1$$

$$f(Y_n) = 2Y_n.$$

Toute algèbre de Lie rigide dont le nilradical est isomorphe à n est de la forme $g = n \oplus t$ avec $dim t = 1$.

Proposition 11 *Soit $g = n \oplus t$ une algèbre de Lie décomposable dont le nilradical n est isomorphe à $C_n(\lambda_1, \cdots, \lambda_t)$. Alors g n'est pas rigide.*

En effet, considérons le 2-cocycle de g défini par

$$\phi(X, Y_1) = Y_n$$

les autres produits étant nuls, X étant une base de t. Si μ_0 est la loi de g, alors $\mu = \mu_0 + \epsilon\phi$ est une loi d'algèbre de Lie non isomorphe à μ_0 pour tout ϵ infiniment petit. Il s'en suit que la loi μ_0 et donc que l'algèbre de Lie g ne sont pas rigides.

References

[1] Ancochea Bermudez J.M. *On the Rigidity of Solvable Lie Algebras.* In Deformation Theory of Algebras and Structures and Applications. Kluwer Academic Publishers, 403-445, 1988

[2] Ancochea Bermudez J.M., Goze M. Le rang du système linéaire des racines d'une algèbre de lie résoluble rigide. *Communications in Algebra.* 20(1992), 875-887.

[3] Ancochea Bermudez J.M., Goze M. Algèbres de Lie rigides dont le nilradical est filiforme. *Note aux C.R.A.Sc.Paris*, 312(1991), 21-24.

[4] Carles R. Weight systems for complex nilpotent Lie algebras. *Preprint Universit de Poitiers* ,96 (1996).

[5] Carles R. Sur les srtuctures d'algèbres de Lie. *Ann. Institut Fourier*,33, 65-82, 1984

[6] Goze M., Khakimdjanov Yu. *Nilpotent Lie Algebras.* Kluwer Academic Publishers, MIA 361, 1996.

[7] Goze M., Khakimdjanov Yu. Sur les algèbres de Lie nilpotentes admettant un tore de dérivations, *Manuscripta math.*, 84(1994), 115-224.

[8] Nijenhuis A., Richardson R.W. Cohomology and Deformation of Algebraic Structures. *Bull.A.M.S.*, 70(1964), 406-411.

[9] Seeley G. Classification of seven dimensional nilpotent Lie Algebras. *T.A.M.S.* 335:2(1993), 479-496.

FAMILY OF p-FILIFORM LIE ALGEBRAS[3]

J.M.CABEZAS, J.R.GÓMEZ, A.JIMENEZ-MERCHÁN

J.M.C.: Departamento de Matemática Aplicada. Universidad del País Vasco. c) Nieves Cano 12. 01006 Vitoria(Spain).

J.R.G.,A.J.-M.: Departamento de Matemática Aplicada I. Universidad de Sevilla. Avenida de Reina Mercedes s.n. 41012 Sevilla(Spain).

Abstract. This paper, an approach to the classification of nilpotent Lie algebras g of any dimension with a Goze's invariant $(\dim g - p, 1, \ldots, 1)$, provides an explicit classification for $p \geq \dim g - 3$.

1 Introduction

Levi's theorem decomposes the classification of Lie algebras into the classification of semisimple and solvable Lie algebras. The classification of the former was solved in 1914 and the classification of solvable ones is actually reduced, modulo the study of derivations, to that of the nilpotent Lie algebras.

Up to now, the study of nilpotent Lie algebras has only dealt with the filiforms, except for some simple situations or for the classification of some specific dimensions. This paper, however, approaches the classification of large families of nilpotent Lie algebras of any dimension.

The lists of nilpotent Lie algebras of dimensions ≤ 5 over an abelian field were first published by J. Dixmier in [4]. Later, V.V. Morosov [8] provided the first complete list for dimension 6, and M. Vergne [9] provided another by means of cohomological methods,but since then only partial advances were made

[3]This work has been partially sponsored by the PAICYT project from the Government / Council of Andalucía.

Y. Khakimdjanov et al (eds.), Algebra and Operator Theory, 93-102.

before the introduction by M. Goze and J.M. Ancochea of an invariant of great importance: the characteristic sequence. This Goze's invariant made possible the classification of nilpotent Lie algebras of dimension 7 in [2] and of filiforms of dimension 8 in [1]. Subsequently, J.R. Gómez and F.J. Echarte have classified those of dimension 9 in [5] and L. Boza, F.J. Echarte and J. Núñez did the same for dimension 10 in [3].

Recently, in [6], J.R. Gómez, M. Goze and Y. Khakimdjanov have worked out techniques that have made possible the classification of subfamilies of filiform Lie algebras, whatever their dimension. Further developments of these techniques have already led to the classification of complex filiform algebras of dimensions ≤ 11 by J.R. Gómez, A. Jiménez-Merchán and Y. Khakimdjanov in [7].

This paper attempts the classification of large families of nilpotent Lie algebras, whatever their dimension, and, consequently, the description of their laws. To this end, it is necessary to introduce the concept of p-filiform Lie algebras, that is, a wider class including filiforms and defined by the characteristic sequence $(\dim \boldsymbol{g} - p, 1, \ldots, 1)$.

2 Definitions and preliminary results

Although most of the definitions and results of this section are well-known, the development of the present study makes it wise to state some of them.

A *Lie algebra* $\boldsymbol{g}$ on a field $\mathbf{F}$ is a vector space on $\mathbf{F}$ where multiplication has been defined, i.e., it is provided with a bilinear

$$\begin{array}{ccc} \boldsymbol{g} \times \boldsymbol{g} & \longrightarrow & \boldsymbol{g} \\ (X,Y) & \longrightarrow & [X,Y] \end{array}$$

which verifies the identities

1) $[X,X] = 0 \quad \forall X \in \boldsymbol{g}$
2) $[X,[Y,Z]] + [Y,[Z,X]] + [Z,[X,Y]] = 0 \quad \forall X,Y,Z \in \boldsymbol{g}$ (Jacobi identity).

The *descending central sequence* of a Lie algebra $\boldsymbol{g}$ is defined by

$$(\mathcal{C}^k(\boldsymbol{g})), \quad k \in \mathbf{N} \cup \{0\}, \quad \text{where} \quad \mathcal{C}^0(\boldsymbol{g}) = \boldsymbol{g} \quad \text{y} \quad \mathcal{C}^i(\boldsymbol{g}) = [\boldsymbol{g}, \mathcal{C}^{i-1}(\boldsymbol{g})], \quad i \in \mathbf{N}.$$

Provided there exists a $k \in \mathbf{N}$ such that $\mathcal{C}^k(\boldsymbol{g}) = \{0\}$, the Lie algebra $\boldsymbol{g}$ is said to be *nilpotent* and the smallest integer verifying this equation is called the *nilindex* of $\boldsymbol{g}$. In this case, $\boldsymbol{g}$ is said to be a *k-nilpotent* algebra.

A Lie algebra $\boldsymbol{g}$, with $\dim \boldsymbol{g} = n$, is called *filiform* if it verifies

$$\dim \mathcal{C}^i(\boldsymbol{g}) = n - i - 1 \quad 1 \leq i \leq n-1.$$

These algebras have $n-1$ as their maximal nilindex. The Lie algebras with a nilindex $n-2$ are called *quasifiliform* and those with a nilindex 1 are called *abelian*.

Let $\boldsymbol{g}$ be a nilpotent Lie algebra of dimension n. For all $X \in \boldsymbol{g} - [\boldsymbol{g},\boldsymbol{g}]$, then

$$c(X) = (c_1(X), c_2(X), \ldots, 1)$$

is the sequence, in decreasing order, of the dimensions of the characteristic subspaces of the nilpotent operator adX, where the adjoint operator of an element $X \in \boldsymbol{g}$, adX, is defined by

$$\begin{array}{rccl} adX: & \boldsymbol{g} & \longrightarrow & \boldsymbol{g} \\ & Y & \longrightarrow & adX(Y) = [X,Y] \end{array}$$

$c(\boldsymbol{g}) = sup\{c(X) : \quad X \in \boldsymbol{g} - [\boldsymbol{g},\boldsymbol{g}]\}$ is called the *characteristic sequence* or *Goze's invariant* of the nilpotent Lie algebra $\boldsymbol{g}$.

The filiform, quasifiliform and abelian Lie algebras of dimension n have as their Goze's invariant $(n-1,1), (n-2,1,1)$ and $(1,1,\ldots,1)$, respectively. In this study *p-filiform* Lie algebras is the name given to a large family of Lie algebras comprising the filiforms as a particular group.

Definition 1. A nilpotent Lie algebra $\boldsymbol{g}$, of dimension n, is called *p-filiform* if its Goze's invariant is $(n-p, 1, \ldots, 1)$.

The filiform, quasifiliform and abelian Lie algebras are 1-filiform, 2-filiform and $(n-1)$-filiform, respectively, provided n is the dimension of the algebra. From the definition it follows that all p-filiform algebras $\boldsymbol{g}$, of dimension n, have nilpotency $n-p$ (the converse is not true), that $1 \leq p \leq n-1$ and that an *adapted basis* of $\boldsymbol{g}$, denoted $\{X_0, X_1, \ldots, X_{n-p}, Y_1, \ldots, Y_{p-1}\}$, can always be found, and this is such that $X_0 \in \boldsymbol{g} - [\boldsymbol{g},\boldsymbol{g}]$ is a *characteristic vector*, i.e., it is verified that

$$\begin{array}{lcll} [X_0, X_i] & = & X_{i+1} & 1 \leq i \leq n-p-1 \\ [X_0, X_{n-p}] & = & 0 & \\ [X_0, Y_j] & = & 0 & 1 \leq j \leq p-1. \end{array}$$

3 Classification

This section gives a classification, up to isomorphism, of p-filiform algebras of dimension n for $p \geq n-3$. In order to do this, the proofs will largely use theorems on nilpotency, Goze's invariant and p-filiformity.

3.1 $(n-1)$-filiform algebras

$(n-1)$-filiform Lie algebras are actually the abelian Lie algebras of dimension n, and therefore, there exists only one for each dimension.

3.2 $(n-2)$-filiform algebras

Theorem 1. *In dimension $n, n \geq 3$, there are exactly $E(\frac{n-1}{2})$ nilpotent, real or complex, pairwise non-isomorphic Lie algebras, denoted as $\boldsymbol{g}_n^i, 1 \leq i \leq E(\frac{n-1}{2})$, with Goze's invariant $(2,1,1,\ldots,1)$ and whose laws are given in an adapted basis $\{X_0, X_1, X_2, Y_1, Y_2, \ldots, Y_{n-3}\}$ by*

$$\boldsymbol{g}_n^q : \begin{array}{lcll} [X_0, X_1] & = & X_2 & \\ [Y_{2k-1}, Y_{2k}] & = & X_2 & 1 \leq k \leq q-1 \end{array}$$

for $1 \leq q \leq E(\frac{n-1}{2})$.

Proof. All these algebras are nilpotent and $(2,1,1,\ldots,1)$ is their Goze's invariant. In addition, they are pairwise non-isomorphic, because the dimensions of the center are

$$\dim \mathcal{Z}(\boldsymbol{g}_n^q) = n - 2q, \qquad 1 \leq q \leq E(\frac{n-1}{2})$$

Therefore, it only remains to prove that any nilpotent Lie algebra $\boldsymbol{g}$ of dimension n having $(2,1,1,\ldots,1)$ as its Goze's invariant is isomorphic to one of the previous ones.

If $X_0 \notin [\boldsymbol{g}, \boldsymbol{g}]$ is a characteristic vector and $\{X_0, X_1, X_2, Y_1, Y_2, \ldots, Y_{n-3}\}$ is an adapted basis of $\boldsymbol{g}$, then it follows that

$$\left\{ \begin{array}{lcll} [X_0, X_1] & = & X_2 & \\ [X_0, X_2] & = & 0 & \\ [X_0, Y_i] & = & 0 & 1 \leq i \leq n-3. \end{array} \right\}.$$

Provided that Jacobi identities are verified for any triad of vectors, one of which be X_0, and considering that $Y_i \notin Im\ adX_0$, $1 \leq i \leq n-3$, then, from the application of any of the apparent conditions concerning nilpotency, it follows that the non-zero bracket products of $\boldsymbol{g}$, up to antisymmetry, are

$$\left\{ \begin{array}{lcll} [X_0, X_1] & = & X_2 & \\ [X_1, X_2] & = & \beta X_2 + \sum_{k=1}^{n-3} \alpha_k Y_k & \\ [X_1, Y_i] & = & a_i X_2 + \sum_{k=1}^{n-3} d_i^k Y_k & 1 \leq i \leq n-3 \\ [Y_i, Y_j] & = & b_{ij} X_2 + \sum_{k=1}^{n-3} c_{ij}^k Y_k & 1 \leq i < j \leq n-3. \end{array} \right.$$

Since $(2,1,1,\ldots,1)$ is the Goze's invariant of $\boldsymbol{g}$ and $X_1 \notin [\boldsymbol{g}, \boldsymbol{g}]$, there cannot exist in adX_1 any minor of order 2 other than zero, this result leading to

$$\alpha_k = 0 \quad \text{and} \quad d_i^k = 0\ , \quad 1 \leq k, i \leq n-3$$

and, therefore, considering nilpotency, $\beta = 0$.

For all $A \in \mathbf{F}$ ($\mathbf{F} = \mathbf{R}$ or $\mathbf{C}$) it follows that $X_0 + AY_i \notin [\boldsymbol{g}, \boldsymbol{g}]$, $1 \leq i \leq n-4$, and therefore, as the previous reasoning showed, if A is correctly chosen, it is obtained that

$$c_{ij}^k = 0, \qquad 1 \leq i < j \leq n \qquad 1 \leq k \leq n-3.$$

Thus, the law of $\boldsymbol{g}$, once the basis has been changed by the relations

$$\begin{cases} X_t^* &= X_t & 0 \leq t \leq 2 \\ Y_i^* &= Y_i + a_i X_0 & 1 \leq i \leq n-3, \end{cases}$$

adopts the form of

$$\begin{cases} [X_0, X_1] &= X_2 \\ [Y_i, Y_j] &= b_{ij} X_2 & 1 \leq i < j \leq n-3. \end{cases}$$

If $b_{ij} = 0, \quad 1 \leq i < j \leq n-3$, then $\boldsymbol{g}_n^1$.

If there exists any $b_{ij} \neq 0$, $\quad 1 \leq i < j \leq n-3$, it can be easily assumed that $b_{12} \neq 0$ implementing an easy change of basis. This provided, it can also be assumed that $b_{12} = 1, \quad b_{1k} = 0 = b_{2k}$, $\quad 3 \leq k \leq n-3$. To obtain an isomorphic algebra with the law

$$\begin{cases} [X_0, X_1] &= X_2 \\ [Y_1, Y_2] &= X_2 \\ [Y_i, Y_j] &= b_{ij} X_2 & 3 \leq i < j \leq n-3 \end{cases}$$

is suffices to apply the change of basis defined by

$$\begin{cases} X_t^* &= X_t & 0 \leq t \leq 3 \\ Y_1^* &= \frac{1}{b_{12}} Y_1 \\ Y_2^* &= Y_2 \\ Y_k^* &= Y_k + \frac{b_{2k}}{b_{12}} Y_1 - \frac{b_{1k}}{b_{12}} Y_2 & 3 \leq k \leq n-3. \end{cases}$$

If all $b_{ij} = 0, \quad 3 \leq i < j \leq n-3$, then it would be $\boldsymbol{g}_n^2$, and in the case of $b_{ij} \neq 0$, it can be presumed $b_{34} = 1, b_{3k} = 0 b_{4k}$, $5 \leq k \leq n-3$, by a change of basis similar to that already mentioned.

By continuing this process, it is proved that the algebra $\boldsymbol{g}$ is isomorphic to some of $\boldsymbol{g}_n^i$, where $\quad 1 \leq i \leq E(\frac{n-1}{2})$. It can be observed that all of them are disjointed if the dimension is an odd number. These algebras are the direct sum of the Heisenberg algebras of dimension $\leq n$, with the necessary copies of the basis field not to alter the dimension.

3.3 $(n-3)$-filiform algebras

Theorem 2. *In dimension $n, n \geq 5$, there are exactly $n-2$ nilpotent, real or complex, pairwise non-isomorphic Lie algebras, denoted as $\boldsymbol{g}_n^i$, $\quad 1 \leq i \leq n-2$,* *with Goze's invariant* $(3,1,1,\ldots,1)$ *and whose laws are given in an adapted basis* $\{X_0, X_1, X_2, X_3, Y_1, Y_2, \ldots, Y_{n-4}\}$ *by*

$$\begin{array}{llllll} \boldsymbol{g}_n^{2q-1}: & [X_0, X_i] & = & X_{i+1} & 1 \leq i \leq 2 & 1 \leq q \leq E(\frac{n-2}{2}) \\ & [Y_{2k-1}, Y_{2k}] & = & X_3 & 1 \leq k \leq q-1 & \end{array}$$

$$\begin{array}{llllll} \boldsymbol{g}_n^{2s}: & [X_0, X_i] & = & X_{i+1} & 1 \leq i \leq 2 & 1 \leq s \leq E(\frac{n-3}{2}) \\ & [X_1, Y_{n-4}] & = & X_3 & & \\ & [Y_{2k-1}, Y_{2k}] & = & X_3 & 1 \leq k \leq s-1 & \end{array}$$

$$\begin{array}{lllll} \boldsymbol{g}_n^{n-2}: & [X_0, X_i] & = & X_{i+1} & 1 \leq i \leq 2 \\ & [X_1, X_2] & = & Y_{n-4} & \end{array}$$

Proof. All previous algebras are nilpotent with $(3,1,1,\ldots,1)$ as Goze's invariant. In addition, they are pairwise non-isomorphic because the dimensions of the center and of the derived algebras are

$$\begin{array}{llll} \dim \mathcal{Z}(\boldsymbol{g}_n^i) & = & n-i-2 & 1 \leq i \leq n-3 \\ \dim \mathcal{Z}(\boldsymbol{g}_n^{n-2}) & = & n-3 & \\ \dim [\boldsymbol{g}_n^i, \boldsymbol{g}_n^i] & = & 2 & 1 \leq i \leq n-3 \\ \dim [\boldsymbol{g}_n^{n-2}, \boldsymbol{g}_n^{n-2}] & = & 3. & \end{array}$$

It remains to be proved that any nilpotent Lie algebra $\boldsymbol{g}$ of dimension n and Goze's invariant $(3,1,1,\ldots,1)$ is isomorphic to one of the preceding ones.

If $X_0 \notin [\boldsymbol{g}, \boldsymbol{g}]$ is a characteristic vector and $\{X_0, X_1, X_2, X_3, Y_1, Y_2, \ldots, Y_{n-4}\}$ an adapted basis of $\boldsymbol{g}$, then it follows that

$$\left\{ \begin{array}{llll} [X_0, X_i] & = & X_{i+1} & 1 \leq i \leq 2 \\ [X_0, X_3] & = & 0 & \\ [X_0, Y_j] & = & 0 & 1 \leq j \leq n-4. \end{array} \right.$$

From the Jacobi identities in which X_0 occurs a preliminary expression of the other bracket products rendered by the elements of the basis can be obtained. Taking into account some simple conditions of nilpotency and $Y_j \notin Im\, adX_0$, $1 \leq j \leq n-4$, the expression of the above mentioned brackets is dramatically simplified. This approach renders the following family of laws:

$$\begin{cases} [X_0, X_i] &= X_{i+1} & 1 \le i \le 2 \\ [X_1, X_2] &= \beta X_3 + \sum_{k=1}^{n-4} \alpha_k Y_k & \\ [X_1, Y_i] &= a_{2i} X_2 + a_{1i} X_3 + \sum_{k=1}^{n-4} d_i^k Y_k & 1 \le i \le n-4 \\ [X_2, Y_i] &= a_{2i} X_3 & 1 \le i \le n-4 \\ [Y_i, Y_j] &= b_{ij} X_3 + \sum_{k=1}^{n-4} c_{ij}^k Y_k & 1 \le i < j \le n-4. \end{cases}$$

For all $A \in \mathbf{F}$ ($\mathbf{F} = \mathbf{R}$ or $\mathbf{C}$) it follows that $X_0 + AY_i \notin [\boldsymbol{g}, \boldsymbol{g}]$, $1 \le i \le n-5$, and since the Goze's invariant is $(3, 1, 1, \ldots, 1)$, there cannot exist a non-zero minor of order 3. Appropriate choices of A together with a process of finite induction lead to

$$c_{ij}^k = 0 \,, \qquad 1 \le i < j \le n-4 \,, \qquad 1 \le k \le n-4.$$

From $ad(AX_0 + X_1)$ it can be concluded that

$$d_i^k = 0 \qquad \text{and} \qquad a_{1i}\alpha_k = 0, \qquad 1 \le i, k \le n-4,$$

and from the change of basis defined by

$$\begin{cases} X_1^* &= X_1 - \beta X_0 & \\ X_j^* &= X_j & j = 0, 2, 3 \\ Y_k^* &= Y_k & 1 \le k \le n-4 \end{cases}$$

it can be assumed that $\beta = 0$.

From the others Jacobi identities and the adjoint nilpotency arises the conclusion that

$$\begin{cases} \sum_{k=2}^{n-4} \alpha_k b_{1k} &= 0 & \\ \sum_{k=i+1}^{n-4} \alpha_k b_{ik} &= \sum_{r=1}^{i-1} \alpha_r b_{ri} & 2 \le i \le n-5 \\ \sum_{k=1}^{n-5} \alpha_k b_{k,n-4} &= 0 & \\ \sum_{k=1}^{n-4} \alpha_k a_{2k} &= 0. & \end{cases}$$

This result together with the change of basis defined by

$$\begin{cases} X_t^* &= X_t & 0 \le t \le 3 \\ Y_i^* &= Y_i + a_{2i}X_0 & 1 \le i \le n-4 \end{cases}$$

lead to $a_{2i} = 0,\ 1 \le i \le n-4$.

Thus, it is proved that all Lie algebra $\boldsymbol{g}$ with Goze's invariant $(3, 1, 1, \ldots, 1)$, are isomorphic to some other showing the law

$$\begin{cases} [X_0, X_i] &= X_{i+1} & 1 \le i \le 2 & \\ [X_1, X_2] &= \sum_{k=1}^{n-4} \alpha_k Y_k & & \\ [X_1, Y_i] &= a_i X_3 & 1 \le i \le n-4 & (a_i = a_{1i}) \\ [Y_i, Y_j] &= b_{ij} X_3 & 1 \le i < j \le n-4. & \end{cases}$$

Since the dimension of $[\boldsymbol{g}, \boldsymbol{g}]$ is respectively 2 or 3, whether all α_k be zero or not, two cases may be distinguished:

3.3.1 $\alpha_k = 0 \quad 1 \le k \le n-4$

$\boldsymbol{g}$ is then isomorphic to an algebra of law

$$\begin{cases} [X_0, X_i] &= X_{i+1} & 1 \le i \le 2 \\ [X_1, Y_i] &= a_i X_3 & 1 \le i \le n-4 \\ [Y_i, Y_j] &= b_{ij} X_3 & 1 \le i < j \le n-4. \end{cases}$$

If $b_{ij} = 0, \quad 1 \le i < j \le n-4$, and $\quad a_i = 0,\ 1 \le i \le n-4$, then $\boldsymbol{g}_n^1$.

If $b_{ij} = 0, \quad 1 \le i < j \le n-4$, and if there exists any $a_i \ne 0,\ 1 \le i \le n-4$, it can be assumed that $a_{n-4} \ne 0$ by means of an obvious change of basis. In this case it can be also assumed that $a_{n-4} = 1$ and $a_i = 0\ ,\ 1 \le i \le n-5$. This result arises from the change of basis given by

$$\begin{cases} X_t^* &= X_t & 0 \le t \le 3 \\ Y_{n-4}^* &= \frac{1}{a_{n-4}} Y_{n-4} & \\ Y_i^* &= -a_i Y_{n-4} + a_{n-4} Y_i & 1 \le i \le n-5 \end{cases}$$

whereby $\boldsymbol{g}_n^2$ appears.

If there exists any $b_{ij} \ne 0\ ,\ 1 \le i < j \le n-4$, then it can be supposed that $b_{12} \ne 0$. It can also be assumed that $b_{12} = 1, b_{1k} = 0 = b_{2k} \quad 3 \le k \le n-4$ and

$a_1 = 0 = a_2$, with no further operation than the change of basis defined by

$$\begin{cases} X_1^* = X_1 - \frac{a_2}{b_{12}} Y_1 + \frac{a_1}{b_{12}} Y_2 & \\ X_t^* = X_t & t = 0, 2, 3 \\ Y_1^* = \frac{1}{b_{12}} Y_1 & \\ Y_2^* = Y_2 & \\ Y_k^* = Y_k + \frac{b_{2k}}{b_{12}} Y_1 - \frac{b_{1k}}{b_{12}} Y_2 & 3 \le k \le n-4. \end{cases}$$

Thus, the algebra with the law

$$\begin{cases} [X_0, X_i] = X_{i+1} & 1 \le i \le 2 \\ [Y_1, Y_2] = X_3 & \\ [X_1, Y_i] = a_i X_3 & 3 \le i \le n-4 \\ [Y_i, Y_j] = b_{ij} X_3 & 3 \le i < j \le n-4 \end{cases}$$

is obtained.

If $b_{ij} = 0$, $3 \le i < j \le n-4$, and $a_i = 0$, $3 \le i \le n-4$, then $\boldsymbol{g}_n^3$.

If $b_{ij} = 0$, $3 \le i < j \le n-4$ and there exists any $a_i \ne 0$, $3 \le i \le n-4$, it can be assumed $a_{n-4} = 1$ and $a_i = 0$, $3 \le i \le n-5$, with the help of changes of basis similar to some of those already seen, thus obtaining $\boldsymbol{g}_n^4$.

In this way there appear in succession the algebras $\boldsymbol{g}_n^i$, $1 \le i \le n-3$, which made up the two families stated in the theorem.

3.3.2 $\alpha_k \ne 0$ for any $1 \le k \le n-4$

Since $a_i \alpha_k = 0$, $1 \le i \le n-4$, it follows that $a_i = 0$, $1 \le i \le n-4$. Moreover, it can be assumed that $\alpha_{n-4} \ne 0$.

The change of basis defined by

$$\begin{cases} X_t^* = X_t & 0 \le t \le 3 \\ Y_k^* = Y_k & 1 \le k \le n-5 \\ Y_{n-4}^* = \sum_{k=1}^{n-4} \alpha_k Y_k & \end{cases}$$

reveals that $\boldsymbol{g}$ is isomorphic to an algebra with the law

$$\begin{cases} [X_0, X_i] = X_{i+1} & 1 \le i \le 2 \\ [X_1, X_2] = Y_{n-4} & \\ [Y_i, Y_j] = b_{ij} X_3 & 1 \le i < j \le n-5. \end{cases}$$

Since the Goze's invariant is $(3, 1, 1, \ldots, 1)$ and considering $ad(X_1 + Y_i)$, $2 \le i \le n-5$, it can be concluded that

$$b_{ij} = 0 \quad 1 \le i < j \le n-5,$$

thus obtaining the law

$$\begin{cases} [X_0, X_i] &= X_{i+1} \qquad 1 \le i \le 2 \\ [X_1, X_2] &= Y_{n-4} \end{cases}$$

corresponding to $\boldsymbol{g}_n^{n-2}$.

Note: In the case of algebras of dimension 4, the Goze's invariant is $(3,1)$, i.e., they are filiforms. For this dimension only one algebra, the model, is obtained.

References

[1] J.M. Ancochea, M. Goze, Classification des algèbres de Lie filiformes de dimension 8, *Archiv. Math.*, 50(1988), 511-525.

[2] J.M. Ancochea, M. Goze, Classification des algèbres de Lie nilpotentes de dimension 7, *Archiv. Math.*, 52(1989), 175-185.

[3] L. Boza, F.J. Echarte, J. Núñez. Classification of complex filiform Lie algebras of dimension 10, *Algebra, Groups and Geometries,* 11:3(1994), 253-276.

[4] J. Dixmier, Sur les reprèsentations unitaires des groupes de Lie nilpotentes III, *Canadian J. Math.*, 10(1958), 321-348.

[5] J.R. Gómez, F.J. Echarte, Classification of complex filiform nilpotent Lie algebras of dimension 9, *Rendiconti Cagliari,* 61:1(1991), 21-29.

[6] J.R. Gómez, M. Goze, Y. Khakimdjanov, On the k-abelian filiform Lie algebras, *Communications in algebra,* 25:2(1997), 431-450.

[7] J.R. Gómez, A. Jiménez-Merchán, Y. Khakimdjanov, Low-Dimensional Filiform Lie Algebras, to appear in *Journal of Pure and Applied Algebra.*

[8] V.V. Morosov, Clasification of nilpotent Lie algebras, (in Russian) *Isvestia Vys Ucheb. Zav.* 4(1958), 161-171.

[9] M. Vergne, Cohomologie des algèbres de Lie nilpotentes. Application à l'étude de la varieté des algèbres de Lie nilpotentes, *Bull. Soc. Math. France,* 98(1970), 81-116.

E-mail: mapcamaj@vc.ehu.es, jrgomez@cica.es, ajimenez@cica.es.

THE FUNCTIONAL REPRESENTATION OF COMMUTATIVE SYMMETRIC OPERATOR ALGEBRAS IN PONTRYAGIN SPACE[4]

V.I.CHILIN, S.Sh.MASHARIPOVA

Departement of Mathematics,
Tashkent State University,
700095, Tashkent, Uzbekistan

Introduction

Symmetric operator algebras are the most powerful tools in the investigations in the quantum field theory. Together with well-known C^* and W^*-algebras acting in the Hilbert space, recently operator algebras acting in the spaces with indefinite metric began to be used in some models of quantized fields. Particularly, these were the operator algebras in Pontryagin space [2]. Unlike deeply developed theory of C^* and W^*-algebras, the theory of symmetric operator algebras in a π_1 space has just begun to develop.

Some results were obtained in [4, 8-13], connected with different algebraic and topologic properties of these algebras, and besides, it was established that many facts of the theory of C^*-algebras are not carried out for the general algebras (they were algebras having an invariant degenerated subspace). At the same time each non-degenerated symmetric operator algebra (i.e. not general algebra) is topologically isomorphic to usual C^*-algebra [3, 7]. Thus, the most interesting objects for investigations are general algebras.

In [13] the system of models describing all general uniformly closed symmetric algebras acting in a Pontryagin π_1 space was constructed. For commutative algebras the same system of models was obtained in [6, 11]. Improved variant of these models was given in [1].

[4]The work is supported by State Committee of Sciences and Technology of Uzbekistan, Grant No 41/96.

Y. Khakimdjanov et al (eds.), Algebra and Operator Theory, 103-110.

In this paper we obtain the functional representation of commutative symmetric operator algebras on a Pontryagin π_1 space, and construct the functional calculus for Borel functions. We use Shul'man's models [13] and terminology from [13, 14].

1. Preliminary.

Let H be a Pontryagin π_1 space with indefinite inner product (,) and with fixed usual norm, $B(H)$ be the algebra of all bounded continious linear operators on H. For each neutral space $L \subset H$ we consider:

$$I_L = \{T \in B(H) : T(L) \subset L, T^*(L) \subset L\}.$$

Since L is a 1-dimensional space, then L is subspace for each $T \in I_L$. We shall denote by λ_T the corresponding eigenvalue. We'll select a neutral subspace $M \subset H$ which is skew-links with L; then $K = L^\perp \bigcap M^\perp$ is the Hilbert space with respect to the usual inner product. For each $T \in I_L$ we'll define $T_1 \in B(K)$: $T_1 x = \pi T x$, where $x \in K$, π is the orthogonal projection of H on K. We fix vectors $\xi_0 \in L, \eta_0 \in M$ such that $(\xi_0, \eta_0) = 1$. For every $T \in I_L$ we put:

$$x_T = T\eta_0, \quad \gamma_T = (T\eta_0, \eta_0), \quad y_T = x_T^*, \quad \mu_T = \lambda_T^*.$$

We denote by N a 2-dimentional space of π_1 type, generated by the vectors ξ_0 and η_0. Then $H = N \oplus K$.

For each set $\{A, x, y, \lambda, \mu, \gamma\}$, consisting of an operator $A \subset B(H)$, vectors $x, y \in K$ and numbers $\lambda, \mu, \gamma \in \mathbf{C}$ we define an operator $T =< A, x, y, \lambda, \mu, \gamma >\in B(N \oplus K)$, such that $T\xi_0 = \lambda\xi_0$, $T\eta_0 = \mu\eta_0 + \lambda + \gamma\xi_0$, $Th = Ah + (h, y)\xi_0$, where $h \in K$. Then $T \in I_L$, and every operator $T \in I_L$ has the form $T =< A, x, y, \lambda, \mu, \gamma >$ (see [13]). Moreover, the uniform topology in I_L is the topology of the norm $|||\cdot|||$:

$$||| < A, x, y, \lambda, \mu, \gamma > ||| = \|A\|_{B(K)} + \|x\|_K + \|y\|_K + |\lambda| + |\mu| + |\gamma|.$$

It is easy to see that an operator $T \in I_L$ is self-adjoint if and only if T has the representation $T =< A, x, x, \lambda, \overline{\lambda}, \gamma >$, where $A = A^*$, $\gamma = \overline{\gamma}$.

Let M be an arbitrary uniformly closed symmetric subalgebra of $B(H)$. If M has no nonzero invariant neutral subspace, then the algebra M is said to be non-degenerated algebra. The structure of those algebras is well-known (see [3,7]): each non-degenerated algebra is topologically isomorphic to a C^*-algebra. If the algebra M is not non-degenerated algebra, i.e. M has a non-trivial invariant neutral subspace, then M is called the general algebra. Substantial classification of general algebras in Pontruyagin π_1 space was offered by V.S.Shul'man in [13]. He divided these algebras into 6 classes. We remind these definitions.

Let L be a non-zero neutral M-invariant space for a general algebra M Then $M \subset I_L$. The symmetric ideal $chM = \{T \in M : T(L^\perp) \subset L\} \subset M$ is called the L-caracteristic of M. If $chM = \{0\}$, for some neutral M-invariant subspace L, then M is called the algebra of class 0. In this case there are such Hilbert space K and C^*-algebra $\mathcal{U}$ of operators on K not containing $\mathbf{I}$ that M is unitarily equivalent to algebra $M^0(K,\mathcal{U}) = \{< A+\lambda, 0, 0, \lambda, \lambda, 0 >: A \in \mathcal{U}, \lambda \in \mathbf{C}\}$, acting on $N \oplus K$.

If the algebra M is not of class 0, and $chM \subset M_0 = \{T \in M : \lambda_T = \mu_T = 0\}$ for some neutral M-invariant subspace L, then M is called the algebra of class 1. For its description we need the notion of quasi-vector.

Let $\mathcal{U}$ be an arbitrary *-subalgebra of $B(K)$. A linear map $p : \mathcal{U} \to K$ is called the quasi-vector, if $p(TS) = Tp(S)$, $\forall T, S \in \mathcal{U}$. For a quasi-vector p we consider a map $\hat{p} : \mathcal{U} \to K \oplus K$: $\hat{p}(T) = p(T) \oplus p(T^*)$, $\forall T \in \mathcal{U}$. The quasi-vector p is called semiclosed quasi-vector, if $\hat{p}$ is the closed map from $(B(K), u)$ into $K \oplus K$, where u is a uniform topology in $B(K)$. In [13] Shul'man showed that every algebra of class 1 is unitarily equivalent to the algebra

$$M^1(K, K_1, K_2, V, p) = \{< A + \lambda\mathbf{I}, x_1 + y + p(A), x_2 + Vy+$$

$$+p(A^*), \lambda, \lambda, \gamma >: A \in \mathcal{U}, x_1, x_2 \in K_F, y \in D_V, \lambda, \gamma \in \mathbf{C}\},$$

where K is a Hilbert space, $\mathcal{U}$ is an algebra of operators on K, p is a semiclosed quasivector, K_1 is a closed $\mathcal{U}$-invariant subspace of K which is orthogonal to $p(\mathcal{U})$, K_2 is a closed subspace of $Ker\ \mathcal{U}$, which is orthogonal to K_1, V is a closed antilinear involutory operator on K_2 defined on the domain D_V.

If an algebra M is not contained in classes 0 and 1, and the functional $T \to \lambda_T$ is a real-valued functional under some choice of neutral M-invariant subspace L and $chM \bigcap M_0 = \{0\}$ (respectively, $chM \bigcap M_0 \neq \{0\}$), then M is said to be the algebra of class $2a$ (respectively $2b$). In this case M is unitarily equivalent to the algebra

$$M^{2a}(K,\mathcal{U}) = \{< A, 0, 0, \lambda, \lambda, 0 >: A \in \mathcal{U}, \lambda \in \mathbf{C}\}$$

(respectively

$$M^{2b}(K, K_1, \mathcal{U}) = \{< A, x_1, x_2, \lambda, \lambda, \gamma >: A \in \mathcal{U}, x_1, x_2 \in K_1, \lambda, \gamma \in \mathbf{C}\})$$

If an algebra M is not contained in classes 0, 1, 2a, 2b, and a functional $T \to \lambda_T$ is not Hermitian for some neutral M-invariant subspace L, and $chM \bigcap M_0 = \{0\}$ (respectively $chM \bigcap M_0 \neq \{0\}$) then M is called the algebra of class $3a$ (respectively $3b$). In this case M in unitarily equivalent to the algebra.

$M^{3a}(K,\mathcal{U}) = \{< A, 0, 0, \lambda, \mu, 0 >: A \in \mathcal{U}, \lambda, \mu \in \mathbf{C}\}$

(respectively

$$M^{3b}(K, K_1, \mathcal{U}) = \{< A, x_1, x_2, \lambda, \mu, \gamma >: A \in \mathcal{U}, x_1, x_2 \in K_1, \lambda, \mu, \gamma \in \mathbf{C}\ \})$$

where K is a Hilbert space, $\mathcal{U}$ is a C^*-algebra of operator on K containing $\mathbf{1}$, K_1 is a $\mathcal{U}$-invariant closed subspace of K [13].

2.The functional representation of commutative symmetric operator algebras.

Let N be a commutative symmetric w-closed algebra of operators on H possessing a positive eigen-vector. If M is a non-degenerated algebra, then, as we mentioned above, M is topologically *-isomorphic to a C^*-algebra. Let us suppose that M is contained in class 0. Then the algebra M admits a representation $M = \{< A + \lambda, 0, 0, \lambda, \lambda, 0 >: A \in \mathcal{U},\ \lambda \in \mathbf{C}\ \}$, where $\mathcal{U}$ is a commutative C^*-algebra without $\mathbf{1}$, acting on $K = \{\xi_0, \eta_0\}^\perp$. We'll consider the C^*-algebra $\hat{\mathcal{U}} = \mathcal{U} + \mathbf{C} \cdot \mathbf{1}$. The spectrum $\sigma(\hat{\mathcal{U}})$ of $\hat{\mathcal{U}}$ coincides with $\sigma(\hat{\mathcal{U}}) = \sigma(\mathcal{U}) \bigcup \{w_0\}$, where $w_0(\mathbf{1}) = 1$, $w_0|_{\mathcal{U}} = 0$. We'll denote by Φ the Gelfand's homomorphism from $\hat{\mathcal{U}}$ onto the C^*-algebra $C(\sigma(\hat{\mathcal{U}}))$ of all continuous complex functions on the compact $\sigma(\hat{\mathcal{U}})$. For every $f \in C(\sigma(\hat{\mathcal{U}}))$ put

$$\Psi(f) =< \Phi^{-1}(f), 0, f(w_0), f(w_0), 0 > .$$

It is easy to see that Ψ is a topological *-isomorphism from $C(\sigma(\hat{\mathcal{U}}))$ onto M.

Now we assume that M is an algebra of class 2; then M may be represented in the form:

$$M = \{< A, 0, 0, \mu, \mu, 0 >: A \in \mathcal{U}, \mu \in \mathbf{C}\ \},$$

where $\mathcal{U}$ is a commutative C^*-algebra with $\mathbf{1}$, acting on $K = \{\xi_0, \eta_0\}^\perp$. We extend each character h on $\mathcal{U}$ to a character $\hat{h}$ on M:

$$\hat{h}(< A, 0, 0, \mu, \mu, 0 >) = h(A).$$

Further consider on M a continuous character $\lambda(< A, 0, 0, \mu, \mu, 0 >) = \mu$. Let $\sigma = \{\hat{h} : h \in \sigma(\mathcal{U})\}$ and $X = \sigma \bigcup \{\lambda\}$. We'll define on X the topology τ:

$$G \in \tau \Longleftrightarrow (G \subset \tau_0 \quad \text{or} \quad G = \{\lambda\} \bigcup G_1,\ G_1 \subset \tau_0),$$

where τ_0 is the Gelfand topology on $\sigma(\mathcal{U}) \cong \sigma$.

If is clear that (X, τ) is a compact. Let Φ be the Gelfand's *-isomorphism from $\mathcal{U}$ onto $C(\sigma(\mathcal{U}), \tau_0)$. For each $f \in C(X, \tau)$ put

$$\Psi(f) =< \Phi^{-1}(f_u), 0, 0, f(\lambda), f(\lambda), 0 >,$$

where $f_u = f|_{\sigma(\mathcal{U})}$. Then Ψ is a topological *-isomorphism from (X,τ) onto M. Thus we obtain the following result.

Theorem 1 *Let M be a commutative symmetric u-closed operator algebra acting on a Pontryagin π_1 space and possessing a positive eigen-vector. Then M is topologically *-isomorphic to a commutative C^*-algebra.*

Let T be a self-adjoint operator from $B(H)$, $\mathcal{P}(T)$ (respectively $\mathcal{P}(T,\mathbf{1})$) a symmetric subalgebra of $B(H)$ generated by T (respectively generated by T and $\mathbf{1}$). Let F be the closure of $F \subset B(H)$ in the uniform topology, $\overline{F}^w$ - the closure of F in the weak topology.

Theorem 1 implies

Corollary 1. *Let T be a self-adjoint operator from $B(H)$, possessing a positive eigen-vector, and let $M = \overline{\mathcal{P}}(T,\mathbf{1})^u$. Then there exists a compact $X \subset \mathbf{R}$ and a topological *-isomorphism Ψ from $C(X)$ onto M, such that, $U(f_0) = \mathbf{1}$, $\Phi(f_1) = T$, where $f_0(t) = 1$, $f_1(t) = t$. This isomorphism can be extended to a (continuous) *-homomorphism $\hat{\Psi}$ from the *-algebra $W(X)$ af all complex Borel functions on X to $\overline{M}^w$, and besides, if $f_u, f \in W(X), f_u(t) \to f(t)$, $\forall t \in X$, and*

$$Sup\{|f_u(t)| : t \in X,\ n \geq 1\} < \infty,$$

then $\hat{\Psi}(f_u)$ converge to $U(f)$ in the weak topology.

For description of commutative algebras of class $3a$ we'll introduce a special class of functional C^*-algebras. Let X be an arbitrary compact, $\lambda, \mu \in X$, $\lambda \neq \mu$; let $\{\lambda\}$ and $\{\mu\}$ be putting closed sets in X. On the algebra $C(X)$ we'll consider an involution $\#$:

$$f^{\#}(t) = \overline{f}(t), \quad t \neq \lambda, \mu;$$

$$f^{\#}(\lambda) = \overline{f}(\mu);$$

$$f^{\#}(\mu) = \overline{f}(\lambda).$$

It is clear that $C(X,\#)$ is a B^*-algebra with respect to the uniform topology (we'll refer to these algebras as functional (λ,μ) - B^*-algebras). Using the representation of the algebra of class $3a$ in the form $M^{3a}(K,\mathcal{U})$, we can obtain following

Theorem 2 *Every commutative symmetric operator algebra of the class $3a$ is topologically *-isomorphic to some functional (λ,μ) - B^*-algebra.*

Corollary 2 *Let $T = T^* \in B(H)$, and $M = \overline{\mathcal{P}(T,\mathbf{1})}^u$ be an algebra of the class $3a$ (i.e. $T\xi_0 = \lambda\xi_0$, $\lambda \neq \overline{\lambda}$ for some neutral vector $\xi_0 \in H$). Then there exist a compact $X_0 \subset \mathbf{R}$ and a topological *-isomorphism Ψ from the* *functional* $(\lambda,\overline{\lambda})$ - B^**-algebra* $C(X,\#)$, *where* $X = X_0\bigcup\{\lambda\}\bigcup\{\overline{\lambda}\}$, *such that* $\Psi(f_0) = \mathbf{1}$, $\Psi(f_1) = T$ *(f_1 and f_2 as in Corollary 1). This isomorphism has an extension (continuous) to a* **-homomorphism* $\hat{\Psi} : W(X,\#) \to \overline{M}^w$, *possessing the property as in Corrolary 1.*

From the results of [1, 5, 9] it follows that each commutative symmetric u-closed algebra of operators on a Pontryagin π_1 space is either a non-degenerated algebra, or it is the algebra of the classes 0, 1, $2a$ and $3a$. Thus we have to consider only the functional representation of commutative algebras of class 1.

Let T be a self-adjoint operator from $B(H)$, such that the algebra $M = \overline{\mathcal{P}(T,\mathbf{1})}^u$ is an algebra of class 1. Then T possesses a representation: $T = < A+\lambda, x+y, x+y, \lambda, \lambda, \gamma >$, where $A = A^* \in B(K)$, $\lambda \in \sigma(A)$, $\gamma \in \mathbf{R}$, $y \in Ker A$, $x \perp Ker A$. Now we define a linear map $p : \mathcal{P}(A) \to K$, such that $p(A) = x$, $p(A^n) = A^{n-1}p(A)$ for each positive integer n. Let us denote by $\mathcal{U}$ the set of all $B \in B(K)$ such that there exists a sequence $\{B_n\} \subset \mathcal{P}(A)$ with properties: $\|B - B_n\|_{B(K)} \to 0$, and the sequences $\{p(B_n)\}$ and $\{p(B_n^*)\}$ are converging in H. For each $B \in \mathcal{U}$ we put: $p(A) = \lim p(B_n)$. It is easy to check that $\mathcal{U}$ is a subalgebra of $B(K)$ and the map $p : \mathcal{U} \to K$ is the semiclosed quasi-vector; besides,

$$M = \{< B+\alpha, p(B)+\beta y, p(B^*)+\overline{\beta}y, \alpha, \alpha, \delta >:\ B \in \mathcal{U}, \alpha, \beta, \delta \in \mathbf{C}\ \}.$$

Let $T_0 = < A, x, x, 0, 0, 0 >$, $M_0 = \overline{\mathcal{P}(T_0)}^u$, $M_1 = \{< 0, \beta y, \overline{\beta}y, 0, 0, \delta >:\ \beta, \delta \in \mathbf{C}\}$. It is clear that $M = M_0 + M_1 + \mathbf{C}\cdot\mathbf{1}$, and moreover M_1 is a 2-dimensional ideal in M, consisting of nilpotent operators with order not more than 3. We'll give a functional description of M_0. We'll consider the set F of all $f \in C(\sigma(A))$ such that the sequences $\{p(q_n(A))\}$, $\{p(\overline{q}_n(A))\}$ are converging in K, $q_n(0) = 0$ and the sequence of numbers $\{(p(\hat{q}_n(A)), p(A))\}$ is also converging, where $\hat{q}_n(t) = q_n(t)/t - q_n'(0)$. For $f \in F$ we put: $p(f(A)) = \lim p(q_n(A))$, $p(\overline{f}(A)) = \lim p(\overline{q}_n(A))$ and $\|f\|_F = \|f(A)\|_{B(K)} + \|p(f(A))\|_K + \|p(\overline{f}(A))\|_K + \gamma_f$, where $\gamma_f = |\lim(p(\hat{q}_n(A)), p(A)|$. One may show that $(F, \|\ \|_F)$ is a B^*-algebra.

For each function $f \in F$ we define an operator

$$f(T_0) = < f(A), p(f(A)), p(\overline{f}(A)), 0, 0, \gamma_f >$$

from $B(H)$. We assume: $\Psi(f) = f(T_0)$.

Theorem 3. *Ψ is a* **-isomorphism from F onto M_0, and besides,* $\Psi(f_0) = T_0$, *for* $f_0 \cong t$ *and* $|||\Psi(f)||| = \|f\|_F$ *for all* $f \in F$.

Theorem 3 implies that the operator S is contained in $M = \overline{\mathcal{P}(T,\mathbf{1})}^u$ if and only if there exist a functional $f \in F$, a nilpotent operator $S_1 \in M_1$, and $\alpha \in \mathbf{C}$ such that $S = f(T_0) + S_1 + \alpha\mathbf{1}$.

References

[1] A.Y.Bendersky, S.N.Litvinov, V.I.Chilin. A description of commutative symmetric operator algebras in a Pontryagin space π_1. *J.Operator Theory,* 37 (1997), 201-222.

[2] K.Yn.Dadashyan, S.S.Horuzhy. On field algebras in the quantum theory with indefinite metric. *Theor. Math. Phys.* 54:1(1983), 57-77; 62:1(1985), 30-44, 70:2(1987), 181-191; 72:3(1987), 340-351. (Russian).

[3] V.I.Liberson, V.S.Shul'man. Non-degenerate algebras of operator in spaces with indefinite metric. *Izv.Akad.Nauk USSR, ser. Math.*, 37(1973), 533-537. (Russian).

[4] S.N.Litvinov. Bicyclic WJ^*-algebras in a Pontryagin space of type π_1. *Func. Anal Appl.*, 26:3(1992), 46-54. (Russian).

[5] A.I.Loginov. On commutative symmetric operator algebras in a Pontryagin space. *Izv. Akad. Nauk USSR, Ser. Math.*, 33:3(1969), 549-569. (Russian).

[6] A.I.Loginov. Complete commutative symmetric operator algebras in a Pontryagin space π_1. *Math. Sbornik,* 84(1971), 575-582. (Russian).

[7] A.I.Loginov. V.S.Shul'man. Non-reduced J-symmetric algebras of operators in space with indefinite metric. *Dokl. Akad. Nauk SSSR,* 240:1(1978), 21-23. (Russian).

[8] S.Sh.Masharipova. A description of weak closure of symmetric algebras of operators in a Pontryagin space. *Uzbek Math. Journal,* 3(1991), 36-42. (Russian).

[9] S.Sh.Masharipova, V.I.Chilin. Functional calculus for commutative symmetric algebras of operators in a Pontryagin space of type π_1. *Dokl. Akad Nauk of Republik Uzbekistan,* 4(1994), 10-12. (Russian).

[10] S.Sh.Masharipova, V.I.Chilin. Kaplansky's density theorem for symmetric algebras of operator in a Pontryagin space of type π_1. *Uzbek Math. Journal,* 2(1996), 68-75. (Russian).

[11] M.A.Naymark. Kommutative symmetricshe Operatoren algebren in Pontryaginschen Raumen π_k. *Math. Annalen,* 162(1965), 147-171. (German).

[12] U.A.Shtrauss. Functional representation of an algebra generated by a self-adjoint operator in a Pontryagin space. *Funct. Anal. Appl.*, 20:1(1986), 91-92. (Russian).

[13] V.S.Shul'man. Banach symmetric algebras of operators in a space of type π_1. *Math. Sbornik*, 89(1972), 264-279. (Russian).

[14] T.Ya.Azizov, I.S.Iohvidov. Foundations of theory of linear operators in spaces with indefinite metric."Nauka", Moskow, 1986. (Russian).

E-Mail: chilin@chilin.silk.org

CONTINUOUS DECOMPOSITION OF REAL VON NEUMANN ALGEBRAS OF TYPE III

Sh.USMANOV

Institute of Mathematics,
Uzbekistan Academy of Sciences
F. Hodjaev str., 29, 700143 Tashkent, Uzbekistan.

Abstract: Let $\mathfrak{R}$ be a real von Neumann algebra. We consider Takesaki's duality for enveloping von Neumann algebra $\mathcal{U}(\mathfrak{R}) = \mathfrak{R} + i\mathfrak{R}$ and the action of modular automorphism σ^{φ} for α-invariant weight on $\mathcal{U}(\mathfrak{R})$, where α is an involutory *-antiautomorphism of $\mathcal{U}(\mathfrak{R})$, generating $\mathfrak{R}$. We obtain the theorem of continuous decomposition of real type III factor into the crossed product of real von Neumann of type II_∞ by one-parameter group of automorphisms.

1. Introduction.

In [1] Takesaki obtained the theorem of duality for action of abelian locally compact groups on von Neumann algebras and used this duality for construction of well-known continuous decomposition of type III factors as the crossed product of von Neumann algebras of type II_∞ on action of group $\mathbf{R}$. We want to obtain similar results for real factors of type III and to apply its to real type III_0 factors.

Let $\mathfrak{R}$ be a real von Neumann algebra, i.e. $\mathfrak{R}$ be a real weakly closed *-algebra of bounded linear operators on a complex Hilbert space H, such that $\mathbf{I} \in \mathfrak{R}$ and $\mathfrak{R} \bigcap i\mathfrak{R} = \{0\}$. $\mathfrak{R}$ is called a real factor of type III if its enveloping von Neumann algebra $\mathcal{U}(\mathfrak{R}) = \mathfrak{R} + i\mathfrak{R}$ is a factor of type III.

This paper is devoted to the structure of real type III factors. We obtain decomposition theorem for real type III factors and give isomorphism criterions

Y. Khakimdjanov et al (eds.), Algebra and Operator Theory, 111-116.

for such factors.

2. Preliminaries.

In this section, we give some basic facts from the theory of complex and real von Neumann algebras.

Let $B(H)$ be the algebra of all bounded linear operators on a complex Hilbert space H. A weakly closed real *-subalgebra $\mathfrak{R}$ in $B(H)$ is called a real von Neumann algebra, if it contains the identity operator $\mathbf{I}$, and $\mathfrak{R}\bigcap i\mathfrak{R} = \{0\}$, where $i\mathfrak{R} = \{ix : x \in \mathfrak{R}\}$ [2-4]. The smallest (complex) von Neumann algebra $\mathcal{U}(R)$ containing R coincides with its complexfication $\mathfrak{R} + i\mathfrak{R}$, i.e. $\mathcal{U}(\mathfrak{R}) = \mathfrak{R} + i\mathfrak{R}$.Moreover $\mathfrak{R}$ generates a natural involutory *-antiautomorphism α of $\mathcal{U}(\mathfrak{R})$, namely, $\alpha(x+iy) = x^* + iy^*$, where $x+iy \in \mathcal{U}(\mathfrak{R})$, $x, y \in \mathfrak{R}$.

It is clear that $\mathfrak{R} = \{x \in \mathcal{U}(\mathfrak{R}) : \alpha(x) = x^*\}$. Conversely, given a von Neumann algebra $\mathcal{U}$ and any involutory $*$-antiautomorphism α on $\mathcal{U}$, the set $\{x \in \mathcal{U} : \alpha(x) = x^*\}$ is a real von Neumann algebra.

Von Neumann algebra M (or real von Neumann algebra $\mathfrak{R}$) is a factor, if $Z(M)$ $(= \{x \in M : xy = yx, \forall y \in M\}) = \{\lambda \cdot \mathbf{I}\}$, $\lambda \in \mathbf{C}$ (respectively, if $Z(\mathfrak{R}) = \{\lambda \cdot \mathbf{I}\}, \lambda \in \mathbf{R}$). We say, that a real von Neumann algebra $\mathfrak{R}$ is of type I_{fin}, I_∞, II_1, II_∞, III_λ $(0 \le \lambda \le 1)$, if $\mathcal{U}(\mathfrak{R})$ has the corresponding type in usual classification of W^* - algebras (see., for example, [5, 6]). Other useful references are [5-9].

3. Takesaki's duality theorem and antiautomorphisms.

Let M be a σ-finite continuous infinite von Neumann algebra, G - an abelian locally compact group, $\sigma = \{\sigma_g\}_{g\in G}$ - a strongly continuous action of G on M. Let $N = M \times_\sigma G$ be the crossed product of M and G by σ; then N is the von Neumann algebra of linear operators on the Hilbert space $\ell^2(H, G)$, generated by operators $\{\pi_\sigma(x), x \in M\}$ and $\{\lambda_\sigma(t), t \in G\}$, where

$$\pi_\sigma(x)\xi(s) = (\sigma_{-s}(x)\xi)(s);$$

$$\lambda_\sigma(t)\xi(s) = \xi(s-t), \quad t, s \in G, \xi(s) \in \ell^2(H, G).$$

We consider also an involutory *-antiautomorphism α on M. We say that: (1) α is commuting with σ, if $\alpha \cdot \sigma_g = \sigma_g \cdot \alpha$ for all $g \in G$; (2) α is anticommuting with σ, if $\alpha \cdot \sigma_g = \sigma_{-g} \cdot \alpha$ for all $g \in G$.

Theorem 1

*(1) If α is commuting with σ, then there exists a unique involutory *-antiautomorphism*

$\overline{\alpha}$ on $N = M \times_\sigma G$ such that

$$\overline{\alpha}|_{\pi(M)} = \pi_\sigma \cdot \alpha, \quad \text{and} \quad \overline{\alpha}(\lambda_\sigma(t)) = \lambda_\sigma^*(t),$$

for all $t \in G$.

*(2) If α is anticommuting with σ, then there exists a unique involutory *-antiautomorphism $\hat{\alpha}$ on $N = M \times_\sigma G$ such that*

$$\hat{\alpha}|_{\pi(M)} = \pi_\sigma \cdot \alpha, \quad \text{and} \quad \hat{\alpha}(\lambda_\sigma(t)) = \lambda_\sigma(t),$$

for all $t \in G$.

Sketch of Proof. It is known, that $M \times_\sigma G \cong (M \otimes B(\ell^2(G)))^{\sigma \otimes Ad\rho} = \{x \in M \otimes B(\ell^2(G))$: $(\sigma_t \otimes Ad\rho_t)(x) = x, \forall t \in G\}$, where ρ_t is the right translation of G on $\ell^2(G)$:

$$\rho_t f(s) = f(s+t), \quad t, s \in G, \quad f(s) \in \ell^2(G)$$

(see, for example [10, Theorem 3.11]).
We consider on $\ell^2(G)$ an antilinear operator J_1:

$$J_1 f(s) = \overline{f}(s), \quad f(s) \in \ell^2(G), \quad s \in G.$$

Evidently, $J_1^2 = \mathbf{1}$. Put $\beta_1(x) = J_1 x^* J_1$, where $x \in B(\ell^2(G))$, then β_1 is in involutory *-antiautomorphism on $B(\ell^2(G))$ (see [9]), and we can consider the restriction of antiautomorphism $\alpha \otimes \beta_1$ on $(M \otimes B(\ell^2(G)))^{\sigma \otimes Ad\rho}$:

$$\overline{\alpha} = \alpha \otimes \beta_1|_{(M \otimes B(\ell^2(G)))^{\sigma \otimes Ad\rho}}.$$

Similarly, let J_2 be an antilinear operator on $\ell^2(G)$:

$$J_2 f(s) = \overline{f}(-s), \quad f(s) \in \ell^2(G), \quad s \in G.$$

Then $J_2^2 = \mathbf{1}$. Put $\beta_2(x) = J_2 x^* J_2$, where $x \in B(\ell^2(G))$. Then

$$\hat{\alpha} = \alpha \otimes \beta_2|_{(M \otimes B(\ell^2(G)))^{\sigma \otimes Ad\rho}}. \quad \Box$$

Let $\hat{G}$ be the dual group of G; we define the unitary operator v_p on $\ell^2(H, G)$, $p \in \hat{G}$:

$$(v_p) f(s) = < \overline{s, p} > f(s), \quad \text{where } s \in G, p \in \hat{G}, f(s) \in \ell^2(H, G),$$

$< s, p >$ is the duality between G and $\hat{G}$. The action $\hat{\sigma}_p = Adv_p$ of $\hat{G}$ on $M \times_\sigma G$ is called dual action (see [1]).

By Takesaki's duality theorem (see [1], Theorem 4.5) the von Neumann algebra $(M \times_\sigma G) \times_{\hat{\sigma}} G$ is isomorphic to $M \otimes B(\ell^2(G))$, and the bidual action $\hat{\hat{\alpha}}$

of G on $(M \times_\sigma G) \times_{\hat{\sigma}} \hat{G}$ is transformed under this isomorphism to the action $\sigma \otimes Ad\rho$ on $M \otimes B(\ell^2(G))$.

Theorem 2

(1) If α is commuting with σ, then $\overline{\alpha}$ is anticommuting with $\hat{\sigma}$, and the anti-automorphism $\hat{\overline{\alpha}}$ of $(M \times_\sigma G) \times_{\hat{\sigma}} \hat{G}$ is transformed to $\alpha \otimes \beta_1$;

(2) If α is anticommuting with σ, then $\hat{\alpha}$ is commuting with $\hat{\sigma}$, and the anti-automorphism $\overline{\hat{\alpha}}$ of $(M \times_\sigma G) \times_{\hat{\sigma}} \hat{G}$ is transformed to $\alpha \otimes \beta_2$;

4. Continuous decomposition of real type III factors.

Let M be a σ - finite type III factor, α - an involutory *-antiautomorphism of M, $\mathfrak{R} = \{x \in M : \alpha(x) = x^*\}$ the real type III factor generated by α. By [2, 11] there exists a faithful normal semifinite α-invariant weight φ on M_+. Let σ^φ be the modular automorphism group of M, associated with φ. Then

$$\sigma_t^\varphi \cdot \alpha = \alpha \cdot \sigma_{-t}^\varphi,$$

$\forall t \in \mathbf{R}$ (see [7, 8]).
We consider on $N = M \times_{\sigma^\varphi} \mathbf{R}$ the dual action $\hat{\sigma}^\varphi$ of the group $\hat{I\!R}$ with respect to the duality $< t, s >= e^{its}$. Let τ be the canonical faithful normal semifinite trace on N_+, such that

$$\tau \cdot \sigma_t^\varphi = e^{-t}\tau$$

(see [1]).

Proposition 3.

(1) $\hat{\alpha}$ is commuting with $\hat{\sigma}^\varphi$;

(2) $\tau \cdot \hat{\alpha} = \tau$.

The following theorem gives a continuous decomposition of real type III factors (by analogy with Takesaki's classification of factors of type III).

Theorem 4

*(1) Let N be a σ-finite von Neumann algebra of type II_∞, α - an involutory *-antiautomorphism of N, τ - a faithful normal semifinite α-invariant trace on N, $\theta = \{\theta_s\}_{s \in I\!R}$ a strongly continuous action of the additive group $I\!R$ on N, commuting with α, acting ergodically on the centre of N and such that*

$$\tau \cdot \theta_s = e^{-s}\tau, \qquad \forall s \in I\!R.$$

Then $\mathfrak{R} = \{x \in N \times_\theta I\!R : \overline{\alpha}(x) = x^*\}$ is a real type III factor.

(2) If $\mathfrak{R}$ is a σ-finite real type III factor, then exists a triple (N, θ, α) which satisfies the conditions of statement (1) of Theorem and such that

$$\mathfrak{R} = \{x \in N \times_\theta I\!R : \overline{\alpha}(x) = x^*\}$$

where $\overline{\alpha}$ is the extension of α on $N \times_\theta \mathbb{R}$ (see Theorem 1.)

*(3) Let $(N_1, \theta^{(1)}, \alpha_1)$ and $(N_2, \theta^{(2)}, \alpha_2)$ be triples which satisfy the statement (1) of Theorem. Then the real type III factors $\mathfrak{R}_1 = \{x \in N \times_{\theta^{(1)}} \mathbb{R} : \overline{\alpha}_1(x) = x^*\}$ and $\mathfrak{R}_2 = \{x \in N \times_{\theta^{(2)}} \mathbb{R} : \overline{\alpha}_2(x) = x^*\}$ are isomorphic if and only if there exist an *-isomorphism $\Phi : N_1 \to N_2$ and a family of unitaries $\{u_t\}_{t\in\mathbb{R}} \subset \mathfrak{R}_1$ such that*

$$\Phi \cdot \alpha_1 = \alpha_2 \cdot \Phi;$$

$$U_{t+s} = u_t \theta_t^{(1)}(u_s);$$

$$\Phi^{-1} \cdot \theta_t^{(2)} \cdot \Phi(x) = u_t \theta_t^{(1)}(x) u_t^*,$$

where $x \in N_1$, $t, s \in \mathbb{R}$.

Remark In statement (1) of Theorem 4 we have two cases:

(1) if $\alpha|_{Z(N)} = id$, then $mod(\overline{\alpha}) = id$ (in the sense of Giordano's fundamental homomorphism , see [7, 8]).

(2) if $\alpha|_{Z(N)} \neq id$, then $mod(\overline{\alpha}) \neq id$.

Application to real type III_0 factors.

Using the Theorem 2 and the Theorem 4, we can obtain a discrete decomposition of real type III_0 factors.

Theorem 5

Let $\mathfrak{R}$ be a σ-finite real type III_0 factor. Then either

*(1) there exist a type II_∞ real von Neumann algebra Q with a continuous centre and with a faithful normal semifinite trace τ; an *-automorphism θ of Q having the property $\tau \cdot \theta \leq \lambda_0 \theta$, $0 \leq \lambda_0 \leq 1$ and acting ergodically on the centre of Q, such that $\mathfrak{R}$ is isomorphic to the real crossed product of Q by $\{\theta^n\}_{n\in Z}$,*

or

*(2) there exist a type II_∞ complex von Neumann algebra N with a continuous centre and with a faituful normal semifinite trace τ; an antilinear *-automorphism δ of N having the property $\tau \cdot \leq \lambda_0 \delta$, $0 \leq \lambda_0 \leq 1$, such that $\mathfrak{R}$ is isomorphic to the crossed product of N by $\{\delta^n\}_{n\in Z}$.*

References

[1] Takesaki M. Duality for crossed products and the structure of von Neumann algebras of type III. *Acta Math.*, 131 (1973), 249-310.

[2] Ayupov Sh.A., Rakhimov A.A., Usmanov Sh.M. Jordan, Real and Lie structures in operator algebras. *Mathematics and Its Appl.*, N 418. Kluwer Academic Publisher, 1997, 235 pp.

[3] Stormer E. Real structure in the hyperfinite factor *Duke Math. J.*, 47:1(1980), 145 - 153.

[4] Ayupov Sh.A. Classification and representation of ordered Jordan algebras, "Fan" , Tashkent , 1986. 121 p. (Russian).

[5] Connes A. Une classification des facteurs de type III *Ann. Sc. Ec. Norm. Sup.*, 6(1973), 133-252.

[6] Takesaki M. Theory of operator algebras, I. Berlin: Springer. 1979. VIII + 415 p.

[7] Giordano T. Antiautomorphismes involutifs des facteurs de von Neumann injectifs. These , Universite de Neuchatel, 1981. p. 106.

[8] Giordano T. Antiautomorphismes involutifs des facteurs de von Neumann injectifs. II *J. Funct. Anal.*, 51:3(1983), 326 - 360.

[9] Hanche-Olsen H., Stormer E. Jordan operator algebras, London: Pitman Ltd, 1984. VIII+183 p.

[10] Van Daele A. Continuous crossed products and type III von Neumann algebras. *London Math. Soc.*, Lecture Notes 31 (Cambridge University Press, London, 1978).

[11] Stacey P.J. Real structure in σ - finite factors of type III_λ where $0 < \lambda < 1$. *Proc. London Math. Soc.*, 47:3(1983), 275-284.

E-mail: root@im.tashkent.su aratex@hal.freenet.uz

ESPACES VECTORIELS DIFFERENTIELS

Kamola KHAKIMDJANOVA

Institute of Mathematics,
Uzbekistan Academy of Sciences
F. Hodjaev str., 29, 700143 Tashkent, Uzbekistan.

Introduction

La notion de l'espace vectoriel différentiel a été introduite pour étudier les algèbres différentielles (J.F.Ritt [8] et E.R.Kolchin [5]). Cette théorie, qui a pour but la généralisation de la géométrie algébrique, dans le cadre de laquelle une équation algébrique peut être considérer comme une équation différentielle, permet une approche algébrique de l'étude des équations différentielles [2], [6], [7]. Dans les nombreux travaux consacrés à l'étude des algèbres différentielles, il y a une lacune importante: une notion de base comme la notion d'espace vectoriel différentiel n'est pas bien étudiée. Dans notre travail nous considérons certains aspects de cette étude, qui montre que les propriétés d'un espace vectoriel différentiel peuvent être très différentes de celles d'un espace vectoriel. Nous bornerons notre étude aux espaces vectoriels différentiels de dimension finie.

1 Généralité

1.1 Corps différentiels

Soit K un corps . Un operateur $d : K \longrightarrow K$ est dit *différentiel,* s'il verifie les conditions suivants:

$$d(a+b) = d(a) + d(b), \ \forall a, b \in K$$

$$d(ab) = d(a)b + ad(b), \ \forall a, b \in K$$

Y. Khakimdjanov et al (eds.), Algebra and Operator Theory, 117-126.

Définition 1.1 *Un corps K muni d'un opérateur différentiel d est appelé corps différentiel.*
Exemple 1. Un corps arbitraire est un corp différentiel pour $d = 0$.
Exemple 2. Soit $K(x)$ le corps des fractions rationnelles, c'est-à-dire des fonctions de la forme $\frac{f(x)}{g(x)}$, ou $f(x)$ et $g(x)$ sont les polynomes réels, $g(x) \neq 0$. $K(x)$ est un corp différentiel par rapport à l'opérateur $d = \frac{d}{dx}$.

Définition 1.2 *Un élément a de corp différentiel (K, d) est appelé constante, si $d(a) = 0$.*

Soit (K, d) un corps différentiel. Le sous ensemble $K_0 \subset K$ constitué des éléments constants de K est evidemment un corps. Ce corps est appelé corps des constantes. Pour le corps différentiel de l'exemple 1 on a $K_0 = K$. Pour l'exemple 2, on a $K_0 = R$.

1.2 Espaces vectoriels différentiels

Définition 1.3 *Soit (K, d_1) un corps différentiel et soit V un espace vectoriel. Soit $d_2 : V \longrightarrow V$ un opérateur verifiant les conditions suivantes:*

$$d_2(x + y) = d_2(x) + d_2(y), \forall x, y \in V$$

$$d_2(ax) = d_1(a)x + ad_2(x), \forall a \in K, \forall x \in V.$$

Alors (V, d_2) est appelé espace vectoriel différentiel sur (K, d_1)
Remarque. Pour simplifier l'ecriture, nous désignons souvent par d les opérateurs d_1 et d_2.

1.3 Applications linéaires différentielles

Soient (V_1, d) et (V_2, d) deux espaces vectoriels différentiels sur le même corp différentiel (K, d) .

Définition 1.4 *Une application $f : V_1 \longrightarrow V_2$ est appelée linéaire différentielle, si*

$$f(x + y) = f(x) + f(y), \forall x, y \in V_1$$

$$f(\alpha x) = \alpha f(x), \forall \alpha \in K, \forall x \in V_1$$

$$f(dx) = d(f(x)), \forall x \in V_1.$$

Remarque. Si $V_1 = V_2$, une telle application linéaire est dite transformation linéaire différentielle.

Définition 1.5 *Les espaces vectoriels différentiels (V_1, d) et (V_2, d) sur un corps différentiel (K, d) sont dit isomorphes, s'il existe une application linéaire différentielle bijective $f : V_1 \longrightarrow V_2$.*

1.4 Matrice d'une transformation linéaire différentielle

Soit (V, d) un espace vectoriel différentiel de dimension n sur un corps (K, d) et soit $f : V \longrightarrow V$ une transformation linéaire différentielle. Choisissons une base $\{e_1, e_2, ..., e_n\}$ de V et soit la matrice $A = (a_{ij})$ de f verifiant $f(e_i) = \sum a_{ki}e_k$. Considerons également la matrice $D = (d_{ij})$ de d définie par $de_i = \sum d_{ki}e_k$.

Théorème 1 *Soient (V, d) un espace vectoriel différentiel sur (K, d), $f : V \longrightarrow V$ une transformation linéaire différentielle et soient A et D respectivement les matrices de f et de d dans la base $\{e_1, ..., e_n\}$ de V. On a alors*

$$d(A) = AD - DA,$$

oú $d(A) = (da_{ij})$.

Démonstration. Pour un vecteur $x = \sum \alpha_i e_i$ de V, légalité $f(d(x)) = d(f(x))$ s'écrit:

$$f(\sum e_i d\alpha_i + \alpha_i de_i) = d(\sum \alpha_i f(e_i))$$

et donc

$$\sum(fe_i d\alpha_i + \alpha_i f de_i) = \sum(d\alpha_i fe_i + \alpha_i df(e_i)).$$

Mais

$$fde_i = f\sum d_{ki}e_k = \sum de_i Ae_k = \sum d_{ki}a_{sk}e_s$$

$$d(fe_i) = d\sum(a_{ki}e_k) = \sum(e_k da_{ki} + a_{ki}de_k) = \sum e_k da_{ki} + \sum a_{ki}d_{sk}e_s$$

D'ou

$$\sum d_{ki}a_{sk} = da_{si} + \sum a_{ki}d_{sk}$$

1.5 Changement de bases

Soit (V, d) un espace vectoriel différentiel sur (K, d) et soient $\{e_1, ..., e_n\}$ et $\{e'_1, ..., e'_n\}$ deux bases de V. Designons par $T = (t_{ij})$ la matrice de passage définie par $e_i = \sum t_{si}e'_s$ et par $D = (d_{ij})$ et par $D' = (d'ij)$ les matrices de d dans les deux bases respectives. Alors

$$de_i = \sum d_{si}es = \sum d_{si}t_{ks}e'_k$$

et

$$d\sum(t_{si}e'_s) = \sum d(t_{ki})e'_k + \sum t_{si}d'_{ks}e'_k.$$

D'ou

$$d(T) = TD - D'T$$

et

$$D' = TDT^{-1} - d(T)T^{-1}$$

1.6 Somme directe des espaces vectoriels différentiels

Soient $(V_1, d_1), (V_2, d_2), ..., (V_k, d_k)$ des espaces vectoriels différentiels sur (K, d)

Alors, nous pouvons construire un espace vectoriel différentiel V, appelé somme directe des (V_i, d_i) de la faon suivante:

$$V = \{(v_1, v_2, ..., v_k) : v_i \in V_i\},$$

$$d(v_1, v_2, ..., v_k) = (d_1 v_1, ..., d_k v_k).$$

Soient $D_1, D_2, ..., D_k$ les matrices respectives des opérateurs différentiels $d_1, d_2, ..., d_k$ dans les base fixés des espaces $V_1, V_2, ..., V_k$. Si on considère la base $\{e_1, e_2, ..., e_{n_1+n_2+...+n_k}\}$ de V, obtenue à partir des bases fixées des $V_1, ..., V_k$ par plongement, la matrice de d s'écrit:

$$D = \begin{pmatrix} D_1 & 0 & \cdots & 0 \\ 0 & D_2 & \cdots & 0 \\ \cdots & & & \\ 0 & 0 & \cdots & D_k \end{pmatrix}$$

1.7 Espace dual différentiel

Soit (V, d) espace vectoriel différentiel sur (K, d). Dans l'espace dual V^* des fonctions linéaires dans V on peut définir l'opérateur différentiel d^* en posant

$$(d^* f)(v) = d(f(v)) - f(dv),$$

où $f \in V^*$. Une vérification immédiate montre, que

$$d^*(\alpha f) = (d\alpha)f + \alpha d^* f, \ \forall \alpha \in K, f \in V^*.$$

Nous obtenons ainsi une structure d'espace vectoriel différentiel dans V^* , appelée espace dual différentiel. Soit $(e_1, e_2, ..., e_n)$ une base de V et $(e^1, e^2, ..., e^n)$ la base dual de V^*. Désignons par D et D^* les matrices de d et de d^* dans ces bases respectives. Alors

$$D^* = -{}^t D.$$

En effet, pour un vecteur $v = \sum \alpha_m e_m \in V$ nous avons

$$d^*(e^k)(v) = d(e^k(v)) - e^k(dv) = d(\alpha_k) - d(\alpha_k) - \sum \alpha_m e^k \sum d_{sm} e_s =$$

$$- \sum \alpha_m d_{km} = (- \sum d_{ki} e^i)(v),$$

d'où

$$d^*(e^k) = \sum d^*_{ik} e^i = - \sum d_{ki} e^i.$$

1.8 Produit tensoriel des espaces vectoriels différentiels

Soient (V, d) et (W, d) deux espaces vectoriels différentiels sur (K, d). Sur le produit tensoriel $V \otimes W$ (au sense ordinaire) on peut introduire l'opérateur différentiel en posant

$$d(x \otimes y) = dx \otimes y + x \otimes dy$$

Il est clair que

$$d(\alpha(x \otimes y)) = d\alpha(x \otimes y) + \alpha d(x \otimes y)$$

et $(V \otimes W, d)$ est un espace vectoriel différentiel sur (K, d).

2 Certains propriétés des espaces vectoriels différentiels

Dans le paragraphe on met en évidence, des différences structurelles sur certaines propriétés des espaces vectoriels différentiels par rapport aux espaces vectoriels ordinaires. Cela nous conduit à définir différement certaines notions comme la dimension.

2.1 Certain propriétés d'isomorphisme

On connais bien, que deux espaces vectoriels de même dimension sur le même corps sont isomorphes. Ce n'est pas le cas s'il s'agit des espaces vectoriels différentiels.

Théorème 2 *Il existe les espaces vectoriels différentiels de même dimension (sur le même corps différentiel) qui ne sont pas isomorphes.*

Démonstration. Soit $(K, d) = R(x)$ le corps différentiel des fonctions rationnelles (exemple 2 de 1.1). Considerons les deux espaces de dimension 1 suivants:

$$V_1 = \{Ke_1 : d(e_1) = \alpha e_1\}$$

$$V_2 = \{Ke_2 : d(e_2) = \beta e_2\}$$

Ici α et β sont des nombres réels. Nous allons montrer que nous pouvons choisir ces nombres de manière à ce que les espaces V_1 et V_2 ne soient pas isomorphes.

Supposons $V_1 \cong V_2$. Alors il existe une application linéaire $A : V_1 \longrightarrow V_2$ verifiant l'egalité $dA = Ad$. Soit $Ae_1 = ae_2$. Comme $e_1 \neq 0$, on a $a \neq 0$. Alors

$$dae_2 + ade_2 = a\alpha e_2.$$

Cela est possible si et seulement si

$$da + a(\beta - \alpha) = 0.$$

Si nous choisissons les nombres α et β pour que $\alpha-\beta = 1$, alors $da = a$. Comme l'équation différentielle

$$\frac{da(x)}{dx} = a(x)$$

n'admet pas des solutions non nulles appartenant à $R(x)$ notre supposition conduit à une contradiction. D'ou le théorème.

2.2 Certains propriétés des sous espaces différentiels

On sait bien que tout espace vectoriel de dimension n possède des sous espaces de dimensions k quel que soit un entier $1 \leq k \leq n$. Ce n'est pas le cas dans un espace vectoriel différentiel.

Théorème 3 *Il exist un espace vectoriel différentiel de dimension 2, qui ne possède pas de sous espace différentiel de dimension 1.*

Démonstration. Soit (K, d) le corps différentiel des fonctions rationnelles, et soit (V, d) un espace vectoriel différentiel sur (K, d) de dimension 2. Pour une base $\{e_1, e_2\}$ de V et pour un vecteur $x = \begin{pmatrix} x_1 \\ x_2 \end{pmatrix} = x_1e_1 + x_2e_2 \in V$ on a

$$d : \begin{pmatrix} x_1 \\ x_2 \end{pmatrix} \longrightarrow \begin{pmatrix} dx_1 \\ dx_2 \end{pmatrix} + \begin{pmatrix} \alpha & \beta \\ \gamma & \delta \end{pmatrix} \begin{pmatrix} x_1 \\ x_2 \end{pmatrix}$$

En effet, si la matrice de d dans la base $\{(e_1, e_2\}$ est $\begin{pmatrix} \alpha & \beta \\ \gamma & \delta \end{pmatrix}$, on a

$$de_1 = \alpha e_1 + \gamma e_2$$

$$de_2 = \beta e_1 + \delta e_2$$

et

$$d(x) = d(x_1e_1 + x_2e_2) = d(x_1e_1) + d(x_2e_2) = dx_1e_1 + x_1de_1 + dx_2e_2 + x_2de_2.$$

D'où la formule cherchée.

Si pour un vecteur x avec les coordonnées x_1 et x_2, $d(x) \in Kx$, c'est-à-dire si Kx est un sous espace de V de dimension 1, alors pour tout element $b \in K$ nous avons $d(bx) = db.x + b.dx \in Kx$. Cela signifie que nous pouvons supposer $x_1 = 1$ ou $x_2 = 1$. Soit $x_1 = 1$ et $dx = tx$. Alors

$$\begin{aligned} d(1) + \alpha + \beta x_2 &= t.1 \\ dx_2 + \gamma + \delta x_2 &= tx_2 \end{aligned}$$

d'où

$$\begin{aligned} t &= \alpha + \beta x \\ dx_2 + \gamma + \delta x_2 &= (\alpha + \beta x_2)x_2 \end{aligned}$$

Si nous choisissons $\alpha = \beta = \delta = -\gamma = 1$, nous obténons l'équation différentielle $dx_2 - x_2^2 - 1 = 0$, qui n'admet pas des solution dans l'ensemble des fonctions rationnelles.

Pour une démonstration complète il nous reste de montrer que le sous ensemble des vecteurs $\begin{pmatrix} 0 \\ x \end{pmatrix}$ ne sera pas un sous espase différentiel de (V, d) pour l'application d construite. Soit $y = \begin{pmatrix} 0 \\ y_2 \end{pmatrix} \in V$. Alors

$$d(y) = \begin{pmatrix} 0 \\ dy_2 \end{pmatrix} + \begin{pmatrix} 1 & 1 \\ -1 & 1 \end{pmatrix} \begin{pmatrix} 0 \\ y_2 \end{pmatrix} = \begin{pmatrix} y_2 \\ y_2 + dy_2 \end{pmatrix}.$$

Si $y_2 \neq 0$, d(y) n'appartient pas au sous ensemble consideré. D'où le théorème.

Le théorème suivant montre, que la méthode appliquée pour démontrer le théorème 3 ne peut pas automatiquement être utilisée pour la dimension 3 ainsi que pour une dimension impaire.

Théorème 4 *Soit (V, d) un espace vectoriel de dimension $n = 2k + 1, k \geq 1$, sur le corps différentiel des fonctions rationnelles $K = (R(x), d/dx)$ et soit dans une base de V la matrice $D = (d_{ij})$ n'admet que les éléments constants. Alors il existe toujour un sous espace différentiel de dimension 1 dans V*

Démonstration. Comme l'ordre de la matrice réelle D est impaire il existe au moins un vecteur propre $v = \beta_1 e_1 + ... + \beta_n e_n$, où $\beta_1, ..., \beta_n \in R$ corréspondant à la valeur propre λ . Alors

$$d : \begin{pmatrix} \beta_1 \\ \beta_2 \\ \vdots \\ \beta_n \end{pmatrix} \longrightarrow \begin{pmatrix} d\beta_1 \\ d\beta_2 \\ \vdots \\ d\beta_n \end{pmatrix} + D. \begin{pmatrix} \beta_1 \\ \beta_2 \\ \vdots \\ \beta_n \end{pmatrix} = 0 + \lambda. \begin{pmatrix} \beta_1 \\ \beta_2 \\ \vdots \\ \beta_n \end{pmatrix}$$

et pour $\forall a \in K$ nous avons

$$d(av_0) = da.v_0 + a.dv_0 = (da + a\lambda).v_0 \in Kv_0$$

D'ou le théorème.

On peut poser une question naturelle: est-que le théorème 4 est valable, si n est pair? Le théorème suivant montre que la reponse est non.

Théorème 5 *Pour tout nombre pair $n = 2k, k \geq 1$, il existe un espace vectoriel différentiel de dimension n , qui n'admèt pas un sous espace différentiel de dimension 1.*

Démonstration. Soit (V_1, d_1) un espace vectoriel différentiel de dimension 2, vérifiant le théorème 3. Considérons la somme dirécte

$$V = \underbrace{V_1 \oplus V_1 \oplus \cdots \oplus V_1}_{k}$$

et montrons , que V vérifie le théorème . Supposons le contraire, soit $W = Kv, v \neq 0$, un sous espace différentiel de dimension 1. Alors $v = (v_1, v_2, \cdots, v_k)$, $v_i \in V_1$. Comme $v \neq 0$ au moins un des v_i est non nul. Nous pouvons supposer, que $v_1 \neq 0$. Dans ce cas $W_0 = Kv_1$ est un sous espace vectoriel différentiel de dimension 1 de V_1. En effét, comme

$$d(v) = \lambda v = \lambda(v_1, v_2, \cdots, v_k),_q uad\lambda \in K,$$

on a $d(v_1) = \lambda v_1$. Cela est contraire au chois de V_1. D'où le théorème.

3 Dimension différentielle

La définition de dimension d'un espace vectoriel peut être donnée de manière suivante: c'est la longeur maximale des drapeaux des sous espaces, c'est - à - dire des suites des sous espaces non nuls stictement croissants

$$V_1 \subset V_2 \subset \cdots \subset V_n$$

Cette définition se généralise facilement au cas des espaces vectoriels différentiels.

Définition 3.1 *On appelle dimension différentielle d'un espace vectoriel différentiel (V, d) la longeur maximale des drapeaux des sous espaces différentiels de V.*

Les théorèmes du chapitre 2 montrent, que la dimension différentiel et la dimension ordinaire d'un espace différentiel peuvent être différents. Le but de ce chapitre est de montrer, que cela est possible dans toutes les dimensions n, c'est-à-dire montrer le théorème suivante:

Théorème 6 *Soit $n \geq 2$ un nombre naturel. Alors il existe un espace différentiel (V, d) de dimension n tel que sa dimension différentiel est strictement inférieure à n*

Pour démontrer ce théorème il est suffit de montrer le lemme suivant
Lemme *Il existe un espace vectoriel différentiel de dimension 3 qui n'admet pas des sous espaces différentiels de dimension 1.*

En effét, le resonnement du théorème 5 montre, qu'à partir des espaces vectoriels différentiels des dimensions 2 et 3 n'admetant pas les sous espaces différentiels de dimension 1, nous pouvons construire les espace dans une dimension $n \geq 2$ quelconque.

Démonstration du lemme. Soit (V, d) un espace vectoriel différentiel de dimension 3 sur le corp $K = (R(x), d/dx)$, et soient (e_1, e_2, e_3) une base de V,

$D = (d_{ij})$ la matrice de d dans cette base. Pour qu'un sous espace $Kv, v = (y_1, y_2, y_3)$ soit un sous espace différentiel (ici, comme dans la démonstration du théorème 3, nous pouvons supposer, que $y_1 = 1$), les équations suivantes doivent se vérifier:

$$\begin{aligned} d_{11} + d_{12}y_2 + d_{13}y_3 &= \lambda \\ dy_2 + d_{21} + d_{22}y_2 + d_{23}y_3 &= \lambda y_2 \\ dy_3 + d_{31} + d_{32}y_2 + d33y_3 &= \lambda y_3 \end{aligned}$$

oú $\lambda \in K$.

Choisissons les éléments d_{ij} de K comme suit:

$$d_{11} = 1, d_{12} = d_{13} = 0$$

$$d_{21} = 0, d_{22} = 3, d_{23} = 1$$

$$d_{31} = \frac{1}{x}, d_{32} = -2, d_{33} = 1$$

Alors les équations s'écrit:

$$\frac{dy_2}{dx} + 2y_2 + y_3 = 0$$

$$\frac{dy_3}{dx} - 2y_2 = -\frac{1}{x}$$

Ce système se reduit à l'équation linéaire du deuxième ordre suivante:

$$y''_2 + 2y'_2 + 2y_2 = \frac{1}{x}$$

Solution générale de cette équation s'ecrit:

$$y_2 = C_1(x)e^{-x}\cos x + C_2(x)e^{-x}\sin x + Ae^{-x}\cos x + Be^{-x}\sin x$$

Où A et B sont les constants réels arbitraires,

$$C_1(x) = \int \frac{1}{x}e^{x}\cos x dx, C_2(x) = \int \frac{1}{x}e^{x}\sin x dx$$

Il est claire que les fonctions $C_1(x)$ et $C_2(x)$ ne sont pas les fonctions rationnelles , c'est-à-dire, aucune des solutions de notre système différentiel n'appartient pas à K. D'oú le lemme.

References

[1] P.Blum. Rational functions on differential closed sets. *Amer. J. Math.*, 94(1972), 676 - 684.

[2] E.Gourin. On irreducible systems of algebraic differential equations. *Bull. Amer. Math. Soc.*, 39(1933), 593 - 595.

[3] J.Johnson. A notion of Krull dimension for differential rings. *Comment. Math. Helv.* 44(1969), 207 -216.

[4] I.Kaplansky. An Introduction to Differential Algebra. Hermann, Paris, 1957.

[5] E.R.Kolchin. Differential Algebra and Algebraic Groups. Academic Press, New York and London, 1973.

[6] E.R.Kolchin. The notion of dimension in the theory of algebraic differential equations. *Bull. Amer. Math. Soc.* 70(1964), 570 - 573.

[7] J.F.Ritt. Differential equations from the algebraic standpoint. *Amer. Math. Soc. Coll. Pub.*, vol. 14, New York, 1932.

[8] J.F.Ritt. Differential algebra. *Amer. Math. Soc. Coll. Pub.*, vol. 33, New York, 1950.

COMPLETUDE DE L'EQUATION D'EULER

Sh.BROMBERG, A.MEDINA

Sh.B.: en année sabatique au Département des Mathématiques de l'Université de Montpellier II
Departamento de Matemáticas
Universidad Autónoma Metropolitana-IZTAPALAPA
México, D. F. MEXICO

A.M.: Département de Mathématiques
Université de Montpellier II, case 051
34095 Montpellier cedex 5 FRANCE

Abstract The main purpose of this work is to study the relationship between the completness of the flow of Euler Equation and the existence of idempotent elements for the related Levi–Civita product. We find new cases of completness.

1 Introduction. Résumé

Soit $\mathcal{G}$ une algèbre de Lie réelle de dimension finie, $\mathcal{G}^*$ l'espace dual de $\mathcal{G}$ et $\Phi : \mathcal{G} \longrightarrow \mathcal{G}^*$ un isomorphisme symétrique d'espaces vectoriels. L'équation différentielle

$$\dot{\xi} = -ad^*_{\Phi^{-1}(\xi)}\,\xi. \tag{1}$$

où ad^* désigne la représentation co–adjointe de l'algèbre $\mathcal{G}$ est appellée l'*équation d'Euler* associée à $(\mathcal{G}, \Phi)$. Dans ce travail nous étudions la complétude du flot du champ de vecteurs associé à cette équation.

Soit G un groupe de Lie connexe d'algèbre de Lie $\mathcal{G} = \mathrm{L}(G)$. La donnée de Φ correspond à la donnée d'une métrique pseudo–riemanniene g_Φ invariante à gauche sur G. L'étude de la complétude de l'équation revient à l'étude

Y. Khakimdjanov et al (eds.), Algebra and Operator Theory, 127-144.

de la complétude du flot géodésique de (G, g_Φ). Si l'équation est complète les mouvements (i.e. les géodésiques relatives à la connexion de Levi–Civita) du *solide généralisé* (G, g_Φ) au sens de V. Arnold sont définis pour toute valeur du paramètre. Des définitions, il résulte que les solutions de l'équation d'Euler sont tracées sur les orbites de la représentation co–adjointe de G. Or ces orbites sont sympléctiques: elles sont en fait les feuilles symplectiques d'une structure de Poisson sur la variété $\mathcal{G}^*$. Le champ restriction du champ d'Euler à chaque orbite est hamiltonien, de hamiltonien la restriction de l'énergie

$$E : \xi \longmapsto \xi\left(\Phi^{-1}(\xi)\right)$$

à l'orbite correspondante ([Ar]. La complète intégrabilité de ces systèmes a été l'objet d'une abondante et importante littérature.

Les zéros du champ d'Euler correspondent à ce qu'on appelle en Mécanique des *rotations stationnaires* (c.f. ibid). Notre lemme 2.1 prouve que si (1) est complète il y a toujours des rotations stationnaires.

Les directions de non complétude des géodésiques ont leur intérêt dans la théorie des *trous noirs* dans les espaces–temps ([H-E]).

Voici une description succinte de nos principaux résultats.

Le théorème 2.1 assure que toute équation d'Euler sur une algèbre de Lie nilpotente *orthogonale* (au sens de Medina–Revoy, [M-R]) est complète.

Soit $\mathbf{r}(X,Y)$ le produit de Levi–Civita déterminé par $< X, Y >:= \Phi(X)Y$:

$$< \mathbf{r}\,(X,Y), Z >= \frac{1}{2}(< [X,Y], Z > - < [Y,Z], X > + < [Z,X], Y >).$$

Si $\mathbf{r}$ est à puissances associatives alors (1) est complète si et seulement si $\mathbf{r}$ est un produit nilpotent (théorème 2.2).

Si $\mathbf{r}$ est plat, c'est–à–dire si $X \mapsto \mathbf{r}(X, \cdot)$ est un homomorphisme d'algèbres de Lie, alors le flot d'Euler associé à $(\mathcal{G}, \Phi)$ est complet si et seulement si $\mathcal{G}$ est unimodulaire (théorème 2.3)

Les résultats du paragraphe deux laissent penser que le flot d'Euler est complet si et seulement si le produit $\mathbf{r}$ n'admet pas d'idempotents non nuls. Au paragraphe trois, nous montrons que ce résultat est vrai pour des algèbres de Lie unimodulaires de dimension trois (proposition 3.2). Cependant il existe des algèbres de Lie de dimension trois non unimodulaires munies de produits de Levi–Civita n'admettant pas d'idempotents et telles que (1) n'est pas complète (3.3).

2 Résultats généraux

Dans la suite $\mathcal{G}$ est une algèbre de Lie réelle de dimension finie, $\mathcal{G}^*$ est l'espace dual de $\mathcal{G}$ et ad^* désigne la représentation co–adjointe de l'algèbre $\mathcal{G}$.

Définition. L'équation différentielle

$$\dot{\xi} = -ad^*_{\Phi^{-1}(\xi)}\,\xi.$$

où $\Phi : \mathcal{G} \longrightarrow \mathcal{G}^*$ est un isomorphisme symétrique d'espaces vectoriels est appellée l'*équation d'Euler* associée à $(\mathcal{G}, \Phi)$.

Nous dirons que (1) est *complète* si le flot du champ de vecteurs associé est complet. A priori, la complétude dépend de la nature de $\mathcal{G}$ et de Φ.

Soit G un groupe de Lie connexe d'algèbre de Lie $\mathcal{G}$. Des définitions, il est clair que les courbes intégrales de (1) sont tracées sur les orbites de la représentation co-adjointe de G.

Par ailleurs, on constate directement que la fonction *énergie* $E : \mathcal{G}^* \longrightarrow \mathbb{R}$ définie par $E(\xi) := \xi(\Phi^{-1}\xi)$ est une intégrale première de (1).

Ces deux remarques impliquent les résultats connus suivants:

si $\mathcal{G}$ est compacte, c'est-à-dire si la forme de Killing de $\mathcal{G}$ est définie négative, ou si Φ est d'indice zéro, l'équation (1) est complète.

En effet, dans les deux cas les solutions de (1) sont tracées sur des compacts; or tout champ de vecteurs lisse sur une variété compacte est complet.

Evidemment, (1) est complète si $\mathcal{G}$ est abelienne.
Le résultat suivant est surement connu:

L'équation (1) est complète si $\mathcal{G}$ est 2-nilpotente c'est-à-dire si l'idéal dérivé $[\mathcal{G}, \mathcal{G}]$ est inclus dans le centre $\mathcal{Z}(\mathcal{G})$ de $\mathcal{G}$.

Ceci résulte du fait que dans ce cas le champ d'Euler est linéaire.

Transportons (1) à $\mathcal{G}$ au moyen de Φ; nous obtenons l'équation

$$\dot{X} = -\mathbf{r}\,(X, X). \tag{2}$$

Voici des conditions nécessaires simples pour que le champ d'Euler soit complet.

Lemme 2.1

(i) Si X_0 est un idempotent non nul de $\mathbf{r}$, c'est-à-dire si $\mathbf{r}(X_0, X_0) = X_0$, alors X_0 est une direction d'incomplétude de l'équation (1) et la solution passant par X_0 à l'instant $t = 0$ est donnée par

$$t \longmapsto \lambda(t)\, X_0$$

où $\dot{\lambda} = -\lambda^2$.

(ii) Si le flot d'Euler est complet, il laisse un point fixe.

Démonstration. Soit b un produit bilinéaire sur un espace vectoriel réel de dimension finie V. D'après ([K-Y]), il existe X non nul dans V, tel que les vecteurs X et $b(X, X)$ sont dépendants entre–eux.

Si le flot est complet il n'existe pas $X \neq 0$ tel que $\mathbf{r}(X, X) = X$, d'après la première assertion du lemme. Par conséquent il existe $Y \neq 0$ vérifiant $\mathbf{r}(Y, Y) = 0$.

□

Supposons $\mathcal{G}$ *orthogonale* au sens de Medina–Revoy ([M-R]). Ceci veut dire que $\mathcal{G}$ est munie d'une forme bilinéaire $(\,|\,)$ symétrique non–dégénérée qui est associative, i.e. elle vérifie l'identité

$$(ad_X Y \mid Z) = (x \mid ad_Y Z).$$

Alors Φ s'écrit comme

$$< X, Y >= (u(X) \mid Y)$$

où u est un isomorphisme $(\,|\,)$–symétrique de l'espace $\mathcal{G}$. On a, en outre,

$$\mathbf{r}\,(X,Y) = \frac{1}{2}[X,Y] + \frac{1}{2}u^{-1}\left([X,u(Y)] + [Y,u(X)]\right)$$

et par suite (2) devient

$$u(\dot{X}) = -\mathbf{r}\,(X, X). \tag{3}$$

D'autre part nous avons

Lemme 2.2 *Si $\mathcal{G}$ est une algèbre de Lie nilpotente et v est un endomorphisme de l'espace $\mathcal{G}$, les solutions de l'équation différentielle sur $\mathcal{G}$*

$$\dot{X} = [v(X), X] \tag{4}$$

sont des fonctions polynomiales.

Démostration. Considérons le produit sur l'espace $\mathcal{G}$ défini par

$$X \star Y = [v(X), Y] =: L_X(Y) = R_Y(X).$$

Désignons par $\mathcal{A}$ la sous–algèbre de l'algèbre associative End($\mathcal{G}$), des endomorphismes de l'espace $\mathcal{G}$, engendrée par l'ensemble $\{L_X ; X \in \mathcal{G}\}$. Il est évident que $\mathcal{A}$ est nilpotente et que l'algèbre de Lie sous–jacente $\mathcal{A}^-$ est aussi nilpotente. Ceci implique ([B, p. 60, prop. 5]) l'existence d'un drapeau

$$\{0\} = \mathcal{G}_0 \subset \mathcal{G}_1 \subset \ldots \subset \mathcal{G}_n = \mathcal{G}$$

de sous–espaces de $\mathcal{G}$ tels que $f(\mathcal{G}_{i+1}) \subset \mathcal{G}_i$ pour tout $0 \leq i \leq n - i$. Soit $\mathcal{B} = (e_1, e_2, \ldots, e_n)$ une base de $\mathcal{G}$ adaptée au drapeau: $\mathcal{G}_i = \text{Vect}(e_1, \ldots, e_i)$.

Si $I \ni t \mapsto X(t) = \sum_{i=1}^{n} X_i(t)e_i$ est une solution de (4), le fait que $X(t) \star X(t) \in \mathcal{G}_{n-1}$ implique que $\dot{X}_n = 0$ ou ce qui est de même que $t \mapsto X_n(t)$ est constante (I est connexe). A son tour ce dernier fait et ce qui précède entraînent que $t \mapsto X_{n-1}(t)$ est polynomiale d'ordre 1. Une itération du même argument montre que $t \mapsto X(t)$ est polynomiale. □

Une application du lemme à l'équation d'Euler nous fournit le résultat non encore publié suivant ([T]).

Théorème 2.1 (A. Tsemo) *Si $\mathcal{G}$ est une algèbre de Lie orthogonale nilpotente alors toute équation d'Euler sur $\mathcal{G}$ est complète.*

Démostration. Si dans l'équation (3) nous posons $u(X) = Y$, nous obtenons

$$\dot{Y} = -[u^{-1}(\dot{Y}), Y]$$

et donc l'equation s'écrit comme celle du lemme, d'où le résultat. □

Munissons l'espace $\mathcal{G}$ de la structure d'algèbre fournie par le produit de Levi–Civita $X \cdot Y := \mathbf{r}(X,Y)$ déterminé par Φ. Si la sous–algèbre $\mathbb{R}[X]$ de $\mathcal{G}$ engendrée par un élément quelconque $X \in \mathcal{G}$ est associative, nous dirons que $\mathbf{r}$ est *à puissances associatives*. Nous avons

Théorème 2.2 *On suppose $\mathbf{r}$ à puissances associatives. Alors*

(i) La solution de l'équation d'Euler

$$\dot{X}(t) = -\mathbf{r}(X(t), X(t)) = -X(t)^2$$

qui passe par $X_0 = X(0)$ s'exprime comme

$$X(t) = \sum_{k=0}^{\infty} (-1)^k X_0^{k+1} t^k$$

(ii) Le champ associé à $(\mathcal{G}, \Phi)$ est complet si et seulement si $\mathbf{r}$ est un produit nilpotent.

Le théorème est une conséquence du lemme suivant qui résulte d'une simple récurrence sur k.

Lemme 2.3 *Si le produit de Levi–Civita $\mathbf{r}(= \mathbf{r}_\Phi)$ sur $\mathcal{G}$ est à puissances associatives toute solution $t \mapsto X(t)$ de (2) vérifie les conditions suivantes pour tout $k \in \mathbb{N} \setminus \{0\}$*

(i)

$$\frac{d}{dt} X(t)^k = -kX(t)^{k+1}$$

(ii)

$$X^{(k)}(t) = (-1)^k k!\, X(t)^{k+1}.$$

Démostration du théorème. Du lemme, il est clair que la solution passant par X_0 s'exprime comme

$$X(t) = \sum_{k=0}^{\infty}(-1)^k X_0^{k+1} t^k.$$

Par conséquent si $\mathbf{r}$ est un produit nilpotent, la solution $t \longrightarrow X(t)$ est polynomiale de sorte que le champ est complet.

Réciproquement, supposons le champ complet et $\mathbf{r}$ non nilpotent. D'après un résultat de A. Albert (voir par exemple [Sc, p. 32])) il existe dans l'algèbre $(\mathcal{G}, \mathbf{r})$ un idempotent non nul e. Ceci implique la non complétude de (2) compte tenu du lemme 2.1

Posons $L_X(Y) := \mathbf{r}(X,Y)$. Si $X \longrightarrow L_X$ est une représentation de l'algèbre de Lie $\mathcal{G}$, on dit que $\mathbf{r}$ est *plat.* Dans ce cas nous avons le résultat suivant d à A. Aubert et A. Medina que nous fournissons sans preuve (voir [Au-M]).

Théorème 2.3 *Si le produit de Levi–Civita* $\mathbf{r}$ *est plat alors le champ d'Euler associé à* $(\mathcal{G}, \Phi)$ *est complet si et seulement si l'algèbre de Lie* $\mathcal{G}$ *est unimodulaire.*

3 Cas de dimension ≤ 3

Proposition 3.1 *Si l'algèbre de Lie* $\mathcal{G}$ *est de dimension deux, l'équation d'Euler associée à* $(\mathcal{G}, \Phi)$ *est complète si et seulement si* $\mathbf{r}_\Phi$ *ne possède pas d'idempotents non nuls.*

Démonstration. Compte tenu du lemme 2.1 (i), il suffit de prouver que (1) est complète si $\mathbf{r}$ n'admet pas d'idempotents non nuls.

Soit E_1, E_2 une base de $\mathcal{G}$ telle que

$$[E_1, E_2] = \alpha\, E_2.$$

Lorsque $\alpha = 0$ ont est dans le cas abélien. Les autres cas peuvent s'identifier au cas $\alpha = 1$ et on est dans l'algèbre de Lie du groupe affine de la droite. Posons pour $X \in \mathcal{G}$ et $\xi \in \mathcal{G}^*$,

$$X = x_1\, E_1 + x_2\, E_2 \qquad \xi = \xi_1\, E_1^* + \xi_2\, E_2^*.$$

Les matrices de ad_X, de ad_X^* et de l'isomorphisme Φ^{-1} s'expriment, respectivement, dans les bases (E_1, E_2) et sa duale (E_1^*, E_2^*) par

$$ad_X = \begin{pmatrix} 0 & 0 \\ -\alpha x_2 & \alpha x_1 \end{pmatrix}, \; -ad_X^* = \begin{pmatrix} 0 & -\alpha x_2 \\ 0 & \alpha x_1 \end{pmatrix},$$

$$\Phi^{-1} = \begin{pmatrix} \lambda & a \\ a & \mu \end{pmatrix}.$$

L'équation d'Euler s'écrit dans ces coordonnées

$$\begin{aligned} \dot{\xi}_1 &= -\alpha x_2 \xi_2 \\ \dot{\xi}_2 &= \alpha x_1 \xi_2, \end{aligned}$$

où $X = \Phi^{-1}(\xi)$. Il est clair que r a d'idempotents non nuls si, et seulement si, l'équation d'Euler en a. Lorsque $\alpha = 0$, l'équation devient $\dot{\xi} = 0$ laquelle ne possède pas d'idempotents non nuls et est, evidemment, complète. Si $\alpha = 1$, et l'équation d'Euler n'a pas d'idempotents alors $\lambda \neq 0$. En effet, si $\lambda = 0$ alors $a \neq 0$ et $(1/a, \mu/(2a^2))$ est idempotent. Par conséquent $< E_2, E_2 > \neq 0$ et on peut prendre $e_1 \neq 0$ orthogonal à E_2 tel que $[e_1, E_2] = E_2$. Dans cette base (et sa duale) les matrices représentatives de Φ (et de Φ^{-1}) sont diagonales et l'équation d'Euler s'exprime comme

$$\begin{aligned} \dot{\xi}_1 &= -\overline{\lambda} \xi_2^2 \\ \dot{\xi}_2 &= \overline{\mu} \xi_1 \xi_2. \end{aligned}$$

L'énergie y est donnée par la forme quadratique

$$E(\xi_1, \xi_2) = \overline{\lambda} \xi_1^2 + \overline{\mu} \xi_2^2,$$

et l'équation d'Euler possède un idempotent si et seulement si E possède un vecteur isotrope.

Par conséquent, si l'équation d'Euler n'a pas d'idempotents, E est une forme quadratique définie et les solutions de l'équation d'Euler, étant tracées sur un compact, sont complètes. □

Dans la suite nous allons utiliser la description des algèbres de Lie de dimension trois faite par J. Milnor dans [Mi]. Le crochet de Lie peut s'exprimer au moyen du produit vectoriel, $\times$, associé à un produit scalaire défini positif, $(\ |\ )$, et à une orientation choisie comme,

$$[X, Y] = L(X \times Y)$$

où $L \in \mathrm{End}(\mathcal{G})$.

3.1 cas unimodulaire

Proposition 3.2 *On suppose $\mathcal{G}$ de dimension trois et unimodulaire. Le champ d'Euler correspondant à $(\mathcal{G}, \Phi)$ est complet si et seulement si l'équation d'Euler ne possède pas d'idempotents non nuls.*

Démonstration. Comme auparavant, il faut seulement démontrer que l'équation d'Euler est complète si elle n'a pas d'idempotents.

Comme trace$(ad_X) = 0$ pour tout X dans $\mathcal{G}$, l'endomorphisme L est symétrique par rapport à $(\ |\ )$. Il existe, donc, une base de $\mathcal{G}$, $\mathcal{B} = (E_1, E_2, E_3)$, positivement orientée formée de vecteurs propres de L, c'est–à–dire, telle que:

$$[E_2, E_3] = \alpha_1\, E_1, \quad [E_3, E_1] = \alpha_2\, E_2, \quad [E_1, E_2] = \alpha_3\, E_3. \tag{5}$$

Les signes des valeurs propres caractérisent l'algèbre de Lie, selon le tableau suivant

	signes de $\alpha_1, \alpha_2, \alpha_3$	Groupe de Lie associé	type
1	0,0,0	$\mathbb{R} \oplus \mathbb{R} \oplus \mathbb{R}$	abélien
2	+,0,0	Heisenberg	nilpotent
3	+,-,0	E(1, 1)	résoluble
4	+,+,0	E(2)	résoluble
5	+,+,+	SO(2)	compact
6	+,+,-	SL(2, $\mathbb{R}$)	simple

Hormis les cas connus (abélien et compact), nous allons aborder chacun de ces cas dans l'ordre du tableau. Tout d'abord fixons un peu les notations. Dorénavant, $\mathcal{B} = (E_1, E_2, E_3)$ est une base de l'algèbre $\mathcal{G}$ et $\mathcal{B}^* = (E_1^*, E_2^*, E_3^*)$, est la base de $\mathcal{G}^*$ duale de $\mathcal{B}$. On exprime $X \in \mathcal{G}$ et $\xi \in \mathcal{G}^*$ par

$$X = x_1 E_1 + x_2 E_2 + x_3 E_3 \qquad \xi = \xi_1 E_1^* + \xi_2 E_2^* + \xi_3 E_3^*.$$

D'après (5), les matrices de ad_X et de ad_X^* s'écrivent dans les bases $\mathcal{B}$ et $\mathcal{B}^*$ comme

$$ad_X = \begin{pmatrix} 0 & -\alpha_1\, x_3 & \alpha_1\, x_2 \\ \alpha_2\, x_3 & 0 & -\alpha_2\, x_1 \\ -\alpha_3\, x_2 & \alpha_3\, x_1 & 0 \end{pmatrix}, \; -ad_X^* = \begin{pmatrix} 0 & \alpha_2\, x_3 & -\alpha_3\, x_2 \\ -\alpha_1\, x_3 & 0 & \alpha_3\, x_1 \\ \alpha_1\, x_2 & -\alpha_2\, x_1 & 0 \end{pmatrix}$$

Par conséquent, l'équation d'Euler dans ces coordonnées prend la forme

$$\begin{aligned} \dot{\xi}_1 &= \alpha_2\, x_3\, \xi_2 - \alpha_3\, x_2 \xi_3 \\ \dot{\xi}_2 &= -\alpha_1\, x_3\, \xi_1 + \alpha_3\, x_1 \xi_3 \\ \dot{\xi}_3 &= \alpha_1\, x_2\, \xi_1 - \alpha_2\, x_1 \xi_2 \end{aligned}$$

où

$$x_1\, E_1 + x_2\, E_2 + x_3\, E_3 = \Phi^{-1}(\xi_1\, E_1^* + \xi_2\, E_2^* + \xi_3\, E_3^*).$$

1. Dans le cas de l'algèbre de Heisenberg, deux des valeurs propres de L son nulles, disons $\alpha_1 = \alpha_2 = 0$, et l'équation devient:

$$\begin{aligned} \dot{\xi}_1 &= -\alpha_3 x_2 \xi_3 \\ \dot{\xi}_2 &= \alpha_3 x_1 \xi_3 \\ \dot{\xi}_3 &= 0 \end{aligned}$$

Cette équation n'a pas d'idempotents non nuls (un idempotent doit satisfaire $\xi_3 = 0$ et par conséquent ξ_1 et ξ_2 seront aussi nuls). Elle est complète car $\xi_3 =$ cte implique que l'équation est, en fait, une équation linéaire.

2. Dans le cas $\mathcal{G} = \mathrm{L}(E(2))$, une des valeurs propres est nulle, disons $\alpha_1 = 0$ et $\alpha_2 \alpha_3 > 0$. L'équation d'Euler devient

$$\begin{aligned} \dot{\xi}_1 &= \alpha_2 x_3 \xi_2 - \alpha_3 x_2 \xi_3 \\ \dot{\xi}_2 &= \alpha_3 x_1 \xi_3 \\ \dot{\xi}_3 &= -\alpha_2 x_1 \xi_2 \end{aligned}$$

L'équation admet comme intégrales premières:

$$x_1 \xi_1 + x_2 \xi_2 + x_3 \xi_3 \qquad \text{(l'énergie)}$$

et $$\alpha_2 \xi_2^2 + \alpha_3 \xi_3^2.$$

Cette équation n'a pas non plus, d'idempotents. Les solutions de l'équation d'Euler sont tracées sur l'intersection des surfaces de niveau de l'énergie avec un cylindre de base elliptique. Un ensemble de ce type étant compact, les solutions sont complètes.

3. Dans le cas $\mathcal{G} = \mathrm{L}E(1,1)$, une des valeurs propres nulle, disons $\alpha_1 = 0$ et $\alpha_2 \alpha_3 < 0$. On peut obtenir une base $\mathcal{B} = (E_1, E_2, E_3)$, de $\mathcal{G}$ telle que

$$[E_1, E_2] = E_2 \quad [E_1, E_3] = -E_3 \quad [E_2, E_3] = 0.$$

Si la matrice de Φ^{-1} s'écrit

$$M(\Phi^{-1})_{\mathcal{B}^*,\mathcal{B}} = \begin{pmatrix} \lambda & a & b \\ a & \mu & c \\ b & c & \nu \end{pmatrix},$$

l'équation d'Euler devient

$$\begin{aligned} \dot{\xi}_1 &= -x_2 \xi_2 + x_3 \xi_3 \\ \dot{\xi}_2 &= x_1 \xi_2 \\ \dot{\xi}_3 &= -x_1 \xi_3. \end{aligned}$$

Elle admet comme intégrales premières:

$$x_1\,\xi_1 + x_2\,\xi_2 + x_3\,\xi_3 \qquad \text{(l'énergie)}$$

et $\qquad \xi_2\,\xi_3.$

Si l'équation d'Euler n'a pas d'idempotents, alors $\lambda \neq 0$. En effet, si $\lambda = 0$,

$$(\frac{\mu}{2a^2}, \frac{1}{a}, 0) \qquad (\frac{\nu}{2b^2}, 0, \frac{1}{b})$$

sont des idempotents (le premier lorsque $a \neq 0$ et le second lorsque $b \neq 0$.) Or dire que $\lambda \neq 0$ équivaut à dire que la restriction de Φ au sous–espace $\mathcal{G}' = \mathrm{Vect}(E_2, E_3)$ est un isomorphisme entre $\mathcal{G}'$ et $(\mathcal{G}')^*$. Par conséquent il existe $e_1 \perp_\Phi \mathcal{G}'$ tel que $\Phi(e_1)e_1 = \lambda$ et

$$[e_1, E_2] = E_2 \quad [e_1, E_3] = -E_3.$$

La non existence d'idempotents implique que $\lambda\mu > 0$ et $\lambda\nu > 0$. L'énergie, dans ces nouvelles coordonnées s'exprime par:

$$\lambda\xi_1^2 + \mu\xi_2^2 + 2c\xi_2\xi_3 + \nu\xi_3^2$$

et, comme $\xi_2\xi_3$ est intégrale première, il résulte que

$$\lambda\xi_1^2 + \mu\xi_2^2 + \nu\xi_3^2$$

est intégrale première de l'équation. Comme il s'agit d'une forme quadratique d'indice zéro, il s' en suit que les solutions sont complètes.

4. Dans le cas $\mathcal{G} = sl(2, \mathbb{R})$, un isomorphisme symétrique $\Phi : \mathcal{G} \to \mathcal{G}^*$ s'exprime comme

$$\Phi(X)Y = k(uX, Y) := \Phi_u$$

où $u \in \mathrm{GL}(\mathcal{G})$ est k–symétrique. Soit $\mathcal{B} = (E_1, E_2, E_3)$ une base ordonnée k–orthonormale de $\mathcal{G}$, i.e.

$$k(E_i, E_i) = \epsilon_i,\ \epsilon_i^2 = 1, \quad k(E_i, E_j) = 0\ i \neq j.$$

Comme u^{-1} est k–symétrique, on a

$$M_{\mathcal{B}}(u^{-1}) = \begin{pmatrix} \lambda & a & b \\ \epsilon_1\epsilon_2\, a & \mu & c \\ \epsilon_1\epsilon_3\, b & \epsilon_2\epsilon_3\, c & \nu \end{pmatrix},$$

où $a, b, c \in \mathbb{R}$. Le crochet $[\ ,\ ]$ dans$\mathcal{G}$ est donné par le tableau

$[\ ,\ ]$	E_1	E_2	E_3
E_1	0	$\epsilon_3\kappa\, E_3$	$-\epsilon_2\kappa\, E_2$
E_2		0	$\epsilon_1\kappa\, E_1$
E_3			0

Les matrices de ad_X, $-ad_X^*$ et de Φ^{-1} s'expriment dans les bases $\mathcal{B}$ et $\mathcal{B}^*$ comme

$$ad_X = \begin{pmatrix} 0 & -\epsilon_1 x_3 & \epsilon_1 x_2 \\ \epsilon_2 x_3 & 0 & -\epsilon_2 x_1 \\ -\epsilon_3 x_2 & \epsilon_3 x_1 & 0 \end{pmatrix},$$

$$-ad_X^* = \begin{pmatrix} 0 & \epsilon_2 x_3 & -\epsilon_3 x_2 \\ -\epsilon_1 x_3 & 0 & \epsilon_3 x_1 \\ \epsilon_1 x_2 & -\epsilon_2 x_1 & 0 \end{pmatrix}$$

$$\Phi^{-1} = \begin{pmatrix} \epsilon_1 \lambda & \epsilon_2 a & \epsilon_3 b \\ \epsilon_2 a & \epsilon_2 \mu & \epsilon_3 c \\ \epsilon_3 b & \epsilon_3 c & \epsilon_3 \nu \end{pmatrix}.$$

L'équation d'Euler se présente alors comme

$$\begin{aligned} \dot{\xi}_1 &= \epsilon_2 x_3 \xi_2 - \epsilon_3 x_2 \xi_3 \\ \dot{\xi}_2 &= -\epsilon_1 x_3 \xi_1 + \epsilon_3 x_1 \xi_3 \\ \dot{\xi}_3 &= \epsilon_1 x_2 \xi_1 - \epsilon_2 x_1 \xi_2 \end{aligned}$$

PREMIER CAS. Supposons u diagonalisable. Alors il existe une base $\mathcal{B} = (E_1, E_2, E_3)$ k-orthonormale qui diagonalise u. Exactementent un des signes $\epsilon_1, \epsilon_2, \epsilon_3$ est égal à -1, sans perte de généralité, on peut supposer $\epsilon_1 = -1$. La matrice de Φ^{-1} dans les bases $\mathcal{B}^*$, $\mathcal{B}$ est donc

$$\Phi^{-1} = \begin{pmatrix} -\lambda & 0 & 0 \\ 0 & \mu & 0 \\ 0 & 0 & \nu \end{pmatrix},$$

où λ^{-1},μ^{-1} et ν^{-1} sont les valeurs propres de u. L'équation d'Euler est ici,

$$\begin{aligned} \dot{\xi}_1 &= (\nu - \mu)\, \xi_2 \xi_3 = (a + b)\, \xi_2 \xi_3 \\ \dot{\xi}_2 &= (\nu - \lambda)\, \xi_1 \xi_3 = a \xi_1 \xi_3 \\ \dot{\xi}_3 &= (\lambda - \mu)\, \xi_1 \xi_2 = b \xi_1 \xi_2 \end{aligned}$$

avec $a = \nu - \mu, b = \nu - \lambda$. Il est clair que la fonction

$$b\xi_2^2 - a\xi_3^2$$

est une intégrale première. Par conséquent, l'équation d'Euler n'a pas d'idempotents non nuls si et seulement si $ab \leq 0$. Dans ce cas, elle est complète.

DEUXIEME CAS. Supposons que u a deux directions propres, c'est-à-dire, que son polynôme caractéristique a une racine double, $\overline{\mu}$ et une racine simple, $\overline{\lambda}$. Soit E un vecteur propre de u associé à la valeur propre simple et $\mathcal{V} = \{X \in \mathcal{G} : (u - \overline{\mu}\mathrm{Id})^2 X = 0\}$. Alors $E \perp \mathcal{V}$ et si $E_2' \in \mathcal{V}$ est vecteur propre associé à $\overline{\mu}$ on a que $k(E,E) > 0$ et $k(E_2', E_2') = 0$. Choisisons une base E_2, E_3 de $\mathcal{V}$ telle que l'on ait

$$k(E_2, E_2) = 0 = k(E_3, E_3),\ k(E_2, E_3) = 1$$

et

$$A(E_2) = \overline{\mu}\, E_2,\ A(E_3) = E_2 + \overline{\mu}\, E_3.$$

Dans cette base, les matrices du système s'expriment comme

$$u = \begin{pmatrix} \overline{\lambda} & 0 & 0 \\ 0 & \overline{\mu} & 1 \\ 0 & 0 & \overline{\mu} \end{pmatrix}, \Phi_k^{-1} = \begin{pmatrix} 1 & 0 & 0 \\ 0 & 0 & 1 \\ 0 & 1 & 0 \end{pmatrix},$$

$$\Phi^{-1} = \begin{pmatrix} \lambda & 0 & 0 \\ 0 & -\mu^2 & \mu \\ 0 & \mu & 0 \end{pmatrix},$$

Le crochet s'écrit

[,]	E_1	E_2	E_3
E_1	0	$\kappa\, E_2$	$-\kappa\, E_3$
E_2		0	$\kappa\, E_1$
E_3			0

de sorte que ad_X et ad_X^* ont pour matrices

$$ad_X = \begin{pmatrix} 0 & -x_3 & x_2 \\ -x_2 & x_1 & 0 \\ x_3 & 0 & -x_1 \end{pmatrix}, \ -ad_X^* = \begin{pmatrix} 0 & -x_2 & x_3 \\ -x_3 & x_1 & 0 \\ x_2 & 0 & -x_1 \end{pmatrix}.$$

Nous pouvons alors écrire l'équation d'Euler comme

$$\begin{aligned} \dot{\xi}_1 &= -\mu^2\, \xi_2^2 \\ \dot{\xi}_2 &= (\lambda - \mu)\, \xi_1 \xi_2 \\ \dot{\xi}_3 &= -\mu^2\, \xi_1 \xi_2 + (\mu - \lambda)\, \xi_1 \xi_3. \end{aligned}$$

Les deux premières équations du système constituent un système planaire qui est complet si et seulement si

$$\lambda - \mu \geq 0.$$

La non existence d'idempotents du système implique la non existence d'idempotents pour le système planaire en ξ_1, ξ_2. Ce dernier système est donc complet. Or cette sa complétude est équivalente à la complétude de l'équation d'Euler.

TROISIEME CAS. u n'a qu'une seule direction propre. Il y a alors deux possibilités: ou bien k est positive dans cette direction et le polynôme caractéristique de u à une unique racine rélle simple ou bien la direction propre est isotrope et le polynôme caractéristique de u et son polynôme minimal coïncident. Ils ont alors une racine réelle triple.

1. Soit E_1 dans la direction propre de u tel que $k(E_1, E_1) = 1$. Il existe une base k–orthonormée (E_2, E_3) telle que la matrice de la restriction de u à $\mathrm{Vect}(E_2, E_3)$ est de la forme

$$\begin{pmatrix} \varrho & -\varsigma \\ \varsigma & \varrho \end{pmatrix}.$$

Sans perte de généralité nous pouvons supposer $\epsilon_2 = 1$, $\epsilon_3 = -1$. La matrice de Φ^{-1} relative a $\mathcal{B}^*$ et $\mathcal{B} = (E_1, E_2, E_3)$ est alors

$$\begin{pmatrix} \lambda' & 0 & 0 \\ 0 & \varrho' & \varsigma' \\ 0 & \varsigma' & -\varrho' \end{pmatrix},$$

et l'équation d'Euler s'écrit dans ces coordonnées:

$$\begin{aligned} \dot{\xi}_1 &= \varsigma' \xi_2^2 + \varsigma' \xi_3^2 \\ \dot{\xi}_2 &= \xi_1(-\varsigma' \xi_2 + (\varrho' - \lambda')\xi_3) \\ \dot{\xi}_3 &= \xi_1((\varsigma' - \lambda')\xi_2 + \varsigma' \xi_3). \end{aligned}$$

Il est clair que cette équation a toujours des idempotents non nuls.

2. Soit E vecteur propre de u tel que $k(E, E) = 0$. Appelons $\overline{\lambda}$ la valeur propre à laquelle il est associé. Soit (E_1', E_2', E_3') une base k–orthonormée quelconque avec $\epsilon_2 = \epsilon_3 = 1$ et $\epsilon_1 = -1$. Il existe θ tel que $E_1' + \cos\theta\, E_2' + \sin\theta\, E_3'$ est colinéaire à E. Prenons la base k–orthonormée

$$\begin{aligned} E_1 &= E_1' \\ E_2 &= \cos\theta\, E_2' + \sin\theta\, E_3' \\ E_3 &= -\sin\theta\, E_2' + \cos\theta\, E_3'. \end{aligned}$$

Le fait que $E_1 + E_2$ soit vecteur propre de u implique que la matrice de u dans cette base s'exprime comme

$$u = \begin{pmatrix} \overline{\lambda} - a & a & b \\ -a & \overline{\lambda} + a & b \\ -b & b & \overline{\gamma} \end{pmatrix}$$

Le polynôme caracteristique d'une telle matrice est $p(x) = (x - \overline{\lambda})^2(x - \overline{\gamma})$, d'où, d'après nos hypothèses $\overline{\gamma} = \overline{\lambda}$. La matrice de u^{-1} a la même forme:

$$u^{-1} = \begin{pmatrix} \lambda - \alpha & \alpha & \beta \\ -\alpha & \lambda + \alpha & \beta \\ -\beta & \beta & \lambda \end{pmatrix}$$

où $\lambda = 1/\overline{\lambda}$, $\beta = -b/\lambda^2$ et $\alpha = (b^2 - a\overline{\lambda})/\overline{\lambda}^3$. La matrice de Φ^{-1} relative à $\mathcal{B}^*$ et $\mathcal{B}$ se présente alors comme

$$\begin{pmatrix} \lambda - \alpha & \alpha & \beta \\ \alpha & \lambda + \alpha & \beta \\ \beta & \beta & \lambda \end{pmatrix}.$$

Avec ces données l'équation d'Euler s'écrit

$$\begin{aligned} \dot{\xi}_1 &= (\beta\xi_1 + \beta\xi_2 + \lambda\xi_3)\xi_2 - (\alpha\xi_1 + (l + \alpha)\xi_2 + \beta\xi_3)\xi_3 \\ \dot{\xi}_2 &= (\beta\xi_1 + \beta\xi_2 + \lambda\xi_3)\xi_1 + ((-\lambda + \alpha)\xi_1 + \alpha\xi_2 - \beta\xi_3)\xi_3 \\ \dot{\xi}_3 &= -(\alpha\xi_1 + (l + \alpha)\xi_2 + \beta\xi_3)\xi_1 - ((-\lambda + \alpha)\xi_1 + \alpha\xi_2 - \beta\xi_3)\xi_2 \end{aligned}$$

En faisant le changement de coordonnées:

$$\begin{aligned} \eta_1 &= \xi_1 + \xi_2 \\ \eta_2 &= \xi_1 - \xi_2 \\ \eta_3 &= \xi_3 \end{aligned}$$

l'équation d'Euler devient

$$\begin{aligned} \dot{\eta}_1 &= \beta\eta_1^2 \\ \dot{\eta}_2 &= \beta\eta_1\eta_2 - 2\alpha\eta_1\eta_3 - 2\beta\eta_3^2 \\ \dot{\eta}_3 &= -\alpha\eta_1^2 - \beta\eta_1\eta_3 \end{aligned}$$

laquelle a toujours des idempotents.

En résumant, l'équation d'Euler associée à $(\mathrm{sl}(2, \mathbb{R}), \Phi_u)$ est complète si et seulement si on se trouve dans l'un des deux cas suivants:

(a) ou bien u a trois directions propres E_1, E_2 et E_3, nécessairement k-orthogonales dont une, disons E_1, est négative. L'équation d'Euler est complète si et seulement si l'on a

$$\left(\frac{1}{\lambda_1} - \frac{1}{\lambda_2}\right)\left(\frac{1}{\lambda_1} - \frac{1}{\lambda_3}\right) \geq 0.$$

où les λ_i sont les valeurs propres correspondant aux vecteurs propres E_i. propres associées.

(b) ou bien u a deux directions propres E_1 et E_2. Dans ce cas une des directions, disons E_1, est positive et l'autre est isotrope. L'équation est complète si et seulement si

$$\frac{1}{\lambda_1} \leq \frac{1}{\lambda_2}.$$

où les λ_i sont les valeurs propres correspondants aux vecteurs propres E_i.

3.2 cas non–unimodulaire

Soit $\mathcal{G}$ une algèbre de Lie non–unimodulaire de dimension trois. Il existe une base (e_1, e_2, e_3), de $\mathcal{G}$ dans laquelle le crochet s'exprime par

$[,]$	e_1	e_2	e_3
e_1	0	$\alpha e_2 + \beta e_3$	$\gamma e_2 + \delta e_3$
e_2		0	0
e_3			0

avec $\alpha + \delta = 2$ (voir [Mi]). On a $\mathcal{G}' := [\mathcal{G}, \mathcal{G}] \subset \mathcal{V} := \mathrm{Vect}(e_2, e_3)$. Considérons la restriction de ad_{e_1} à $\mathcal{V}$ que l'on notera par T. Supposons que T a deux valeurs propres distinctes α_2 et α_3 et soient $E_2, E_3 \in \mathcal{V}$ des vecteurs propres associés. Dans la base $\mathcal{B} = (E_1 = e_1, E_2, E_3)$, et sa duale les matrices de ad_X et de ad^*_X s'écrivent

$$ad_X = \begin{pmatrix} 0 & 0 & 0 \\ -\alpha_2 x_2 & \alpha_2 x_1 & 0 \\ -\alpha_3 x_3 & 0 & \alpha_3 x_1 \end{pmatrix}, \; -ad^*_X = \begin{pmatrix} 0 & -\alpha_2 x_2 & -\alpha_3 x_3 \\ 0 & \alpha_2 x_1 & 0 \\ 0 & 0 & \alpha_3 x_1 \end{pmatrix}$$

Si la matrice de Φ^{-1} relative a $\mathcal{B}^*$ et $\mathcal{B}$, s'écrit

$$\Phi^{-1} = \begin{pmatrix} \lambda & a & b \\ a & \mu & c \\ b & c & \nu \end{pmatrix}$$

le champ d'Euler prend la forme

$$\begin{aligned} \dot{\xi}_1 &= -\alpha_2 x_2 \xi_2 - \alpha_3 x_3 \xi_3 \\ \dot{\xi}_2 &= \alpha_2 x_1 \xi_2 \\ \dot{\xi}_3 &= \alpha_3 x_1 \xi_3. \end{aligned}$$

Lorsque $\ker T \neq (0)$, c'est–à–dire lorsque une des valeurs propres, disons α_2, est nulle, l'équation d'Euler devient

$$\begin{aligned} \dot{\xi}_1 &= -2 x_3 \xi_3 \\ \dot{\xi}_2 &= 0 \\ \dot{\xi}_3 &= 2 x_1 \xi_3. \end{aligned}$$

L'équation sur le plan $\xi_2 = 0$ est alors

$$\begin{aligned} \dot{\xi}_1 &= -2(b\xi_1 + \nu\xi_3)\xi_3 \\ \dot{\xi}_3 &= 2(\lambda\xi_1 + b\xi_3)\xi_3. \end{aligned}$$

Celle ci est complète si et seulement si l'une des conditions suivantes est vérifiée:

(i) $\lambda = b = 0$

(ii) $\lambda \neq 0$ et $b^2 - \lambda\nu < 0$

Dans le premier cas, l'équation globale est complète puisque ξ_2 est constante et ξ_3 est solution d'une équation linéaire. Dans le deuxième cas, comme $\lambda \neq 0$, nous pouvons choisir $E_1' \perp_k \mathcal{V}$, $E_1' = E_1 + v$ et $ad_{E_1'} = ad_{E_1}$. Par conséquent l'équation d'Euler a la même structure qu'auparavant, tout en ayant

$$\begin{aligned} x_1 &= \lambda\xi_1 \\ x_3 &= c\xi_2 + \nu\xi_3 \end{aligned}$$

L'énergie du système est donnée par

$$\lambda\xi_1^2 + \mu\xi_2^2 + 2c\xi_2\xi_3 + \nu\xi_3^2$$

Compte tenue de ce que $\xi_2 =$ cte et que $\nu \neq 0$, la fonction

$$\lambda\xi_1^2 + \nu(\xi_3 + \frac{2c\xi_2}{\nu})^2$$

est une intégrale première. Comme $\lambda\nu > 0$, la quadrique en question est positive définie et les solutions sont complètes.

En résumé, le champ d'Euler est complet si

ou bien $\ker\Phi\,|_{\mathcal{V}} \neq (0)$ et $E_2 \in \ker\Phi\,|_{\mathcal{V}}$

ou bien $\ker\Phi\,|_{\mathcal{V}} = (0)$ et $(\det\Phi)\ \Phi(E_2)E_2 > 0$.

Si $\alpha_2\alpha_3 \neq 0$, posons $\alpha_2 = 1-\alpha$, $\alpha_3 = 1+\alpha$, $\alpha^2 \neq 1$. Le champ d'Euler dans ce cas est

$$\begin{aligned} \dot{\xi}_1 &= -(1-\alpha)\,x_2\xi_2 - (1+\alpha)\,x_3\xi_3 \\ \dot{\xi}_2 &= (1-\alpha)\,x_1\xi_2 \\ \dot{\xi}_3 &= (1+\alpha)\,x_1\xi_3. \end{aligned}$$

Si $\lambda = 0$ les deux dernières équations constituent un système planaire:

$$\begin{aligned} \dot{\xi}_2 &= (1-\alpha)(a\xi_2 + b\xi_3)\xi_2 \\ \dot{\xi}_3 &= (1+\alpha)(a\xi_2 + b\xi_3)\xi_3. \end{aligned}$$

qui a toujours des idempotents non nuls. Supposons donc que $\lambda \neq 0$ et prenons $e_1 \perp_\Phi \mathrm{Vect}(E_2, E_3)$ avec

$$ad_{e_1} = ad_{E_1}.$$

Dans la base (e_1, E_2, E_3), $a = b = 0$ et l'énergie s'exprime par

$$E(\xi)\xi = \lambda\xi_1^2 + \mu\xi_2^2 + 2c\xi_2\xi_3 + \nu\xi_3^2.$$

Le champ d'Euler s'écrit:

$$\begin{aligned} \dot{\xi}_1 &= \lambda\xi_1^2 - E(\xi)\xi + \alpha\mu\xi_2^2 - \alpha\nu\xi_3^2 \\ \dot{\xi}_2 &= (1-\alpha)\lambda\xi_1\xi_2 \\ \dot{\xi}_3 &= (1+\alpha)\lambda\xi_1\xi_3. \end{aligned}$$

Cette équation n'a pas d'idempotents non nuls si et seulement si $\lambda\mu \geq 0$ et $\lambda\nu \geq 0$. On a le résultat suivant

Proposition 3.3 *Pour tout $\alpha \neq 1, -1$, le champ d'Euler associé à l'isomorphisme $\Phi : \mathcal{G} \longrightarrow \mathcal{G}^*$, défini par*

$$\begin{aligned} \Phi(E_1) &= E_1^* \\ \Phi(E_2) &= E_3^* \\ \Phi(E_3) &= E_2^* \end{aligned}$$

n'a pas d'idempotents non nuls et cependant n'est pas complet.

Démonstration. En effet, dans ces conditions ls champ d'Euler s'exprime comme

$$\begin{aligned} \dot{\xi}_1 &= -2c\xi_2\xi_3 \\ \dot{\xi}_2 &= (1-\alpha)\lambda\xi_1\xi_2 \\ \dot{\xi}_3 &= (1+\alpha)\lambda\xi_1\xi_3, \end{aligned}$$

qui n'a pas d'idempotents non nuls. Comme l'énergie est donnée par $E(\xi)\xi = \lambda\xi_1^2 + 2c\xi_2\xi_3$, la première équation peut aussi s'écrire comme

$$\dot{\xi}_1 = \lambda\xi_1^2 - E(\xi)\xi$$

qui n'est pas complète lorsque l'énergie est non positive.

References

[Ar] V.Arnold. *Méthodes Mathématiques de la Mécanique Classique.* M.I.R., Moscou, 1976.

[Au-M] A.Aubert; A. Medina. Groupes de Lie pseudo–riemanniens plats. *(A paraître).*

[B] N. Bourbaki. *Groupes et Algèbres de Lie, Chapitre I.* Hermann, Paris,1960.

[H-E] S.W. Hawking; G.F.R. Ellis. *The large scale structure of space–time.* Cambridge University Press, London, 1973

[K-Y] J. L. Kaplan; J. A. Yorke. Non associative real algebras and quadratic differential equations. *Nonlinear Analysis* 3:1(1979), 49–51, 1978

[M-R] A. Medina, P. Revoy. Algèbres de Lie et Produit Scalaire Invariant. *Ann. Ec. Norm. Sup., 4ème série,* 18(1985), 553-561.

[Mi] J. Milnor. Curvatures of Left Invariant Metrics On Lie Groups. *Advances in Mathematics*, 21(1976), 293-329.

[Sc] R.D. Schaffer. *An introduction to non–associative algebras.* Academic Press, 1966.

[T] A.Tsemo. Communication privée.

Mathematics Subject Classification: Primary 58F17;Secondary 58F21, 34A12.
Key Words: Euler Equation, Levi–Civita product, non–vanishing idempotent elements.

E-mail: stbs@xanum.uam.mx, medina@math.univ-montp2.fr.

ON INVARIANTS OF SECOND ORDER LINEAR PARTIAL DIFFERENTIAL EQUATIONS IN TWO VARIABLES

Relative to transformations of variables and unknown

U.BEKBAEV

Institute of Mathematics,
Uzbekistan Academy of Sciences
F. Hodjaev str., 29, 700143 Tashkent, Uzbekistan.

1 Introduction

In this paper we deal with second order linear partial differential equations (d.e.) over an abstract differential field $(F, \partial_1, \partial_2)$ of characteristic zero, define an analogue of transformations of "indeterminates" (special transformations of ∂_1 and ∂_2) for such a differential field, consider equivalence of such d.e. relative to these transformations of ∂_1, ∂_2 and multiplications the unknown element of the d.e. by any element of F^*. This equivalence will be reduced to a more simple one (Theorem 1), the corresponding field of invariant differential rational functions will be described (Theorem 2) and separation of "common type orbits" by invariants will be shown (Theorem 3). An analogical result was presented in [3].

As far as I know the algebraization presented as well as the results obtained are new though some (relative) invariants of second order d.e. have arised in equivalence problem of such d.e. [9,10].

As to the nth- order ordinary linear differential equations, their equivalence and invariants have been studied by many authors [4,6,8,11-13]. An algebraic approach to studying their invariants is presented in [1-2].

The results and notations of differential algebra used in this paper can be found in [5, 7].

Y. Khakimdjanov et al (eds.), Algebra and Operator Theory, 145-156.

2 Preliminary results

Definition. A (commutative, with unit) ring F with given mutually commuting additive maps $\partial_i : F \to F$ $(i = \overline{1,m})$ is said to be a *differential ring* if $\partial_i(ab) = \partial_i(a)b + a\partial_i(b)$ is valid for any $i = \overline{1,m}$ and $a, b \in F$.

Here we list some preliminary results without proofs, which will be used to prove the main results of this paper.

Proposition 1. *Let $(F, \partial_1, \partial_2)$ be a differential field and $\lambda \in F$ be any element. Then there exists an extension $(\overline{F}, \partial_1, \partial_2)$ of this field such that the system of equations*

$$\begin{cases} \partial_2 s & = & \partial_1 t \\ s & = & \lambda t \end{cases}$$

has a nontrivial solution.

We need the following two Propositions at $m = 2$.

Proposition 2. *Let $(F, \partial_1, \partial_2, ..., \partial_m)$- be a differential field of characteristic zero. The following three properties are equivalent.*

1. The differential operators $\partial_1, \partial_2, ..., \partial_m$ are linearly independent over F.
2. There is no nonzero ∂- differential polynomial over F which vanishes at all values of indeterminates from F.
3. If

$$p^\partial\{x_{11}, x_{12}, ..., x_{1m}, x_{21}, ..., x_{2m}, ..., x_{m1}, ..., x_{mm}\} = p^\partial\{(x_{ij})_{i,j=\overline{1,m}}\}$$

is a differential polynomial over F such that $p^\partial\{g\} = 0$ at any $g = (g_{ij})_{i,j=\overline{1,m}}$

$$\in GL^\partial(m, F) = \{g = (g_{ij})_{i,j=\overline{1,m}} \in GL(m, F) : \partial_k g_{ij} = \partial_i g_{kj} \quad at\ i, j, k = \overline{1,m}\}$$

then $p^\partial\{(t_{ij})_{i,j=\overline{1,m}}\} = 0$ is also valid, for any indeterminates $(t_{ij})_{i,j=\overline{1,m}}$, for which $\partial_k t_{ij} = \partial_i t_{kj}$ at $i, j, k = \overline{1,m}$

Proposition 3. *Let $(F, \partial_1, \partial_2, ..., \partial_m)$- be a differential field of characteristic zero, $\partial_1, \partial_2, ..., \partial_m$ be linearly independent over F, $(t_{ij})_{i,j=\overline{1,m}}$- be such differential indeterminates that $\partial_k t_{ij} = \partial_i t_{kj}$ at $i, j, k = \overline{1,m}$, and $p^\partial\{x_1, x_2, ..., x_m\}$- be an arbitrary nonzero differential polynomial over F. Then*

$$p^\partial\{T_{11}/T, T_{21}/T, ..., T_{m1}/T\} \neq 0,$$

where $T = \det(t_{ij})_{i,j=\overline{1,m}}$, and T_{k1} is the algebraic cofactor of the element t_{k1} in $(t_{ij})_{i,j=\overline{1,m}}$.

3 Equivalence

Let now $(F, \partial_1, \partial_2)$ be a differential field of characteristic zero, $g \in GL^{\partial}(2, F)$, and $\delta = g^{-1}\partial$, where ∂ is the column vector with the "coordinates" ∂_1, ∂_2 and δ is the column vector with the "coordinates" δ_1, δ_2. It can be checked that (F, δ_1, δ_2) is a differential ring, too i.e. the new derivatives δ_1, δ_2 also commute with each other. The above transformation $\delta = g^{-1}\partial$ is an analogue of transformations of indeterminates in classical functional differential rings for abstract differential rings. It should be noted that $GL^{\partial}(2, F)$ is not a subgroup of $GL(2, F)$ though the union of all $GL^{g^{-1}\partial}(2, F)$, where $g \in GL^{\partial}(2, F)$, can be considered as a "generalized" groupoid .

Under such a transformation and $v = h^{-1}u$,where $h \in F^*$ the following differential equation

$$A\partial_1^2 u + B\partial_1\partial_2 u + C\partial_2^2 u + a\partial_1 u + b\partial_2 u + cu = 0 \tag{1}$$

where A, B, C, a, b are elements of $(F, \partial_1, \partial_2)$, transforms to the analogous one

$$A_1\delta_1^2 v + B_1\delta_1\delta_2 v + C_1\delta_2^2 v + a_1\delta_1 v + b_1\delta_2 v + c_1 v = 0, \tag{2}$$

Moreover the coefficients of these equations, h and components of matrix g are connected in the following way.

$$\left\{\begin{array}{l} A_1 = h(Ag_{11}^2 + Bg_{11}g_{21} + Cg_{21}^2) \\ B_1 = h(2Ag_{11}g_{12} + B(g_{11}g_{22} + g_{12}g_{21}) + 2Cg_{21}g_{22}) \\ C_1 = h(Ag_{12}^2 + Bg_{12}g_{22} + Cg_{22}^2) \\ a_1 = h(A\partial_1 g_{11} + B\partial_1 g_{21} + C\partial_2 g_{21} + ag_{11} + bg_{21}) + 2A\partial_1 hg_{11} + \\ B(\partial_1 hg_{21} + \partial_2 hg_{11}) + 2C\partial_2 hg_{21} \\ b_1 = h(A\partial_1 g_{12} + B\partial_1 g_{22} + C\partial_2 g_{22} + ag_{12} + bg_{22}) + 2A\partial_1 hg_{12} + \\ B(\partial_1 hg_{22} + \partial_2 hg_{12}) + 2C\partial_2 hg_{22} \\ c_1 = A\partial_1^2 h + B\partial_1\partial_2 h + C\partial_2^2 h + a\partial_1 h + b\partial_2 h + ch \end{array}\right. \tag{3}$$

Definition. If $h \in F^*, \delta = g^{-1}\partial$ for some $g \in GL^{\partial}(2, F)$ and the system of equalities (3) is valid, then equations (1) and (2) are said to be *equivalent* relative to g, h and in this case we write $V_1 = \rho^{\partial} < g, h, V >$, where $V_1 = (A_1, B_1, C_1, a_1, b_1, c_1), V = (A, B, C, a, b, c)$.

In future $\rho_i^{\partial} < g, h, V >$ denotes the i-th component of the vector $\rho^{\partial} < g, h, V >$. For convenience even in the abstract differential field case we use

u_x $(u_y,$ $u_\xi,$ $u_\eta)$ to denote $\partial_1 u$ (correspondingly, $\partial_2 u,$ $\delta_1 u,$ $\delta_2 u$).

The following equality

$$\rho^{g_1^{-1}\partial} < g_2, h_2, \rho^{\partial} < g_1, h_1, V >>= \rho^{\partial} < g_1 g_2, h_1 h_2, V > \tag{4}$$

is evident for any $g_1 \in GL^{\partial}(2, F)$, $g_2 \in GL^{g_1^{-1}\partial}(2, F)$ and $h_1, h_2 \in F^*$.

To prove the main result on equivalence we use the following Proposition, too.

Proposition 4. *If* $D<V>\neq 0$ *for equation (1)and* $g^1\in GL^\partial(2,F)$ *is such an element that* $A(g^1_{11})^2+Bg^1_{11}g^1_{21}+C(g^1_{21})^2=A(g^1_{12})^2+Bg^1_{12}g^1_{22}+C(g^1_{22})^2=0$ *(i.e. the diagonal elements of the matrix*

$$(g^1)^t\begin{pmatrix} A & B/2 \\ B/2 & C \end{pmatrix} g^1$$

are zeros) then the equality

$$\begin{pmatrix} \rho_4^\partial <g^1,h,V> \\ \rho_5^\partial <g^1,h,V> \end{pmatrix} = -h(g^1)^t \begin{pmatrix} 2A & B \\ B & 2C \end{pmatrix} \begin{pmatrix} \alpha^\partial <V> - & h^{-1}h_x \\ \beta^\partial <V> - & h^{-1}h_y \end{pmatrix}$$

is valid, where t means transposition, $D<V>=B^2-4AC$,

$$\alpha^\partial <V>=a^\partial\{V\}/D<V>, \quad \beta^\partial <V>=b^\partial\{V\}/D<V>,$$

$$a^\partial\{V\}=AC_x-A_xC+BC_y-B_yC-bB+2aC,$$
$$b^\partial\{V\}=A_xB-AB_x+A_yC-AC_y-aB+2bA.$$

The next theorem deals with equations of the type (1) for which $D<V>\neq 0$ and reduces equivalence of such d.e. to a more simple one.

Theorem 1. *If* $h\in F^*$, $g\in GL^\partial(2,F)$, $\delta=g^{-1}\partial$ *and* $c_1^\partial<V>\in F^*$ $(\partial_2\alpha^\partial<V>=\partial_1\beta^\partial<V>)$ *then equations* (1) *and* (2) *are equivalent relative to* g,h, *if and only if the following equalities*

$$\begin{pmatrix} A_1 & B_1/2 \\ B_1/2 & C_1 \end{pmatrix} = hg^t \begin{pmatrix} A & B/2 \\ B/2 & C \end{pmatrix} g$$

$$\begin{pmatrix} \alpha^\delta <V_1> \\ \beta^\delta <V_1> \end{pmatrix} = g^{-1} \begin{pmatrix} \alpha^\partial <V> -h^{-1}\partial_1 h \\ \beta^\partial <V> -h^{-1}\partial_2 h \end{pmatrix} \tag{5}$$

$$\gamma^\delta <V_1>=h\gamma^\partial <V>$$
$$(\text{ corresp. } \gamma_0^\delta <V_1>=h\gamma_0^\partial <V>),$$

are valid , where
$c^\partial<V>=(\partial_2\alpha^\partial<V>-\partial_1\beta^\partial<V>)^2D<V>$,
$c_1^\partial<V>=-1/2(c^\partial<V>)^{-1}\partial_1 c^\partial<V>$,
$c_2^\partial<V>=-1/2(c^\partial<V>)^{-1}\partial_2 c^\partial<V>$,
$\gamma^\partial<V>=A(\partial_1 c_1^\partial<V>+(c_1^\partial<V>)^2)+B(\partial_2 c_1^\partial<V>+c_1^\partial<V>c_2^\partial<V>)+$

$C(\partial_2 c_2^\partial < V > +(c_2^\partial < V >)^2) + +ac_1^\partial < V > +bc_2^\partial < V > +c,$
$\gamma_0^\partial < V >= A(\partial_1\alpha^\partial < V > +(\alpha^\partial < V >)^2) + B(\partial_2\alpha^\partial < V > +\alpha^\partial < V > \beta^\partial <$
$V >)+$
$C(\partial_2\beta^\partial < V > +(\beta^\partial < V >)^2) + a\alpha^\partial < V > +b\beta^\partial < V > +c.$

Proof of Theorem 1. First of all one should notice that the first equality of (5) is nothing other than a matrix expression of the first three equations of (3). Moreover one can assume the existence of an element $g^1 \in GL^\partial(2, F)$ for which the condition of Proposition 4 is valid. Indeed if $A = C = 0$, $B \neq 0$ then the second order unit matrix E can taken for g^1. If $A \neq 0$ or $C \neq 0$ (let $A \neq 0$) then for the solutions λ_1, λ_2 of the equation

$$A\lambda^2 + B\lambda + C = 0$$

due to Proposition 1 there is such an extension $(F, \partial_1, \partial_2) \subset (\overline{F}, \partial_1, \partial_2)$ and $\{g_{ij}^1 : i, j \in \{1,2\}\} \in \overline{F}$ that

$$\begin{cases} \partial_2 g_{11}^1 = \partial_1 g_{21}^1 \\ g_{11}^1 \quad = \lambda_1 g_{21}^1 \end{cases} \qquad \begin{cases} \partial_2 g_{12}^1 = \partial_1 g_{22}^1 \\ g_{12}^1 = \lambda_2 g_{22}^1 \end{cases}$$

are valid. For such g^1 it is easy to see that $g^1 \in GL^\partial(2, \overline{F})$. Independently which system of equalities (3) or (5) is valid it can be checked that for the function $f_1^\partial < z > = \partial_2\alpha^\partial < z > -\partial_1\beta^\partial < z >$ the following equality $f_1^\delta < V_1 >= \Delta^{-1} f_1^\partial < V >$ (Unlike the discriminant here appears Jacobian's first power), where $\delta = g^{-1}\partial$, $V_1 = \rho^\partial\{g, h, V\}$, $\Delta = det(g) = g_{11}g_{22} - g_{12}g_{21}$, is valid at any $g^1 \in GL^\partial(2, F), h \in F^*$. Therefore in the case $f_1^\partial < V >\in F^*$ one has

$$(f_1^\delta < V_1 >)^{-1}\delta f_1^\delta < V_1 >= g^{-1}((f_1^\partial < V >)^{-1}\partial f_1^\partial < V > -\Delta^{-1}\partial\Delta) \qquad (6)$$

But due to the equality
$D < V_1 >= h^2\Delta^2 D < V >$ we have $D < V_1 >^{-1} \delta D < V_1 >=$

$$g^{-1}(D < V >^{-1} \partial D < V > +2\Delta^{-1}\partial\Delta + 2h^{-1}\partial h) \qquad (7)$$

By the use of (6) and (7) it is easy to see that

$$\begin{pmatrix} c_1^\delta < V_1 > \\ c_2^\delta < V_1 > \end{pmatrix} = g^{-1} \begin{pmatrix} c_1^\partial < V > -h^{-1}\partial_1 h \\ c_2^\partial < V > -h^{-1}\partial_2 h \end{pmatrix} \qquad (8)$$

It should be noted that equality (8) is valid independently which system of equalities (3) or (5) is valid. But because of the equality $\partial_1 c_2^\partial < V >= \partial_2 c_1^\partial < V >$ we may assume that the system

$$\begin{cases} h_1^{-1}\partial_1 h_1 = c_1^\partial < V > \\ h_1^{-1}\partial_2 h_1 = c_2^\partial < V > \end{cases} \qquad (9)$$

has a solution in F. For any solution h_1 of this system and the above g^1 due to Proposition 4 one has

$$\begin{pmatrix} \rho_4^{\partial} < g^1, h_1, V > \\ \rho_5^{\partial} < g^1, h_1, V > \end{pmatrix} = -h_1 (g^1)^t \begin{pmatrix} 2A & B \\ B & 2C \end{pmatrix} \begin{pmatrix} \alpha^{\partial} < V > - & h_1^{-1}\partial_1 h_1 \\ \beta^{\partial} < V > - & h_1^{-1}\partial_2 h_1 \end{pmatrix} \tag{10}$$

Moreover it can be checked that

$$\rho_6^{\partial} < g^1, h_1, V >= h_1 \gamma^{\partial} < V > \tag{11}$$

Let us show now that (3) implies (5). If (3) is valid then it is easy to see that diogonal elements of the matrix

$$(\overline{g}^1)^t \begin{pmatrix} A_1 & B_1/2 \\ B_1/2 & C_1 \end{pmatrix} \overline{g}^1, \tag{12}$$

are zeros and the system of equalities

$$\begin{cases} h_2^{-1}\delta_1 h_2 = c_1^{\delta} < V_1 > \\ h_2^{-1}\delta_2 h_2 = c_2^{\delta} < V_1 > \end{cases} \tag{13}$$

is valid, where $\overline{g}^1 = g^{-1}g^1$, $\overline{h}_1 = h_1 h^{-1}$ and $\delta = g^{-1}\partial$. Therefore one analogously has

$$\begin{pmatrix} \rho_4^{\delta} < \overline{g}^1, \overline{h}_1, V_1 > \\ \rho_5^{\delta} < \overline{g}^1, \overline{h}_1, V_1 > \end{pmatrix} = -\overline{h}_1 (\overline{g}^1)^t \begin{pmatrix} 2A_1 & B_1 \\ B_1 & 2C_1 \end{pmatrix} \begin{pmatrix} \alpha^{\delta} < V_1 > - & \overline{h}_1^{-1}\delta_1 \overline{h}_1 \\ \beta^{\delta} < V_1 > - & \overline{h}_1^{-1}\delta_2 \overline{h}_1 \end{pmatrix} \tag{14}$$

$$\rho_6^{\delta} < \overline{g}^1, \overline{h}_1, V_1 >= \overline{h}_1 \gamma^{\delta} < V_1 > \tag{15}$$

But due to (4)

$$\rho^{\delta} < \overline{g}^1, \overline{h}_1, V_1 >= \rho^{\delta} < \overline{g}^1, \overline{h}_1, \rho^{\partial} < g, h, V >>=$$

$$= \rho^{\partial} < g\overline{g}^1, h\overline{h}_1, V >= \rho^{\partial} < g^1, h_1, V >$$

Comparison coordinates of this vector equality with respect to (10), (11),(14), (15) shows that (3) implies (5).

Now we deduce (3) from (5). Let g^1 be the same matrix and h_1 be the same element as above. Once again all diogonal elements of (12), where $\overline{g}^1 = g^{-1}g^1$, are zeros. Moreover system of equalities (13) is valid too, where $\overline{h}_1 = h_1 h^{-1}$. Therefore we have equalities (14) and (15). In this case system of equalities (5) means that

$$\rho^{\delta} < \overline{g}^1, \overline{h}_1, V_1 >= \rho^{\partial} < g^1, h_1, V > .$$

But due to (4) once again

$$\rho^{\delta} < \overline{g}^1, \overline{h}_1, \rho^{\partial} < g, h, V >>= \rho^{\partial} < g\overline{g}^1, h\overline{h}_1, V >= \rho^{\partial} < g^1, h_1, V >$$

The last two equalities mean that

$$V_1 = \rho^{\partial} < g, h, V >$$

As to the proof of Theorem 1 in $f_1^{\partial} < V >= 0$, i.e. $\partial_2\alpha^{\partial} < V >= \partial_1\beta^{\partial} < V >$ case it is enough to consider the following

$$\begin{cases} h_1^{-1}\partial_1 h_1 = \alpha^{\partial} < V > \\ h_1^{-1}\partial_2 h_1 = \beta^{\partial} < V > \end{cases}$$

system instead of (9) and a proof of theorem 1 in this case can be done as above. This is the proof of Theorem 1.

In order to describe invariants let $z_1, z_2, ..., z_6$ - be differential indeterminates over F , $z = (z_1, z_2, ..., z_6)$, $K = \{a \in F : \partial_1 a = \partial_2 a = 0\}$ - be the field of constants of F, $K < z >$ be the field of all ∂ -differential rational functions in z over K and we assume linear independence of ∂_1, ∂_2 over F.

Definition A differential rational function $f^{\partial} < z >$ over K is called an *invariant* if for any $g \in GL^{\partial}(2, F)$ the equality

$$f^{g^{-1}\partial} < \rho^{\partial} < g, h, z >>= f^{\partial} < z >$$

is valid.

Let K_1 be the set of all such invariant functions. To describe K_1 we need a nonzero element $\varphi^{\partial} < z >$ of $K < z >$ and a matrix $\Phi^{\partial} < z > \in GL^{\partial}(2, K < z >)$ that

$$\Phi^{g^{-1}\partial} < \rho^{\partial} < g, h, z >>= g^{-1}\Phi^{\partial} < z > . \tag{16}$$

$$\varphi^{g^{-1}\partial} < \rho^{\partial} < g, h, z >>= h^{-1}\varphi^{\partial} < z > \tag{17}$$

for any $g \in GL^{\partial}(2, F)$, $h \in F^*$. The existence of such an element and matrix will be shown a little later. Let $\bar{\partial} = (\Phi^{\partial} < z >)^{-1}\partial$. It should be noted that K_1 is not invariant relative to the differential operators ∂_1 and ∂_2 i.e. $(K_1, \partial_1, \partial_2)$ is not a differential field.

Theorem 2. *The field K_1 is a differential field over K relative to the differential operators $\bar{\partial}_1, \bar{\partial}_2$ and as such a differential field it is generated by the components of*

$$\rho^{\partial} < \Phi^{\partial} < z >, \varphi^{\partial} < z >, z >$$

Moreover this differential field has differential transcendence degree 3 over K.

Proof of Theorem 2. Invariance of K_1 relative to $\bar{\partial}_1, \bar{\partial}_2$ is an easy consequence of (16). For any $g \in GL^{\partial}(2, F)$, $h \in F^*$ one has

$$\rho^{g^{-1}\partial} < \Phi^{g^{-1}\partial} < \rho^{\partial} < g, h, z >>, \varphi^{g^{-1}\partial} < \rho^{\partial} < g, h, z >>, \rho^{\partial} < g, h, z >>=$$

$$\rho^{g^{-1}\partial} < g^{-1}\Phi^{\partial} < z >, h^{-1}\varphi^{\partial} < z >, \rho^{\partial} < g, h, z >>$$

and according to (4) the right side of this equality equals

$$\rho^{\partial} < \Phi^{\partial} < z >, \varphi^{\partial} < z >, z >,$$

i.e. the components of $\rho^{\partial} < \Phi^{\partial} < z >, \varphi^{\partial} < z >, z >$ are invariants.

If for the differential rational function $f^{\partial} < z >$ the equality

$$f^{g^{-1}\partial} < \rho^{\partial} < g, h, z >>= f^{\partial} < z >$$

holds for any $g \in GL^{\partial}(2, F)$, $h \in F^*$ then according to Propositions 2 and 3 in particular at $g = \Phi^{\partial} < z >$, $h = \varphi^{\partial} < z >$ on has

$$f^{(\Phi^{\partial}<z>)^{-1}\partial} < \rho^{\partial} < \Phi^{\partial} < z >, \varphi^{\partial} < z >, z >>= f^{\partial} < z >$$

i.e. the K_1 is generated as a $\bar{\partial}$ - differential field over K by the components of $\rho^{\partial} < \Phi^{\partial} < z >, \varphi^{\partial} < z >, z >$.

To prove the equality $\bar{\partial}$ -tr.deg$K_1/K = 3$ we consider the following tower of $\bar{\partial}$ - differential fields

$$K \subset K_1 \subset K_1 < \Phi^{\partial}_{11} < z >,\ \Phi^{\partial}_{12} < z >,\ \Phi^{\partial}_{21} < z >,\ \Phi^{\partial}_{22} < z >, \varphi^{\partial} < z >> .$$

The last field is $K < z >$- the field of all ∂ -differential rational functions in z over K , considered as a $\bar{\partial}$ -differential field (because of

$$\rho^{\bar{\partial}} < (\Phi^{\partial} < z >)^{-1}, (\varphi^{\partial} < z >)^{-1}, \rho^{\partial} < \Phi^{\partial} < z >, \varphi^{\partial} < z >, z >>= z).$$

Therefore $\bar{\partial}$-tr.deg$K < z > /K= 6$ and due to this it is enough to verify the equality $\bar{\partial}$-tr.deg$K < z > /K_1= 3$.

The elements $\Phi^{\partial}_{21} < z >, \Phi^{\partial}_{22} < z >$ are $\bar{\partial}$- differential algebraic over

$$K_1 < \Phi^{\partial}_{11} < z >, \Phi^{\partial}_{12} < z >, \varphi^{\partial} < z >> .$$

Let us verify it for $\Phi^{\partial}_{21} < z >$. According to the equalities

$$\partial_1\Phi^{\partial}_{21} < z >= \partial_2\Phi^{\partial}_{11} < z >,\quad \partial_1\Phi^{\partial}_{22} < z >= \partial_2\Phi^{\partial}_{12} < z >,$$

$$\partial_1 = \Phi^{\partial}_{11} < z >\bar{\partial}_1 + \Phi^{\partial}_{12} < z >\bar{\partial}_2 \text{ and } \partial_2 = \Phi^{\partial}_{21} < z >\bar{\partial}_1 + \Phi^{\partial}_{22} < z >\bar{\partial}_2 \,.$$

one has the following vector form equalities

$$\begin{cases} (\Phi^{\partial}_{11} < z >, \Phi^{\partial}_{12} < z >)\ \bar{\partial}\ \Phi^{\partial}_{21} < z >= (\Phi^{\partial}_{21} < z >, \Phi^{\partial}_{22} < z >)\ \bar{\partial}\ \Phi^{\partial}_{11} < z > \\ (\Phi^{\partial}_{11} < z >, \Phi^{\partial}_{12} < z >)\ \bar{\partial}\ \Phi^{\partial}_{22} < z >= (\Phi^{\partial}_{21} < z >, \Phi^{\partial}_{22} < z >)\ \bar{\partial}\ \Phi^{\partial}_{12} < z > \end{cases} \tag{18}$$

Due to the first equality of system (18)

$$\bar{\partial}\,[(\Phi_{11}^{\partial}<z>\bar{\partial}_1+\Phi_{12}^{\partial}<z>\bar{\partial}_2)\Phi_{21}^{\partial}<z>]=\bar{\partial}\,(\Phi_{21}^{\partial}<z>\bar{\partial}_1\,\Phi_{11}^{\partial}<z>)+$$
$$\bar{\partial}\,\Phi_{22}^{\partial}<z>\bar{\partial}_2\,\Phi_{11}^{\partial}<z>+\Phi_{22}^{\partial}<z>\bar{\partial}\bar{\partial}_2\,\Phi_{11}^{\partial}<z>$$

Multiplying both sides of this equality from the left by

$$(\Phi_{11}^{\partial}<z>,\Phi_{12}^{\partial}<z>)$$

and taking into account the second equality of (18), one has

$$\begin{aligned}&(\Phi_{11}^{\partial}<z>,\Phi_{12}^{\partial}<z>)\;\bar{\partial}\,[(\Phi_{11}^{\partial}<z>\bar{\partial}_1+\Phi_{12}^{\partial}<z>\bar{\partial}_2)\Phi_{21}^{\partial}<z>]=\\&(\Phi_{11}^{\partial}<z>,\Phi_{12}^{\partial}<z>)\;\bar{\partial}\,(\Phi_{21}^{\partial}<z>\bar{\partial}_1\,\Phi_{11}^{\partial}<z>)+\\&(\Phi_{21}^{\partial}<z>,\Phi_{22}^{\partial}<z>)\;\bar{\partial}\,\Phi_{12}^{\partial}<z>\bar{\partial}_2\,\Phi_{11}^{\partial}<z>+\\&\Phi_{22}^{\partial}<z>\,(\Phi_{11}^{\partial}<z>,\Phi_{12}^{\partial}<z>)\;\bar{\partial}\bar{\partial}_2\,\Phi_{11}^{\partial}<z>\end{aligned}\tag{19}$$

If $\bar{\partial}_2\Phi_{11}^{\partial}<z>=0$, then the first equality of (18) shows $\Phi_{21}^{\partial}<z>$ is $\bar{\partial}$-differential algebraic over $K_1<\Phi_{11}^{\partial}<z>,\Phi_{12}^{\partial}<z>,\varphi^{\partial}<z>>$.

If $\bar{\partial}_2\Phi_{11}^{\partial}<z>\neq 0$, then due to the first equality of (18) $\Phi_{22}^{\partial}<z>$ can be expressed by $\Phi_{11}^{\partial}<z>,\Phi_{12}^{\partial}<z>,\Phi_{21}^{\partial}<z>$ and their $\bar{\partial}$- derivatives. Putting this expression for $\Phi_{22}^{\partial}<z>$ in (19) gives us a nonzero $\bar{\partial}$- differential polynomial over $K_1<\Phi_{11}^{\partial}<z>,\Phi_{12}^{\partial}<z>,\varphi^{\partial}<z>>$, which vanishes at $\Phi_{21}^{\partial}<z>$. This shows $\Phi_{21}^{\partial}<z>$ is $\bar{\partial}$- differential algebraic over

$$K_1<\Phi_{11}^{\partial}<z>,\Phi_{12}^{\partial}<z>,\varphi^{\partial}<z>>.$$

Let us now verify $\bar{\partial}$- differential algebraic independence of the system of elements $\Phi_{11}^{\partial}<z>,\Phi_{12}^{\partial}<z>,\varphi^{\partial}<z>$ over K_1. Let $p^{\bar{\partial}}\{t_1,t_2,t_3\}$ be such a nonzero differential polynomial over K_1 that

$$p^{\bar{\partial}}\{\Phi_{11}^{\partial}<z>,\Phi_{12}^{\partial}<z>,\varphi^{\partial}<z>\}=0$$

In this case the equality should remain true under the transformations

$$z\longmapsto\rho^{\partial}<g,h,z>,\quad \partial\longmapsto g^{-1}\partial,$$

where $g\in GL^{\partial}(2,F)$, $h\in F^*$. Relative to such transformations the differential operator $\bar{\partial}$ and coefficients of $p^{\bar{\partial}}\{t_1,t_2,h_3\}$ are invariant . Therefore one has

$$p^{\bar{\partial}}\{G_{11}\Phi_{11}^{\partial}<z>+G_{12}\Phi_{21}^{\partial}<z>,$$
$$G_{11}\Phi_{12}^{\partial}<z>+G_{12}\Phi_{22}^{\partial}<z>,h^{-1}\varphi^{\partial}<z>\}=0,$$

where G_{ij} are corresponding elements of g^{-1}. Due to this equality one can conclude that neither t_3 nor its derivatives take part in $p^{\bar\partial}\{t_1, t_2, t_3\}$ and indeed

$$p^{\bar\partial}\{G_{11}\Phi^{\partial}_{11} < z > +G_{12}\Phi^{\partial}_{21} < z >, G_{11}\Phi^{\partial}_{12} < z > +G_{12}\Phi^{\partial}_{22} < z >\} = 0 \quad (20)$$

Let $(t_{ij})_{i,j=1,2}$ be such $\bar\partial$ - indeterminates that $\bar\partial_1\ t_{21} = \bar\partial_2\ t_{11}$, $\bar\partial_1\ t_{22} = \bar\partial_2\ t_{12}$. It is easy to verify that in this case $\partial_2 g_{1j} = \partial_1 g_{2j}$ at $j = 1, 2$ for the elements of matrix

$$g = \begin{pmatrix} \Phi^{\partial}_{11} < z > & \Phi^{\partial}_{12} < z > \\ \Phi^{\partial}_{21} < z > & \Phi^{\partial}_{22} < z > \end{pmatrix} \begin{pmatrix} t_{11} & t_{12} \\ t_{21} & t_{22} \end{pmatrix}$$

and therefore according to Proposition 3 and equality (20) one has

$$p^{\bar\partial}\{T_{11}/T, T_{21}/T\} = 0,$$

because in this case

$$G_{11}\Phi^{\partial}_{11} < z > +G_{12}\Phi^{\partial}_{21} < z > = T_{11}/T = t_{22}/(t_{11}t_{22} - t_{12}t_{21}),$$

$$G_{11}\Phi^{\partial}_{12} < z > +G_{12}\Phi^{\partial}_{22} < z > = T_{21}/T = -t_{12}/(t_{11}t_{22} - t_{12}t_{21}).$$

The equality obtained $p^{\bar\partial}\{T_{11}/T, T_{21}/T\} = 0$ contradicts Proposition 3. The proof of Theorem 2 is completed.

The next result can be considered as a differential analogue of the well known result in the invariant theory of algebraic groups on separation of common type orbits by algebraic invariant functions.

Let $\chi^{\partial}_1 < z >$ and $\chi^{\partial}_2 < z >$ be such elements of K_1 that $\det \Phi^{\partial} < z > \neq 0$, where

$$\Phi^{\partial}_{11} < z > = \partial_1 \chi^{\partial}_1 < z >, \quad \Phi^{\partial}_{12} < z > = \partial_1 \chi^{\partial}_2 < z >,$$

$$\Phi^{\partial}_{21} < z > = \partial_2 \chi^{\partial}_1 < z >, \quad \Phi^{\partial}_{22} < z > = \partial_2 \chi^{\partial}_2 < z > .$$

In this case the set $X = \{V \in F^6 : \chi^{\partial}_1 < z >, \chi^{\partial}_2 < z >, c^{\partial} < z >$ are defined at V and $c^{\partial} < V > \gamma^{\partial} < V > det\Phi^{\partial} < V > \neq 0\}$ is not empty according to Proposition 2 and by the use of Theorem 1 the following theorem can be proved.

Theorem 3. *Let $g \in GL^{\partial}(2, F)$, $\delta = g^{-1}\partial$. If $V \in X$ for equation (1) and for any function $f^{\partial} < z > \in K_1$ defined over X the equality $f^{\partial} < V > = f^{\delta} < V_1 >$ is valid then equations (1) and (2) are equivalent relative to g and some $h \in F^*$.*

Let us show now existance of corresponding matrix and function with desired properties (16),(17). Invariance of the function $\chi^{\partial}_1 < z > = (c^{\partial} < z >)^{-1} \cdot$

$$\cdot \left(\begin{pmatrix} \alpha^{\partial} < z > +c^{\partial}_1 < z > \\ \beta^{\partial} < z > +c^{\partial}_2 < z > \end{pmatrix}^t \begin{pmatrix} z_1 & z_2/2 \\ z_2/2 & z_3 \end{pmatrix} \begin{pmatrix} \alpha^{\partial} < z > +c^{\partial}_1 < z > \\ \beta^{\partial} < z > +c^{\partial}_2 < z > \end{pmatrix} \right)^2$$

is a consequance of Theorem 1. Therefore the function

$$\chi_2^\partial < z >= (c^\partial < z >)^{-1}((\partial\chi_1^\partial < z >)^t \begin{pmatrix} z_1 & z_2/2 \\ z_2/2 & z_3 \end{pmatrix} \partial\chi_1^\partial < z >)^2$$

is invariant one too. The matrix $\Phi^\partial < z >$ constracted as above by the use of these invariant functions $\chi_1^\partial < z >, \chi_2^\partial < z >$ will be an element of $GL^\partial(2, K < z >)$ with property (16). The last equality of (3) can be used to obtain a function $\varphi^\partial < V >$ with the needed property (17).

References

[1] U.D.Bekbaev. On equivalence and invariants of ordinary linear differential equations, *Differential equations*, 4(1994), 911-912. (Russian)

[2] U.D.Bekbaev. On equivalence and invariants of differential equations of the type $y^{(n)} + a_1 y^{(n-1)} + ... + a_n y = 0$ *Uzbek mathematical journal.*,3(1995), 19-31. (Russian)

[3] U.D.Bekbaev. On equivalence and invariants of second order linear partial differential equations in two variables. *ICM-94. Abstracts of short communications.* Zurich 3-11 August (1994), 163.

[4] L.M.Berkovich. Halphan's problem on equivalence of ordinary linear differential equations, *Usp. mat. nauk*, 41:1 (247)(1986), 183-184 (Russian)

[5] E.R.Kolchin. Differential algebra and algebraic groups. New-York-London, Academic press, 1973.

[6] G.M.Korolev. Differential and integral equations. Gorkiy (1981), 38-41. (Russian)

[7] A.V.Mikhaliov , E.V.Pankratev. Differential and difference algebra. *Itogi nauki i tekhniki. Algebra. Topologia. Geometria*, 25(1987), 67-139. (Russian)

[8] F.Newman. A survey of global properties of linear differential equations of the n-th order, *Lecture Notes Math.*, 964(1982), 548-563.

[9] L.V.Ovsyannikov. Group alalysis of differential equations. Moscow, Nauka, 1978 (Russian)

[10] I.G.Petrovskii. Lectures on partial differential equations. Moscow, FM, 1961, 400 p. (Russian)

[11] V.G.Sirulik , F.A.Brusensov. On equivalence of commutative factorizability and reducibility of linear differential operators. *Izvestia visshix uchebnix zavedeniy. Matematica.* 8(1976), 81-84. (Russian)

[12] A.S.Udalov. Theory of curves in n-dimensional projective spaces. *DAN SSSR*, 146:1, 46-49. (Russian)

[13] E.J.Wilczynski. Projective differential geometry of curves and ruled surfaces. Leipzig: B.G.Teubner. 1906.

E-mail: laru@saturn.silk.org

LATTICE-ORDERED GROUPOIDS AND THEIR PRIME SPECTRUMS

Dj.KHADJIEV, T.M.SHAMILEV

*Dj.K.: Departement of Mathematics,
Tashkent State University,
700095, Tashkent, Uzbekistan*

*T.Sh.: Simferopol Engineering Pedagogical
Institute, Ukraine*

Introduction

Our main results are the following:
1. Let L be a complete ordered groupoid ([1], ch. XIV). We introduce definitions of r -radical and R -radical elements in L and describe some their properties.
2. Let L be a complete ordered groupoid in which every element is ideal. Denote by L_r the lattice of all r-radical elements in L. Then L_r satisfies the infinite $\wedge$ -distributive condition:

$$a \wedge (\vee_{\tau \in T} b_\tau) = \vee_{\tau \in T}(a \wedge b_\tau)$$

for any $a, b_\tau \in L_r, \tau \in T$. Let L be a complete ordered groupoid in which every element is ideal and r-radical. Then $a \cdot b = a \wedge b$ for any $a, b \in L$ and L satisfies the infinite $\wedge$- distributive condition. Analogous statements are hold for R -radical elements.

Let K be a ring. Denote by $L_r(K)$ the lattice of radical ideals of K. Then $L_r(K)$ satisfies the mentioned condition. In particular, $L_r(K)$ is distributive.
3. Complete ordered groupoids are described in which every element is ideal and r-radical.
4. By analogy with commutative rings for any complete ordered groupoid L we define the topological space $Spec(L)$ -prime spectrum (or structural space) of

Y. Khakimdjanov et al (eds.), Algebra and Operator Theory, 157-164.

L.In particular,the topological space $Spec(L)$ is defined for any ring (associative or non-associative) and any T_0 -lattice.
A part of main results of our paper was published in [2-5].
Further we use the terminology and notations of books [1,6].

1. The theorem on infinite $\cap$-distributivity of the lattice of radical elements

An ordered groupoid L is called l_0 *-groupoid* if L is complete as a lattice. Further we shall suppose that L is complete as a lattice. A given binary operation on L we shall call a multiplication and denote by $a \cdot b$ for $a, b \in L$. The greatest element in L denote by 1.
An l_0 -groupoid L shall be call *semi-integral* if every element of L is ideal ([1], ch. XIV). Every integral ([1], ch. XIV) l-groupoid is semi-integral.
An ordered groupoid L is called *l-groupoid* ([1], ch. XIV) if

$$a(b \vee c) = ab \vee ac, (b \vee c)a = ba \vee ca$$

for any $a, b, c \in L$.
An ordered groupoid is called *complete l -groupoid* ([1], ch. XIV) if

$$a(\vee_{\tau \in T} b_\tau) = \vee_{\tau \in T}(ab_\tau), (\vee_{\tau \in T} b_\tau)a = \vee_{\tau \in T}(b_\tau a)$$

for any $a, b_\tau, c \in L$, $\tau \in T$.
The following examples are important classes of l - groupoids:

1. Let G be a groupoid with the zero and the unit. By $L(G)$ denote the set of its ideals. For $A, B \in L(G)$ denote by $A \cdot B$ the intersection of all $C \in L(G)$ such that $a \cdot b \in C$, for any $a \in A, b \in B$. $L(G)$ is a complete integral l -groupoid with respect to the relation of the inclusion of ideals and the operation $A \cdot B$.
2. Let K be a ring (associative or non-associative). Denote by $L(K)$ the set of its ideals. For $A, B \in L(K)$ denote by $A \cdot B$ the intersection of all $C \in L(K)$ such that $a \cdot b \in C$ for any $a \in A$, $b \in B$. $L(K)$ is a complete semi-integral l-groupoid with respect to the relation of the inclusion of ideals and the multiplication $A \cdot B$. $L(K)$ is a complete integral l -groupoid if K have the unit.
3. Let L be a complete lattice. For $a, b \in L$ put $a \cdot b = a \wedge b$. L is an integral l_0-groupoid with the operation $a \cdot b$. L is an integral l- groupoid iff it is distributive.
4. Let G be a group. Denote by $L(G)$ the lattice of its normal subgroups. For $A, B \in L(G)$ denote by $[A, B]$ the normal subgroup generated by $a^{-1}b^{-1}ab, a \in A, b \in B$. $L(G)$ is a complete semi-integral l-groupoid with respect to the relation of the inclusion of subgroups and the operation $[A, B]$.
5. Let G be a set and let ω be a ternary relation on G. Denote by $L(G)$ the lattice of its subsets. For $a, b \in G$ denote by $a \cdot b$ the set of $x \in G$ such that

$(a,b,x) \in \omega$. If it is empty then put $a \cdot b = \emptyset$. For $A, B \in L(G)$ put

$$A \cdot B = \cup_{a \in A, b \in B}(a \cdot b),$$

$L(G)$ is a complete l-groupoid with respect to the operation $A \cdot B$.

Let L be an l_0 -groupoid. An element $p \in L$ is called prime ([1], ch.XIV) if $p \neq 1$ and $a \cdot b \leq p$ implies that $a \leq p$ or $b \leq p$.

An element $p \in L$ is called $\wedge$ *-prime* if $p \neq 1$ and $a \wedge b \leq p$ implies that $a \leq p$ or $b \leq p$.

Proposition 1.1. *Let L be a semi-integral l_0-groupoid. Then:*

(1) $(a \cdot b) \vee (b \cdot a) \leq a \wedge b$ for any $a, b \in L$;

(2) any prime element is $\wedge$ -prime.

Denote by $r(a)$ the intersection of all prime elements p such that $a \leq p$. If this set is empty put $r(a) = 1$. The element $r(a)$ is called r *-radical* of a.

Remark. Other definition of a radical for elements of commutative integral l_0 -groupoids was given in ([1], ch. XIV).

Proposition 1.2. *Let L be an l_0 -groupoid. Then:*

(1)$a \leq r(a)$ for any $a \in L$;

(2) $r(a) = (r(r(a))$ for any $a \in L$;

(3)$a \leq b$ implies that $r(a) \leq r(b)$.

We call an element $a \in L$ r *-radical* if $a = r(a)$. Denote by L_r the set of all r-radical elements in L . We put $1 \in L_r$. For $a, b \in L_r$ we put $a \leq b$ if $a \leq b$ in L.

Proposition 1.3. *L_r is a complete lattice. For $a, b \in L_r$ put $a \circ b = r(a \cdot b)$. Then L_r is an l_0 -groupoid with respect to the relation $a \leq b$ in L_r and the operation $a \circ b$.*

Lattice operations in L denote by $\cap, +$ and in L_r by $\wedge, \vee$.

Proposition 1.4. *Let L be an l_0 -groupoid. Then:*

(1) $r(a \cap b) \leq r(a) \wedge r(b) \leq r(a \cdot b)$ for any $a, b \in L$.

(2) $r(a + b) = r(r(a) + r(b)) = r(r(a) \vee r(b))$ for any $a, b \in L$;

(3) for any $a_\tau \in L_r, \tau \in T$ we have $\cap_{\tau \in T} a_\tau \in L_r$ and $\cap_{\tau \in T} a_\tau = \wedge_{\tau \in T} a_\tau$;

(4) $a \vee b = r(a + b)$ for any $a, b \in L_r$.

Proposition 1.5. *Let L be a semi-integral l_0 -groupoid. Then:*

$$r(a \cdot b) = r(a \cap b) = r(a) \wedge r(b)$$

for any $a, b \in L$.

Remark. The property of semi-integrality was used only in the proposition 1.5.

Proposition 1.6. *Let L be an l_0 -groupoid. Then:*

(1) $(L_r)_r = L_r$ and the multiplication operations on L_r and $(L_r)_r$ coincide;

(2) $a \wedge b \leq a \circ b$ for any $a, b \in L_r$.

Theorem 1.1. *Let L be a semi-integral l_0 -groupoid in which every element is r -radical. Then:*

(1) $a \cdot b = a \wedge b$ for any $a, b \in L$;
(2) L satisfies the infinite $\wedge$ -distributive condition.

For any l_0 -groupoid L the l_0 -groupoid L_r satisfies the conditions of the theorem 1.1.:

Corollary. *For any l_0 -groupoid L the lattice L_r satisfies the infinite $\wedge$ -distributive condition. In particular, L_r is distributive.*

Remark. The statement (1) of the theorem 1.1. is known for the l -groupoid $L(K)$, where K is a hereditary idempotent associative ring with the unit ([7], ch.4, 5).

Further we are going to describe all semi-integral l_0 -groupoids such that every element of them is r -radical.

Proposition 1.7. *Let L be a semi-integral l -groupoid and $1^2 = 1$. Then every maximal element in L is prime.*

Remark. *The lemma 1 in ([1], ch. XIV, n 9) is not true.*

Example. We consider the modular lattice $M = \{0, x, y, z, 1\}$, where $0 < x < 1, 0 < y < 1$, $0 < z < 1$, $x \wedge y = x \wedge z = y \wedge z = 0$, $x \vee y = x \vee z = y \vee z = 1$. M is not distributive.

Consider the operation $a \cdot b = a \wedge b$ on M . Then M is an integral l_0 -groupoid but it is not l -groupoid. Elements x, y, z are maximal in it but they are not $\wedge$ -prime.

But the mentioned lemma is true for any integral l- groupoid. The proposition 1.7. is a generalization of this statement for semi-integral l -groupoids.

Let L be a semi-integral l -groupoid such that $1^2 = 1$. For $a \in L$, $a \neq 1$,denote by $R(a)$ the intersection of maximal elements containing a . If there is no such element put $R(a) = 1$. For $a = 1$ we put $R(a) = 1$.

We call an element $a \in L$ R-radical if $a = R(a)$.

Denote by L_R the set of all R-radical elements in L . In L_R we consider the order induced from L . Then L_R is a complete lattice.

Theorem 1.2. *Let L be a semi-integral l -groupoid such that $1^2 = 1$. Then L_R satisfies the infinite $\wedge$ -distributive condition.*

Further we formulate some corollaries from theorems 1.1. and 1.2.

Let K be a ring. Denote by $L_r(K)$ the lattice of all its radical ideals (example 2).

Corollary 1. *For any ring K the lattice $L_r(K)$ satisfies the infinite $\wedge$ -distributive condition. In particular, $L_r(K)$ is distributive.*

Let L be a complete lattice. For $a \in L$ denote by $r_\wedge(a)$ the intersection of all $\wedge$-prime elements containing a. If there is no such $\wedge$-prime element put $r_\wedge(a) = 1$.

By $L_{r\wedge}$ denote the set of elements $a \in L$ such that $a = r_\wedge(a)$. We consider the order on $L_{r\wedge}$ induced from L . Then $L_{r\wedge}$ is a complete lattice.

Corollary 2. *For any lattice L the lattice $L_{r\wedge}$ satisfies the infinite $\wedge$ -distributive condition. In particular, $L_{r\wedge}$ is distributive.*

Let L be a lattice. Denote by IdL the lattice of all ideals of L . For $A, B \in IdL$ put $A \cdot B = A \cap B$.Then IdL is an integral l_0-monoid.

The following result is known ([6], ch. II, 1, corollary 18). If L is distributive then every element of IdL is r-radical. The inverse statement is also true:

Corollary 3. *Let L be a lattice such that every element of IdL is r -radical. Then L is distributive.*

Corallary 4. *If L is distributive then the lattice IdL satisfies the infinite $\wedge$-distributive condition.*

Now we describe all semi-integral l_0 -groupoids every element of which is r -radical.

A complete distributive lattice L is called T_0 -lattice if every element $a \in L$ is a union of all $\vee$ -indecomposable elements b, $b \leq a$.

The following result is known [8]:

Theorem. *A lattice L is a T_0 -lattice iff it is isomorphic to the lattice of all closed sets of some topological T_0 -space X .*

Let L be a T_0 -lattice. Denote by $L^\star$ the lattice which is dual to L . Then every element $a \in L^\star$ is an intersection of all $\wedge$-indecomposable elements containing a .

For $a, b \in L^\star$ put $a \cdot b = a \wedge b$. Then $L^\star$ is an integral l-monoid with respect to the operation $a \cdot b$. Such l -monoids shall be call a T_0 -monoid.

Theorem 1.3. *Any T_0 -monoid is an integral l -monoid in which every element is r -radical;*

Conversely, let M be a semi-integral l_0 -groupoid in which every element is r -radical. Then M isomorphic to some T_0-monoid.

Let L be a T_1 -lattice ([1], ch.IX) and let $L^\star$ be the dual to L . For $a, b \in L^\star$ put $a \cdot b = a \wedge b$. Then $L^\star$ is an integral l -monoid with respect to the operation $a \cdot b$. Such monoids shall be call a T_1 -monoid.

Theorem 1.4.(1) *Any T_1 -monoid is an integral l -monoid in which every element is R -radical.*

(2) Conversely, let M be a semi-integral l -groupoid in which $1^2 = 1$ and every element in M is R -radical. Then M isomorphic to some T_1 -monoid.

2. Prime spectrums of l_0-groupoids and T_0 -lattices.

Similarly to commutative rings we define prime spectrums (or structural space) of l_0-groupoids.

Let L be a semi-integral l_0-groupoid. Denote by $P(L)$ the set of all prime elements in it.

For $a \in L_r$ denote by $V(a)$ the set of all prime elements containing a .

Proposition 2.1. *The system $\{V(a), a \in L_r\}$ has the following properties:*

(1) $V(0) = P(L)$, where 0 is the smallest element of L_r;

(2) $V(1) = \emptyset$;

(3) $a \le b$ iff $V(b) \le V(a)$;

(4) $V(a \wedge b) = V(a) \cup V(b)$ for any $a, b \in L_r$;

(5) $V(\vee_{\tau\in T} a_\tau) = \cap_{\tau\in T} V(a_\tau)$ for any subset $\{a_\tau, \tau \in T\}$ of L_r ;

(6) $a \ne b$ iff $V(a) \ne V(b)$.

From this proposition we obtain that the system $\{V(a), a \in L_r\}$ is a system of all closed subsets for some topology on the set $P(L)$.

We call it the spectral topology on $P(L)$. This topological space is denoted by $Spec(L)$ and called it the prime spectrum of the l_0 -groupoid L .

Remark. If an l_0 -groupoid L is the l -semigroup of ideals of a commutative associative ring K then $Spec(L)$ is a prime spectrum of the ring K ([10], ch. 1). If K is associative non-commutative then in general $Spec(L)$ is not equal to the set of all primitive ideals with spectral topology ([11], ch.IX). If K is an associative non-commutative ring then $Spec(K)$ is the prime spectrum of K ([12], p.280)

The topological space $Spec(L^*)$ can be defined for any T_0 - lattice L.

Proposition 2.2. *The topological space $Spec(L)$ have following properties:*

(1) for any $x \in P(L)$ $\overline{\{x\}} = V(x)$;

(2) for $x, y \in P(L)$ $y \in \overline{\{x\}}$ iff $V(y) \subseteq V(x)$;

Denote by L_r^* the lattice of all closed subsets in $Spec(L)$. We introduce an order in L_r^* as the inclusion.

Proposition 2.3.

(1) The lattice L_r^* is complete; (2) the map $\varphi : L_r \to L_r^*$, where $\varphi(a) = V(a)$, is an anti-isomorphism of lattices; (3) $V(\wedge_{\tau\in T} a_\tau) = \overline{\cup_{\tau\in T} V(a_\tau)}$ for any subset $\{a_\tau, \tau \in T\}$ of L_r .

Proposition 2.4. *$Spec(L)$ is a topological T_0 -space.*

Let L^* be a T_0 -lattice. We describe topological spaces $Spec(L)$ among all topological T_0 -spaces.

Let X be a topological space. Denote by $l(X)$ the lattice of all open subsets of X .

Proposition 2.5. *If topological spaces X and Y are homeomorphic then the lattices $l(X)$ and $l(Y)$ are isomorphic.*

Remark. This proposition is known [8]. There are non-homeomorphic T_0 -spaces X and Y such that $l(X)$ and $l(Y)$ are isomorphic ([1], ch. IX).

A topological space X is called *balanced* ([9], ch.5, §5) if every $\vee$ -indecomposable closed set of X is a closure of some point of X .

Every topological T_2 -space is balanced. But there is a non-balanced T_1-space. There is a balanced T_1 -space which is not a T_2 -space.

Theorem 2.1. *For any l_0-groupoid L $Spec(L)$ is balanced.*

Proposition 2.6. *Let L_i be a T_0-lattice, $i = 1, 2$. Then topological spaces $Spec(L_1^*)$ and $Spec(L_2^*)$ are homeomorphic iff L_1 and L_2 are isomorphic.*
Let X be a topological T_0 -space. The topological space $Spec(l(X))$ we call the balanced envelope of X. It has the following property:

Any homeomoprhism of a T_0 -space X into a balanced T_0 -space Y has a unique extension to a homeomorphism of $Spec(l(X))$ into Y.

Theorem 2.2. *Let X be T_0-space. Then:*

(1) the map $F(x) = X\backslash\overline{\{x\}}$ is a homeomorphism X of onto some dense subspace of $Spec(l(X))$.

(2) the mentioned homeomorphism is a homeomorphism of X onto $Spec(l(X))$ iff X is balanced.

Remark. From this theorem we obtain the following description of the class of topological spaces $Spec(L)$ among of all topological T_0 -spaces:
Any topological space $Spec(L)$ is balanced. Conversely, any balanced T_0 -space X is naturally homeomorphic to the topological space $Spec(l(X))$.

From the theorem 2.2 we obtain the following corollary too:

Corollary. *Balanced topological T_0 -spaces X and Y are homeomorphic iff the lattices $l(X)$ and $l(Y)$ are isomorphic.*

We formulate two properties of the ordered set of prime elements.

Let L be an l_0 -groupoid and let I be a chain in $P(L)$.Then:

(1) $\cap_{a\in I} a \in P(L)$

(2) for any $a \in P(L)$ there is a minimal element $b \in P(L)$ such that $b \leq a$.

References

[1] Birkhoff G., Lattice theory.- Providence, Rhode Island.-1967

[2] Khadjiev Dj., Shamilev T.M., About conditions of completeness of l - groupoids. Tashkent, Tashkent State University, 1996, 1-6, dep. In UzNI-INTI 27.12.94, 2299 - Uz.94 (in Russian).

[3] Khadjiev Dj., Shamilev T.M., Distributive lattices and T_0 -spaces. - Tashkent, Tashkent State University, 1995, 1-9, dep. In UzNIINTI 07.02.95, N. 2233 - Uz.95 (in Russian).

[4] Khadjiev Dj., About a connection between properties of a ring and its subring of invariants for actions of finite groups. *Dokl. A.N. Resp. Uzbekistan*, 5-6(1995), 6-7 (in Russian).

[5] Khadjiev Dj., Shamilev T.M., Complete l -groupoids and their prime spectrums. *Algebra i logica,* 36:3(1997), 341-355 (in Russian).

[6] Gratzer G., General lattice theory. - Akademie- Verlag, Berlin, 1978.

[7] Andrunakievich V.A., Ryabuhin Y.M., Radicals of algebras and a structural theory. - Moskow, Nauka, 1979 (in Russian).

[8] Drake D., Thron W.J., On the representations of an abstract lattices as the family of closed sets of a topological space. *Trans. Amer. Math. Soc.,* 120:1(1965), 57-71.

[9] Artamonov V.A., Saliy V.N., Skornyakov L.A., Shevrin L.N., Shulgeyfer E.G., Ceneral algebra, V.2. - M, Nauka, 1991.

[10] Atiyah M.F., frs Macdonald I.G., Introduction to commutative algebra. - Addison-Wesley Publishing Company, Reading, Massachusetts, 1969.

[11] Jacobson N., Structure of rings. - AMS, Providence, R.I., 1956.

[12] Rowen L.H., Ring theory, Vol.1, Acad., Press. INC., London, 1988.

E-mail: hadjiev@iaph.silk.glas.apc.org or bobur@simplex.co.uz

SUR UN PROBLEME D'ELIE CARTAN

T.HANGAN

Faculté des Sciences et Techniques,
Université de Haute Alsace
4, rue des Frères Lumière, 68093 Mulhouse cedex
et I.R.M.A.(Unité de recherche associée au CNRS 001),
7 rue René Descartes, 67084 Strasbourg Cedex, France

Abstract A linear complex of lines in the real projective space defines through its focal hyperplanes a totally géodésic distribution of hyperplanes. E. Cartan has shown that conversely, the only non-integrable, totally géodésic distributions of hyperplanes in projective space are of this type. The note gives a new proof of Cartan's result, based on the notion of totally géodesic quasi-connection.

1 Complexe linéaire de droites

Soit $\mathbb{R}P^n = \{[x] = \mathbb{R}x \mid x = (x_0, ..., x_n) \in \mathbb{R}^{n+1} \setminus \{0\}\}$ l'espace projectif réel de dimention n et soit $d_{[x],[y]}$ la droite déterminée par les points $[x]$ et $[y]$ $\in \mathbb{R}P^n$.On note

$$p_{ij} = x_i y_j - x_j y_i \quad ,i,j = 0,1,...,n. \tag{1}$$

les coordonnées plückeriennes de la droite $d_{[x],[y]}$. Elles sont déterminées à un facteur près. On appelle *complexe linéaire de droites*, un ensemble C de droites solutions d'une équation

$$\sum_{i,j=0}^{n} c_{ij} p_{ij} = 0 \tag{2}$$

où $c_{ij} = -c_{ji}$sont des constantes non toutes nulles. Si on fixe le point $[x] \in \mathbb{R}P^n$,

Y. Khakimdjanov et al (eds.), Algebra and Operator Theory, 165-176.

l'équation linéaire en $y_0, ..., y_n$

$$\sum_{i,j=0}^{n} c_{ij}\,(x_i y_j - x_j y_i) = 0 \tag{3}$$

détermine ce qu'on appelle l'*hyperplan focal* $C_{[x]}$ du point $[x]$ par rapport au complexe linéaire (2). Ces équations montrent que $[x] \in C_{[x]}$ et que une droite $d \in C$ appartient à tous les hyperplans focaux de ses points, i.e. $[x] \in d \Rightarrow d \subset C_{[x]}$. On dit aussi que la distribution d'hyperplans$[x] \in \mathbb{R}P^n \mapsto C_{[x]} \in P((\mathbb{R}^{n+1})^*)$ est *totalement géodésique* (abréviation t.g.).

Une définition plus générale s'obtient en considérant l'ensemble $\mathcal{L}_p(\mathbb{R}P^n)$ des variétés linéaires de dimension p de $\mathbb{R}P^n$. On appelle *distribution de variétés linéaires* de dimension p définie dans un ouvert $U \subset \mathbb{R}P^n$, une application $L : U \longrightarrow \mathcal{L}_p(\mathbb{R}P^n)$

$$[x] \in U \mapsto L_{[x]} \in \mathcal{L}_p(\mathbb{R}P^n)$$

telle que pour tout $[x] \in U$ on ait $[x] \in L_{[x]}$. Les droites qui passent par un point $[x]$ et sont contenues dans $L_{[x]}$seront dites de *contact en* $[x]$. On dira alors que la distribution est t.g. si les droites qui sont de contact en un de leur points sont de contact en tous leurs points, i.e.

$$[x] \in d \subset L_{[x]} \Rightarrow d \subset L_{[x']}, \forall [x'] \in d \cap U$$

Etant donnés plusieurs complexes linéaires $C^{(1)}, ..., C^{(q)}$ la distribution de variétés linéaires

$$[x] \in U \subset \mathbb{R}P^n \longmapsto L_{[x]} = C^{(1)}_{[x]} \cap ... \cap C^{(q)}_{[x]}$$

est donc t.g. (U désigne le domaine où les équations (3) des hyperplans focaux sont linéairement indépendantes).

2 Les théorèmes d'Elie Cartan

La notion de distribution t.g. s'étend dans le contexte de la géométrie différentielle aux espaces à connexion affine ou projective ; il suffit de remplacer les variétés linéaires $L_{[x]}$ par des sous espaces linéaires H_x de l'espace $T_x(M)$ tangent en x à l'espace à connexion considéré M et d'exiger que toute géodésique de l'espace soit ou bien partout tangente à la distribution ou bien partout transverse à celle-ci. En plus, on suppose la distribution définie par un système de Pfaff

$$(S) \qquad a^\alpha \equiv \sum_{i=1}^{n} a_i^\alpha(x) dx^i = 0 \qquad , \alpha = 1, ..., q \tag{4}$$

de rang constant q, à coefficients C^∞ ; nous dirons que H est une *distribution différentielle* de l'espace considéré. Dans [1], E. Cartan démontre les deux théorèmes suivants.

Théorème 1 *Dans un ouvert $U \subset \mathbb{R}P^n, n \geq 3$, toute distribution différentielle t.g. H de codimension $q = 1$, non intégrable, provient d'un complexe linéaire de droites C, $(\forall [x] \in U \subset \mathbb{R}P^n,\ H_{[x]} = T_{[x]}(C_{[x]}))$.*

Théorème 2 *Dans un ouvert $U \subset \mathbb{R}P^4$, toute distribution différentielle t.g. H de codimension $q = 2$, définie par un système S qui n'admet aucune combinaison intégrable, provient de deux complexes linéaires C et C' par intersection, $(\forall [x] \in U \subset \mathbb{R}P^n,\ H_{[x]} = T_{[x]}(C_{[x]} \cap C'_{[x]}))$.*

Basé sur ces résultats, E.Cartan propose l'étude des distributions différentielles t.g. dans le cas où *le système S ne possède aucun sous-système complètement intégrable* ; il formule la conjecture selon laquelle la condition imposée à S caractériserait les intersections de complexes linéaires de droites.

Dans ses démonstrations, E.Cartan utilise les équations de structure de l'espace projectif, et la méthode du repère mobile en géométrie projective. Ceci explique le peu d'intérêt suscité par ce problème qui est resté pratiquement au même point jusqu'à ce jour.

3 Quasi-connexion totalement géodésique

Le but de la présente note est d'attaquer le problème des distribution t.g. de l'espace affine par la méthode tensorielle en coordonnées affines; le problème étant à caractère local, en se restreignant à un ouvert affine de l'espace projectif, on ne perd rien à la généralité. En revanche, on gagne en ce qui regarde la géométrisation du problème qui repose sur la notion de quasi-connexion t.g. définie dans la suite.

Soit $H : x \in U \mapsto H_x \subset T_x(U)$ une distribution differentielle de codimension q, pas nécéssairement t.g. définie dans un ouvert $U \subset \mathbb{R}^n$ de l'espace affine $\mathbb{R}^n$. On va noter H° la distribution des annulateurs des espaces H_x ; à $x \in U$ elle associe le sous espace $H_x^\circ \subset (T_x(U))^*$, défini par

$$H_x^\circ = \{f \mid f \in T_x(U)^*, \quad f(H_x) = \{0\}\}$$

Nous allons définir la notion de H-équivalence, noté $\varphi \overset{H}{\sim} \varphi'$, pour deux connexions φ et φ' du fibré H°. Les notations employées seront :

- $\Gamma(H^\circ)$ pour le module des sections $s : U \longrightarrow H^\circ$, $\pi \circ s = id_U$ du fibré $H^\circ \xrightarrow{\pi} U$.

- $\nabla_v^\varphi s$ pour la dérivée covariante par rapport à la connexion φ de la section s dans la direction du vecteur $v \in T_x(U)$.

Définition 1 *Deux connexions φ et φ' du fibré $H^\circ \xrightarrow{\pi} U$ sont dites H-équivalentes si les conditions suivantes sont satisfaites :*

$$\begin{array}{ll} (\mathcal{H}_1) & \nabla_v^\varphi s = \nabla_v^{\varphi'} s \quad \forall v \in H,\ \forall s \in \Gamma(H^\circ) \\ (\mathcal{H}_2) & (\nabla_v^\varphi s)(v) = (\nabla_v^{\varphi'} s)(v) \quad \forall v \in T(U),\ \forall s \in \Gamma(H^\circ). \end{array} \tag{5}$$

Définition 2 *On appelle quasi-connexion t.g. du fibré $H^\circ \xrightarrow{\pi} U$ une famille Φ de connexions deux à deux H-équivalentes.*

Voyons comment s'expriment les conditions (5) en termes de formes de connexion. Supposons pour cela que les 1-formes du système S, voir (4), qui définit le fibré H, constituent un champ de repères pour H° au dessus de U. Une connexion φ du fibré H° au dessus de U sera définie par un champ de 1-formes sur U à valeurs endomorphismes de H° notées

$$\varphi_\beta^\alpha = \sum_{i=1}^{n} \varphi_{\beta i}^\alpha dx^i \quad , \quad \alpha, \beta = 1, ..., q \ ; \tag{6}$$

la différentielle absolue d'une section $s : x \in U \mapsto s(x) = \sum_{\alpha=1}^q s_\alpha(x) a^\alpha$ sera donnée par la formule

$$\nabla^\varphi s = \sum_{\alpha=1}^{q} ds_\alpha a^\alpha + \sum_{\alpha,\beta=1}^{q} s_\alpha \varphi_\beta^\alpha a^\beta = \sum_{\alpha=1}^{q} (ds_\alpha + \sum_{\rho=1}^{q} \varphi_\alpha^\rho s_\rho) a^\alpha \tag{7}$$

Le tenseur δ, différence des connexions φ' et φ, qui s'exprime par la 1-forme matricielle $\delta_\beta^\alpha = \varphi_\beta'^\alpha - \varphi_\beta^\alpha = \delta_{\beta i}^\alpha dx^i$ devra alors satisfaire aux conditions

$$\begin{array}{ll} (\mathcal{H}_1') & \sum_{i=1}^n \delta_{\beta i}^\alpha v^i = 0 \quad , \quad \forall v \in H \\ (\mathcal{H}_2') & \sum_{\beta,i,j} \delta_{\beta i}^\alpha v^i a_j^\beta v^j = 0 \quad \forall v \in T(U) \end{array} \tag{8}$$

et devra donc s'exprimer par la formule

$$\delta_\beta^\alpha = \sum_{i=1}^{n} \delta_{\beta i}^\alpha dx^i = \sum_{\rho=1}^{q} \theta_{\beta\rho}^\alpha a^\rho, \tag{9}$$

où

$$\theta_{\beta\rho}^\alpha = -\theta_{\rho\beta}^\alpha. \tag{10}$$

Les fonctions $\theta_{\beta\rho}^\alpha$, composantes du tenseur différence $\varphi' - \varphi$, seront appelées coefficients de *déviation*.

Les objets différentiels, indépendants de la déviation, qui auront les mêmes expression par rapport à toutes les connexions de la famille Φ seront les invariants différentiels de la quasi-connexion t.g. Φ.

Pour donner un exemple considérons la formule

$$\varphi'^{\alpha}_{\beta} = \varphi^{\alpha}_{\beta} + \delta^{\alpha}_{\beta},$$

qui relie les deux connexions H-équivalentes φ et φ' et soit

$$r'^{\alpha}_{\beta} = d\varphi'^{\alpha}_{\beta} + \sum_{\sigma=1}^{q} \varphi'^{\sigma}_{\beta} \wedge \varphi'^{\alpha}_{\sigma} \quad , \quad r^{\alpha}_{\beta} = d\varphi^{\alpha}_{\beta} + \sum_{\sigma=1}^{q} \varphi^{\sigma}_{\beta} \wedge \varphi^{\alpha}_{\sigma} \tag{11}$$

leur forme de courbure. On a

$$r'^{\alpha}_{\beta} = r^{\alpha}_{\beta} + D^{\varphi}\delta^{\alpha}_{\beta} + \delta^{\sigma}_{\beta} \wedge \delta^{\alpha}_{\sigma} \tag{12}$$

où D^{φ}désigne le symbole de différentiation absolue par rapport à la connexion φ donc

$$D^{\varphi}\delta^{\alpha}_{\beta} = d\delta^{\alpha}_{\beta} - \varphi^{\alpha}_{\sigma} \wedge \delta^{\sigma}_{\beta} + \varphi^{\sigma}_{\beta} \wedge \delta^{\alpha}_{\sigma} \tag{13}$$

En tenant compte de (9) on peut aussi écrire

$$r'^{\alpha}_{\beta} = r^{\alpha}_{\beta} + \sum_{\rho=1}^{q} \left(D^{\varphi}\theta^{\alpha}_{\beta\rho}\right) \wedge a^{\rho} + \sum_{\rho=1}^{q} \theta^{\alpha}_{\beta\rho} D^{\varphi} a^{\rho} + \delta^{\sigma}_{\beta} \wedge \delta^{\alpha}_{\sigma}. \tag{14}$$

Soit $p = u \wedge v$ un bivecteur décomposable où $u, v \in H_x$ sont deux vecteurs qui appartiennent à la distribution H au point $x \in U$; leur produits intérieurs avec les formes a^{ρ} et δ^{α}_{β} sont nuls. Quant au produits intérieurs de p avec les 2-formes $D^{\varphi}a^{\alpha}$, $\alpha = 1, ..., q$, ils représentent les composantes $I^{\alpha}(p)$du tenseur d'intégrabilité I_H de la distribution H. Vu l'antisymétrie des $\theta^{\alpha}_{\beta\rho}$, formule (10),on en déduit l'existence d'un tenseur

$$\sum_{\beta=1}^{q} r'^{\alpha}_{\beta}(p) I^{\beta}(p) = \sum_{\beta=1}^{q} r^{\alpha}_{\beta}(p) I^{\beta}(p) \quad , \quad p = u \wedge v, \tag{15}$$

qui est le même pour toute les connexions de la famille Φ.

4 Quasi-connexion associée à une distribution t.g. H

Soit H une distribution différentielle t.g. définie par le systèmes S dans un ouvert U de l'espace affine $\mathbb{R}^n$. Le vecteur $v \in T_x(U)$ appartient donc à $H_x \subset T_x(U)$ si

$$\sum_{i=1}^{n} a^{\alpha}_{i}(x) v^{i} = 0 \qquad , \alpha = 1, ..., q. \tag{16}$$

La distribution H étant supposée t.g., la droite $t \in \mathbb{R} \longmapsto x + tv \in \mathbb{R}^n$ qui est tangente en x à H est tangente en tous ses points à H et donc le système

$$\sum_{i=1}^{n} a_i^\alpha(x+tv)v^i = 0 \qquad , \alpha = 1, ..., q \tag{17}$$

est une conséquence de (16). La première dérivée de (17) par rapport à t donne

$$\sum_{i,j=1}^{q} \frac{\partial a_i^\alpha}{\partial x^j} v^i v^j = 0 \qquad , \alpha = 1, ..., q. \tag{18}$$

Afin que ces équations soient une conséquence de (16), il faut qu'il existe un systèmes de nq^2 fonctions $\varphi_{\beta i}^\alpha$ de façon qu'on ait

$$\frac{\partial a_i^\alpha}{\partial x_j} + \frac{\partial a_j^\alpha}{\partial x_i} = \sum_{\rho=1}^{q} \varphi_{\rho i}^\alpha a_j^\rho + \sum_{\rho=1}^{q} \varphi_{\rho j}^\alpha a_i^\rho \qquad , \alpha = 1, ..., q. \tag{19}$$

Tous les systèmes déduits de (17) en dérivant successivement par rapport à t se déduisent alors de (19) qui représente l'unique condition infinitésimale de totale géodicité.

Le système (16)&(18) ne détermine pas de façon unique les $\varphi_{\beta j}^\alpha$; si $\varphi_{\beta j}'^\alpha$ est une autre solution de (19) on a

$$\varphi_{\beta j}'^\alpha = \varphi_{\beta j}^\alpha + \sum_{i=1}^{q} \lambda_{\beta\rho}^\alpha a_i^\rho \quad , \quad \alpha, \beta = 1, ..., q, \quad i = 1, ..., n \tag{20}$$

où $\lambda_{\beta\rho}^\alpha = -\lambda_{\rho\beta}^\alpha$ sont $q^2(q-1)/2$ fonctions arbitraires.

La forme matricielle

$$\varphi_\beta^\alpha = \sum_{i=1}^{n} \varphi_{\beta i}^\alpha dx^i \tag{21}$$

est donc bien définie modulo S, car (20) devient

$$\varphi_\beta'^\alpha = \varphi_\beta^\alpha + \sum_{i=1}^{q} \lambda_{\beta\rho}^\alpha a^\rho \quad , \quad \lambda_{\beta\rho}^\alpha + \lambda_{\rho\beta}^\alpha = 0. \tag{22}$$

Ce résultat est exprimé par la

Proposition 1 *A la distribution t.g. H définie dans un ouvert $U \subset \mathbb{R}^n$ par le système de Pfaff S est associée une famille Φ_S de 1-formes φ sur U à valeurs endomorphismes du fibré H° définies (mod S) de façon unique mais qui diffèrent deux à deux en accord avec la formule (22).*

Observons que pour un complexe linéaire de droites C d'équation (2), l'équation de Pfaff correspondante est (on pose $x_0 = 1$)

$$2\sum_{i,j=1}^{n} c_{ij}x^i dx^j + 2\sum_{j=1}^{n} c_{0j}dx^j = 2\sum_{j=1}^{n}\left(\sum_{i=1}^{n} c_{ij}x^i + c_{0j}\right)dx^j = 0,$$

et le premier membre de (19)

$$\frac{\partial a_i}{\partial x^j} + \frac{\partial a_j}{\partial x^i} = c_{ji} + c_{ij} = 0$$

est identiquement nul. Cela veut dire que si H provient par intersection d'un système de q complexes linéaires de droites, il en existe un système S' équivalent à S, pour lequel la famille $\Phi_{S'}$ contient la 1-forme matricielle identiquement nulle; nous dirons dans ce cas que la distribution H est *plate*.

5 Invariance par rapport au changement de repère S

Les formules

$$(S') \qquad a^{\alpha'} = \sum_{\rho=1}^{q} c_\rho^{\alpha'} a^\rho \quad , \alpha = 1, 2 ..., q \tag{23}$$

où $(c_\beta^{\alpha'})$ est une matrice variable, non singulière, définie dans $U \subset \mathbb{R}^n$ définissent un système de formes de Pfaff S' équivalent à S. Entre deux systèmes Φ_S et $\Phi_{S'}$ de formes matricielles associées à la distribution t.g. H, subsiste alors la relation

$$\sum_{\sigma=1}^{q} \varphi_{\sigma'}^{\alpha'} c_\beta^{\sigma'} = \sum_{\tau=1}^{q} c_\tau^{\alpha'} \varphi_\beta^\tau + dc_\beta^{\alpha'} + \sum_{\tau=1}^{q} \theta_{\beta\tau}^{\alpha'} a^\tau \tag{24}$$

où $\theta_{\beta\tau}^{\alpha'} = -\theta_{\tau\beta}^{\alpha'}$ sont $q^2(q-1)/2$ fonctions arbitraires ; elles représentent les coefficients de déviation associés aux deux connexions φ et φ' H-équivalentes.

Décider si H est plate revient à chercher dans quelles conditions, à partir d'une forme matricielle $(\varphi_\beta^\alpha) \in \Phi_S$ on peut trouver une solution des équations (24) où $\varphi_{\sigma'}^{\alpha'} \equiv 0$.

Les inconnues du problèmes sont donc les composantes de la matrice de passage $(c_\beta^{\alpha'})$ et les déviations $\theta_{\beta\gamma}^{\alpha'}$.

Les équations (24) donnent par différentiation extérieure le système

$$\begin{aligned} &c_\beta^{\tau'}(d\varphi_{\tau'}^{\alpha'} + \varphi_{\tau'}^{\sigma'} \wedge \varphi_{\sigma'}^{\alpha'}) - c_\tau^{\alpha'}(d\varphi_\beta^\tau + \varphi_\beta^\sigma \wedge \varphi_\sigma^\tau) \\ &= \theta_{\beta\tau}^{\alpha'} da^\tau + d\theta_{\beta\tau}^{\alpha'} \wedge a^\tau + \theta_{\beta\tau}^{\sigma'} a^\tau \wedge \varphi_{\sigma'}^{\alpha'} - \theta_{\rho\tau}^{\alpha'} a^\tau \wedge \varphi_\beta^\rho. \end{aligned} \tag{25}$$

Les restrictions à $H \wedge H$ des trois derniers termes sont nulles ; on obtient les formules plus simples

$$c_\beta^{\tau'}(d\varphi_{\tau'}^{\alpha'} + \varphi_{\tau'}^{\sigma'} \wedge \varphi_{\sigma'}^{\alpha'})_{|H\wedge H} - c_\tau^{\alpha'}(d\varphi_\beta^\tau + \varphi_\beta^\sigma \wedge \varphi_\sigma^\tau)_{|H\wedge H} = \theta_{\beta\tau}^{\alpha'} da^\tau_{|H\wedge H} \tag{26}$$

qui prouvent que le tenseur d'intégrabilité I_H de la distribution H qui s'exprime par les q 2-formes $da^{\alpha}_{|H\wedge H}$, empêche les 2-formes matricielles

$$r^{\alpha}_{\beta|H\wedge H} = \left(d\varphi^{\alpha}_{\beta} + \varphi^{\sigma}_{\beta} \wedge \varphi^{\alpha}_{\sigma}\right)_{|H\wedge H} \tag{27}$$

de représenter un vrai tenseur de courbure.

En conclusion nous avons trouvé que la distribution differentielle t.g. H définit à travers les équations (19) une quasi-connexion t.g. notée désormais Φ^H.

6 L'équation fondamentale de la quasi-connexion Φ^H

Pour les calculs qui vont suivre, il est convenable d'utiliser des dérivées covariantes. Ainsi, soit $\mathcal{T}^h_k(T\mathbb{R}^n)\otimes\mathcal{T}^l_m(H^\circ)$ le fibré vectoriel au dessus de l'ouvert $U \subset \mathbb{R}^n$ dont la fibre en x est l'espace $\mathcal{T}^h_k(T_x\mathbb{R}^n) \otimes \mathcal{T}^l_m(H^\circ_x)$ où $\mathcal{T}^h_k(V)$ indique l'espace des tenseurs h fois contravariants et k fois covariants de l'espace vectoriel V. Les sections de cet espace fibré seront dérivées en utilisant la connexion affine plate de $\mathbb{R}^n$ en ce qui regarde le facteur $\mathcal{T}^h_k(T\mathbb{R}^n)$, et une connexion fixée $\varphi \in \Phi^H$ en ce qui concerne le facteur $\mathcal{T}^l_m(H^\circ)$; l'opérateur de différentiation sera noté D^φ et la dérivée en direction du vecteur $v \in T_x\mathbb{R}^n$ sera notée D^φ_v. Avec ces conventions, le système (19) devient

$$(D^\varphi a)(u,v) + (D^\varphi a)(v,u) = 0, \quad u,v \in T_xU \tag{28}$$

ou bien

$$(D^\varphi_v a)(u) + (D^\varphi_u a)(v) = 0, \quad u,v \in T_xU \tag{29}$$

ou bien

$$a^{\alpha}_{i|j} + a^{\alpha}_{j|i} = 0 \quad , i,j = 1,...,n; \alpha = 1,...,q \tag{30}$$

où

$$(D^\varphi_v a)^{\alpha}_{ij} = a^{\alpha}_{i|j} := \frac{\partial a^{\alpha}_i}{\partial x_j} - \sum_{\rho=1}^{q} \varphi^{\alpha}_{\rho j} a^{\rho}_i.$$

L'équation fondamentale que nous allons établir par la suite est

$$3\sum_{\rho=1}^{q} r^{\alpha}_{\rho}(u,v) D^\varphi a^\rho(u,v) - \sum_{\rho=1}^{q} (D^\varphi_u r^{\alpha}_{\rho})(u,v) a^\rho(v) + \sum_{\rho=1}^{q} (D^\varphi_v r^{\alpha}_{\rho})(u,v) a^\rho(u) = 0, \tag{31}$$

u,v étant des vecteurs arbitraires tangents à $\mathbb{R}^n$ et $\varphi \in \Phi^H$.

Pour cela considérons la dérivée covariante des équations (30) en direction du champ $\frac{\partial}{\partial x^k}$; on obtient

$$a^{\alpha}_{i|j|k} + a^{\alpha}_{j|i|k} = 0. \tag{32}$$

En permutant cycliquement les indices i, j, k, on écrit encore deux équations

$$a^{\alpha}_{j|k|i} + a^{\alpha}_{k|j|i} = 0 \quad , \quad a^{\alpha}_{k|i|j} + a^{\alpha}_{i|k|j} = 0. \tag{33}$$

Si on retranche la dernière équation de la somme des deux équations qui la précèdent et si on tient compte de la formule de commutation des dérivées covariantes d'ordre deux

$$a^{\alpha}_{j|k|i} - a^{\alpha}_{j|i|k} = -r^{\alpha}_{\rho ik} a^{\rho}_{j} \tag{34}$$

on obtient la formule

$$2a^{\alpha}_{j|k|i} = +a^{\rho}_{j} r^{\alpha}_{\rho ki} - a^{\rho}_{k} r^{\alpha}_{\rho ji} - a^{\rho}_{i} r^{\alpha}_{\rho jk} \tag{35}$$

qui exprime les dérivées covariantes du deuxième ordre des formes a^{α} en fonction des même formes et de la "quasi-courbure r". Par dérivation covariante des formules (35) par rapport à la variable x^l suivie de la permutation des indices i et l et soustraction des résultats, on obtient finalement la formule

$$\begin{aligned} &2r^{\alpha}_{\rho li} a^{\rho}_{j|k} + 2r^{\alpha}_{\rho jk} a^{\rho}_{l|i} + r^{\alpha}_{\rho ji} a^{\rho}_{l|k} + r^{\alpha}_{\rho ik} a^{\rho}_{l|j} + r^{\alpha}_{\rho lk} a^{\rho}_{j|i} + r^{\alpha}_{\rho lj} a^{\rho}_{i|k} = \\ &a^{\rho}_{i} r^{\alpha}_{\rho jk|l} + a^{\rho}_{l} r^{\alpha}_{\rho kj|i} + a^{\rho}_{j} r^{\alpha}_{\rho il|k} + a^{\rho}_{k} r^{\alpha}_{\rho li|j} \end{aligned} \tag{36}$$

où on a posé $r^{\alpha}_{\rho jk|l} := r^{\alpha}_{\rho jk,l} - \varphi^{\alpha}_{\sigma l} r^{\sigma}_{\rho jk} + \varphi^{\sigma}_{\rho l} r^{\alpha}_{\sigma jk}$. Cette dernière équation est équivalente à (31) dont elle provient par polarisation.

La formule fondamentale a une conséquence intéressante, déduite en considérant les vecteurs u, v dans (31) tangents à la distribution H ; le tenseur $r(p)(I_H(p))$,voir formule (15), est nul si la distribution H est t.g. et $\varphi \in \Phi^H$.

7 Démonstration du théorème 1

Si $q = 1$, l'unique indice grec α, β,...sera omis ; la formule (36) dont on désignera par le symbole$[li, jk]$ son premier membre devient

$$\begin{aligned} [li, jk] \equiv\ & 2r_{li} a_{j|k} + 2r_{jk} a_{l|i} + r_{ji} a_{l|k} + r_{ik} a_{l|j} + r_{lk} a_{j|i} \\ & + r_{lj} a_{i|k} + a_j r_{li,k} - a_k r_{li,j} + a_l r_{jk,i} - a_i r_{jk,l} = 0. \end{aligned} \tag{37}$$

L'hypothèse du théorème 1 - la distribution H n'est pas intégrable - s'exprimes en termes de la forme $a = a^1$ par

$$a \wedge da \neq 0. \tag{38}$$

On peut donc supposer que le coefficient $(a \wedge da)_{123}$ du produit $dx^1 \wedge dx^2 \wedge dx^3$ dans la 3-forme $a \wedge da$ est différent de zéro

$$\begin{aligned} (a \wedge da)_{123} = a_1(a_{2,3} - a_{3,2}) + a_2(a_{3,1} - a_{1,3}) + a_3(a_{1,2} - a_{2,1}) = \\ 2[a_1 a_{2|3} + a_2 a_{3|1} + a_3 a_{1|2}] \neq 0. \end{aligned} \tag{39}$$

La démonstration se décompose en 5 étapes, notées E_1,...,E_5 et utilise le système suivant déduit de (37)

$$[li, 123] \equiv [li, 12]a_3 + [li, 23]a_1 + [li, 31]a_2 = 0. \tag{40}$$

En regroupant les termes on trouve

$$\begin{aligned}[li, 123] \equiv\ & 2r_{li}(a_{1|2}a_3 + a_{2|3}a_1 + a_{3|1}a_2) + 2a_{l|i}(r_{12}a_3 + r_{23}a_1 + r_{31}a_2) \\ & +a_3(r_{l2}a_{1|i} + r_{l1}a_{i|2} + r_{1i}a_{l|2} + r_{2i}a_{1|l} + a_l r_{12,i} - a_i r_{12,l}) \\ & +a_1(r_{l3}a_{2|i} + r_{l2}a_{i|3} + r_{2i}a_{l|3} + r_{3i}a_{2|l} + a_l r_{23,i} - a_i r_{23,l}) \\ & +a_2(r_{l1}a_{3|i} + r_{l3}a_{i|1} + r_{3i}a_{l|1} + r_{1i}a_{3|l} + a_l r_{31,i} - a_i r_{31,l}) = 0\end{aligned} \tag{41}$$

E_1.On démontre la formule

$$(r \wedge a)_{123} \equiv r_{12}a_3 + r_{23}a_1 + r_{31}a_2 = 0 \tag{42}$$

Pour la preuve on utilise l'équation (41) pour $l = 1$ et $i = 2$

$$\begin{aligned}[12, 123] \equiv\ & 2r_{12}(a_{1|2}a_3 + a_{2|3}a_1 + a_{3|1}a_2) + 2a_{1|2}(r_{12}a_3 + r_{23}a_1 + r_{31}a_2) \\ & +a_3(2r_{12}a_{1|2} + a_1 r_{12,2} - a_2 r_{12,1}) \\ & +a_1(r_{12}a_{2|3} + r_{32}a_{2|1} + a_1 r_{23,2} - a_2 r_{23,1}) \\ & +a_2(r_{13}a_{2|1} + r_{12}a_{3|1} + a_1 r_{31,2} - a_2 r_{31,1}) = 0.\end{aligned} \tag{43}$$

Une nouvelle combinaison

$$\begin{aligned}& a_3[12, 123] + a_1[23, 231] + a_2[31, 312] \equiv \\ & 6(r_{12}a_3 + r_{23}a_1 + r_{31}a2)(a_{1|2}a_3 + a_{2|3}a_1 + a_{3|1}a_2) = 0\end{aligned}$$

donne le résultat annoncé (42).

E_2. On démontre que

$$r_{12} = r_{23} = r_{31} = 0 \tag{44}$$

En effet en dérivant l'équation (42) par rapport à la variable x^i, on a

$$\begin{aligned}& r_{12,i}a_3 + r_{23,i}a_1 + r_{31,i}a_3 = \\ & -\left(r_{12}(a_{3|i} + \varphi_i a_3) + r_{23}(a_{1|i} + \varphi_i a_1) + r_{31}(a_{2|i} + \varphi_i a_2)\right).\end{aligned} \tag{45}$$

On peut alors éliminer les dérivées $r_{12,2}$...de l'équation (43) qui se réduit à

$$r_{12}(a_{1|2}a_3 + a_{2|3}a_1 + a_{3|1}a_2) = 0$$

d'où (44).

E_3 On démontre la formule

$$r_{i[1}a_{2|3]} \equiv r_{i1}a_{2|3} + r_{i2}a_{3|1} + r_{i3}a_{1|2} = 0 \quad , i > 3. \tag{46}$$

Suite à (42) et (45), l'équation (41) se simplifie et devient

$$\begin{aligned}[li,123]' \equiv\ & 2r_{li}(a_{1|2}a_3 + a_{2|3}a_1 + a_{3|1}a_2) \\ & + a_3(r_{l2}a_{1|i} + r_{l1}a_{i|2} + r_{1i}a_{l|2} + r_{2i}a_{1|l}) \\ & + a_1(r_{l3}a_{2|i} + r_{l2}a_{i|3} + r_{2i}a_{l|3} + r_{3i}a_{2|l}) \\ & + a_2(r_{l1}a_{3|i} + r_{l3}a_{i|1} + r_{3i}a_{l|1} + r_{1i}a_{3|l}) = 0.\end{aligned} \tag{47}$$

Pour $l = 1$ on trouve

$$3r_{1i}(a_{1|2}a_3 + a_{2|3}a_1 + a_{3|1}a_2) - a_1(r_{1i}a_{2|3} + r_{2i}a_{3|1} + r_{3i}a_{1|2}) = 0. \tag{48}$$

En multipliant le premier membre par $a_{2|3}$, en permutant cycliquement les indice $1, 2, 3$ et en faisant la somme des résultats, on trouve l'équation

$$(a_1a_{2|3} + a_2a_{3|1} + a_3a_{1|2})(r_{1i}a_{2|3} + r_{2i}a_{3|1} + r_{3i}a_{1|2}) = 0$$

qui démontre (46).

E_4. On démontre les équations

$$r_{1i} = r_{2i} = r_{3i} = 0 \quad , i > 3. \tag{49}$$

Ces équations résultent de (48)

E_5. Des équations (49) et (47) on déduit enfin les équations

$$r_{li} = 0 \quad , \quad i, l > 3. \tag{50}$$

Le théorème 1 résulte alors du fait que pour $q = 1$, la quasi-connexion associée à la distribution H est une vraie connexion pour le fibré normal $T(U)/H = H^{\perp}$,le dual du fibré H°, et que l'annulation du tenseur de courbure r représente la condition de platitude de la connexion φ.

Autrement dit, des équations (44), (49), (50) on déduit que la 1-forme $\varphi = \varphi_1^1$ est fermée ; on peut alors écrire localement $\varphi = df$ et en choisissant $c_1^{1'} = (\exp f)^{-1}$ on déduit de l'équation (24) que $\varphi' = 0$.

Des démonstrations différentes du théorème 1 ont déjà été données dans [3] et [4]. Dans [2] on trouvera des commentaires de géométrie différentielle globale au sujet de l'article 1.

References

[1] E. Cartan, Sur un problème de géométrie différentielle projective, *Ann. Ec. Norm.* 62(1945).

[2] Th. Hangan, R. Lutz, Champs d'hyperplans totalement géodésiques sur les sphères, *Soc. Math. France, Asterisque* 107-108(1983), 189-200.

[3] Th. Hangan, On Totally Geodesic Distribution of Planes, Coll. Math. *Soc. János Bolyai,* 46, Debrecen 1984.

[4] Th. Hangan, Sur les distribution de plans totalement géodésiques, *Publicationes de la Universidad de Murcia,* 1983.

E-mail: T.Hangan@univ-mulhouse.fr

CLASSIFICATION OF NON-COMMUTATIVE ARENS ALGEBRAS ASSOCIATED WITH SEMI-FINITE TRACES

R.Z.ABDULLAEV

Institute of Mathematics,
Uzbekistan Academy of Sciences
F. Hodjaev str., 29, 700143 Tashkent, Uzbekistan.

Let (Ω, Σ, μ) be a measurable space with a finite measure, $L^p(\mu) = L^p(\Omega, \Sigma, \mu)$ be the Banach space of all μ - measurable complex functions on Ω, the pth degrees of which are integrable $p \in [1, \infty)$. P.Arens had introduced and studied the set $L^w(\mu) = \bigcap_{1 \le p < \infty} L^p(\mu)$. It was demonstrated, in particular, that $L^w(\mu)$ is a metrizable locally convex $\star$ - algebra with respect to the topology t generated by the system of norms $||f||_p = (\int_\Omega |f|^p d\mu)^{1/p}, p \ge 1$. The additional study of the Arens algebras $L^w(\mu)$ was made by S.J.Bhaft [2,3] who has described the algebras $L^w(\mu)$ and examined some classes of homomorphism of these algebras. B.Z. Zakirov [4] showed that $L^w(\mu)$ is a EW^* algebra and gave an example of two measures, μ and ν, on an atomic Boolean algebra for which the algebras $L^w(\mu)$ and $L^w(\nu)$ are not isomorphic. The Arens algebras associated with the non-commutative von Neumann algebras were examined for the first time in the works [4,5].

The present article gives a classification of Arens algebras, associated with a semi-finite von Neumann algebra. It is proved that the isomorphism of the Arens algebras associated with the centre of von Neumann algebra implies the isomorphism of the Arens algebras constructed on the whole von Neumann algebra.

Y. Khakimdjanov et al (eds.), Algebra and Operator Theory, 177-181.

Let M be an arbitrary von Neumann algebra, μ be a faithful normal semi-finite trace on M. Denote the $\star$ - algebra of all μ - measurable operators affilated with M as $K(M,\mu)$ [6]. In the case when M is a commutative von Neumann algebra, i.e. when M is identified with $L^\infty(\Omega,\Sigma,\mu)$, for some space (Ω,Σ,μ) and $\mu(x) = \int\limits_\Omega x d\mu$, $x \in L^\infty(\Omega,\Sigma,\mu)$, the $\star$ - algebra $K(M,\mu)$ may be identified with the $\star$ - algebra of all measurable complex functions on Ω.

Let $p \in [1,\infty)$ and $L^p(M,\mu) = \{x \in K(M,\mu)\ \mu(|x|^p) < \infty\}$, where $|x| = (x^*x)^{1/2}$. The set $L^p(M,\mu)$ is a subspace in $K(M,\mu)$ and the function $||x||_p = (\mu(|x|^p))^{1/p}$ is a Banach norm on $L^p(M,\mu)$ [6].

In [5] it was shown that the set $L^w(M,\mu) = \bigcap\limits_{p\geq 1} L^p(M,\mu)$ is a metrizable locally convex $\star$ - algebra in the topology generated by the system of norms $\{||\cdot||_p\}_{p\geq 1}$. The algebras of this kind were called non-commutative Arens algebras.

In [5] it was also demonstrated that the dual $L^w(M,\mu)^*$ of $L^w(M,\mu)$ may be identified with the space $Lin \bigcap\limits_{1<p\leq\infty} L^p(M,\mu)$, which we denote by $L_0(M,\mu)$. Here $L^\infty(M,\mu)$ is identified with M.

Let μ and ν be faithful normal semi - finite traces on a von Neumann algebra M. We shall denote by $h = \frac{d\mu}{d\nu}$ the Radon - Nikodym derivate of the trace μ with respect to the trace ν, i.e. h is a positive operator from the centre of the algebra $K(M,\mu)$ for which the equality $\mu(x) = \nu(hx)$ is implemented in all $x \in M$.

The set of all $\star$ - automorphisms α of the von Neumann algebra M satisfying the conditions

$$\frac{d\mu}{dv\circ\alpha^{-1}} \in L_0(M, v\circ\alpha^{-1})$$

and

$$\frac{dv\circ\alpha^{-1}}{d\mu} \in L_0(M,\mu)$$

is denoted as $AutM(\mu;v)$.

Now we can formulate the criterion of $\star$ - isomorphism of the noncommutative Arens algebras.

Theorem 1. *Let μ and ν be faithful normal semi-finite traces on a von Neumann algebra M. The following conditions are equivalent:*

1) the Arens algebras $L^w(M,\nu)$ and $L^w(M,\mu)$ are $\star$ - isomorphic.

2) Aut $M(\mu,\nu) \neq \emptyset$.

When examining the isomorphisms of the Boolean algebras with the finite measure in [8], the notion of "passport" for normalised Boolean algebra's was used.

To introduce the notion of the passport for a Boolean algebra with σ - finite measure we need the definition of the homogeneous Boolean algebra and its weight from [8].

Let X be an arbitrary complete Boolean algebra, $e \in X$, $X_e = [0,e] = \{g \in X : g \leq e\}$. Denote by $\tau(X_e)$ the minimal cardinality of the set dence in X_e in the (o) - topology. The infinite complete Boolean algebra X is called homogeneous if $\tau(X_e) = \tau(X_g)$ for any non-zero $e, g \in X$. The cardinality $\tau(X) = \tau(X_{\mathbf{1}})$, where $\mathbf{1}$ is a unit of the Boolean algebra X, is called the weight of the homogeneous Boolean algebra X.

Let μ be a σ - finite essentially positive measure on the Boolean algebra X. Then there is also the positive quasimeasure. Therefore (according to [8], pp.56, 237), X is a Boolean algebra of the countable type and its decomposition into homogeneous components is at most countable. Let $\{X_{s_i}\}$ be the homogeneous components of the Boolean algebra X for which $\tau_{s_i} = \tau(x_{s_i}) < \tau_{s_{i+1}}$, $\mu(s_i) = \infty$, and $\{X_{n_i}\}$ be the homogeneous components of X for which $\tau_{u_i} = \tau(X_{u_i}) < \tau_{u_{i+1}}$, $\mu_i = \mu(u_i) < \infty$, $i = 1, 2, \ldots$.

Then the matrix

$$\begin{pmatrix} \tau_{s_1} & \tau_{s_2} & \ldots \\ \tau_{u_1} & \tau_{u_2} & \ldots \\ \mu_1 & \mu_2 & \ldots \end{pmatrix},$$

is called the passport of the Boolean algebra X with σ - finite measure μ.

In the next theorem the problem of the $\star$ - isomorphism of commutative Arens algebras is reduced to the verification of convergence of the corresponding number series.

Theorem 2. *Let μ and ν be faithful normal σ - finite traces on a non - atomic commutative von Neumann algebra M and let (X,μ) and (X,ν) be the Boolean algebras of the projections of the von Neumann algebra M with σ - finite measures μ and ν and with the passports*

$$\begin{pmatrix} \tau_{s_1} & \tau_{s_2} & \ldots \\ \tau_{u_1} & \tau_{u_2} & \ldots \\ \mu_1 & \mu_2 & \ldots \end{pmatrix} \quad \text{and} \quad \begin{pmatrix} \tau_{t_1} & \tau_{t_2} & \ldots \\ \tau_{v_1} & \tau_{v_2} & \ldots \\ \nu_1 & \nu_2 & \ldots \end{pmatrix}$$

respectively. Then the following conditions are equivalent:

(i) The Arens algebras $L^w(M,\mu)$ and $L^w(M,\nu)$ are $\star$ - isomorphic;

(ii) there are numbers $1 < p_1 < p_2 \leq \infty$, $1 < q_1 < q_2 \leq \infty$ such that

$$\sum_{k=1}^{\infty}(\mu_{n_k}^{p_1}\nu_{n_k}^{1-p_1}+\mu_{z_k}^{p_2}\nu_{z_k}^{1-p_2})<\infty,$$
$$\sum_{k=1}^{\infty}(\nu_{m_k}^{q_1}\mu_{m_k}^{1-q_1}+\nu_{r_k}^{q_2}\mu_{r_k}^{1-q_2})<\infty,$$

where

$$\{n_k\}\cap\{z_k\}=\emptyset,\quad \{n_k\}\cup\{z_k\}=\{1,2,...\}$$

$$\{m_k\}\cap\{r_k\}=\emptyset,\quad \{m_k\}\cup\{r_k\}=\{1,2,...\}$$

in the case $p_2\neq\infty$, $q_2\neq\infty$ *and*

$$\sum_{k=1}^{\infty}(\mu_{n_k}^{p_1}\nu_{n_k}^{1-p_1})+\sup(\mu_{z_k}\nu_{z_k}^{-1})<\infty \text{ if } p_2=\infty,$$
$$\sum_{k=1}^{\infty}(\nu_{m_k}^{q_1}\mu_{m_k}^{1-q_1})+\sup(\nu_{r_k}^{q_2}\mu_{r_k}^{-1})<\infty \text{ if } q_2=\infty.$$

Let now μ and ν be faithful normal semi-finite traces on a von Neumann algebra M. Denote by Z_M the centre of the von Neumann algebra M and by μ' and ν' the restrictions of the traces μ and ν on the centre Z_M respectively.

Proposition. *If the Arens algebras $L^w(M,\mu)$ and $L^w(N,\nu)$ are $\star$ - isomorphic, then the Arens algebras $L^w(Z_M,\mu')$ and $L^w(Z_M,\nu')$ are also $\star$ - isomorphic.*

The converse statement is not always true.

The following theorem can be obtained by using the theorem 1 and the equality $\frac{d\mu}{d\nu}=\frac{d\mu'}{d\nu'}$.

Theorem 3. *The $\star$ - isomorphism of the Arens algebras $L^w(Z_M,\mu')$ and $L^w(Z_M,\nu')$ implies the $\star$ - isomorphism of the Arens algebras $L^w(M,\mu)$ and $L^w(M,\nu)$ if and only if, at least one $\star$ - isomorphism from Aut $Z_M(\mu',\nu')$ extends to a $\star$ - automorphism on M.*

Thus, the problem of isomorphism of noncommutative Arens algebras is "often" reduced to the problem of $\star$ - isomorphism of the corresponding commutative Arens algebras associated with the centre ($\star$ - isomorphism of these algebras has been examined in the theorem 2).

References

[1] Arens R. The space L^w and the convex topological rings. *Bull. Amer. Math. Soc.*, 52(1946), 931-935.

[2] Bhaft S.J. On Arens algebras L^w. *Glas. Math.*, 15:2(1980), 305-312.

[3] Bhaft S.J. On Arens algebras L^w. *II Glas. Math.*, 16:2(1981), 297-306.

[4] Zakirov B.C. Non-commutative Arens algebras (Russian). *Uzbekski Matematicheski Jurnal,* 1(1997), 17-24.

[5] Abdullaev R.Z. Dual of non-commutative Arens algebras. *Uzbekski Matematicheski Jurnal,* 2(1997), 3-7.

[6] Yeadon F.J. Non-commutative L^p - space. *Math. Proc. Cambridge. Phil.Soc.*, 77:1(1995), 91-102.

[7] Schaeffer H.H. Topological vector spaces. Mc. Millan. London. 1966.

[8] Vladimirov D.A. The Boolean algebras (Russian). "Nauka". Moscow. 1969.

ORDER UNIT SPACE OF TYPE I_n WITH BANACH BALL PROPERTY

M.A.BERDIKULOV

Institute of Mathematics,
Uzbekistan Academy of Sciences
F. Hodjaev str., 29, 700143 Tashkent, Uzbekistan.

Alfsen and Shultz have proved in [1] that the state space of a JB-algebra has the Hilbert ball property: for each pair ρ, σ of extreme points, the face generated by ρ and σ is a norm-exposed face affinely isomorphic to the closed unit ball in some Hilbert space ([1], corollary 3.12). Conversely, they have proved, that the order unit space A being in spectral duality with its predual space is a JB-algebra if the state space of A has the Hilbert ball property ([1],theorem 7.2).

In this paper, we define one class of order unit spaces of type I_2-generalized spin factors, the state space of which is isomorphic to a smooth and strictly convex closed unit ball in some Banach space.

We also define Banach ball property for order unit spaces and study order unit spaces, the state space of which has the Banach ball property.

Let (A, e) be an order unit space which is in the spectral duality with a base-norm space (V, K) and $V^* = A$. In this case we can consider the base K, as the state space of A.

The sets of all P-projections on A and all projective units in A are denoted by $\mathcal{P}$ and U, respectively. We recall (see [1]) that a P-projection R is *central* if $R + R' = I$, where I is the identity mapping. We say that an order unit space A is a *factor* (under the given duality with V), if it contains no central P-projection except O and I . We say that the projective unit $u \in U$ is an *atom* if u a minimal (non-zero) element of the logic U. A factor A has *type* I if it contains an atom. It follows from [2] that if u is an atom, then the corresponding projective face consists of exactly, one (extreme) point, which we will henceforth

Y. Khakimdjanov et al (eds.), Algebra and Operator Theory, 183-186.

denote by $\hat{u}$.

An element $u \in U$ is called *finite* if it is the l.u.b. of a finite number of atoms. The minimal number of atoms whose l.u.b. is u, is called *dimension* of u. We say that a factor A has *type* I_n if the dimension of e is equal n. An order unit space A is called a *homogeneous factor of type* I_n if A is factor of *type* I_n and e is the *l.u.b.* of only n orthogonal atoms.

Definition 1. Linear function τ on A is said to be a *trace* if $\tau(Ra+R'a) = \tau(a)$ for every $a \in A$ and $R \in \mathcal{P}$.

Definition 2. We say that a set K (base V) is *smooth* if every extreme point of K has only one supporting hyperplane. A set K is said to be *strictly convex* if every open line segment in K does not contain boundary point of K , in other words, the proper faces of K have the form $\{\sigma\}$ -where σ is an extreme point of K.

Definition 3. A convex set K has the *Banach ball property*, if for every pair ρ, σ of extreme points of K FACE($\{\rho, \sigma\}$)- the minimal face generated by ρ and σ, is norm-exposed face and it is affinely isomorphic to a closed unit ball in some Banach space.

Example. Let E be an Banach space. Denote $A_s = \mathbf{R} + E^*$, then A_s is an order unit space with respect to the order:

$a = \alpha + x \geq 0 \overset{def}{\Longleftrightarrow} \alpha \geq \| x \|$ and norm: $\| a \| = |\alpha| + \| x \|$ for every $a \in A_s$. In this case, A_s is the order unit space of type I_2 and A_s is factor, therefore its state space K coincides with a shift of a closed unit ball E.

If the state space K of A (unit ball E) in the above example is a smooth and strictly convex set then the order unit space A is called a *generalized spin factor*.

So, factor A_s of type I_2 is the generalized spin factor if and only if its state space has the Banach ball property.

Let A be a homogeneous factor of type I_n, (V, K) is a base-norm space. We suppose that A and V are in the spectral duality [2] and $V^* = A$.

Theorem 1. *The state space of A has the Banach ball property if and only if there exists a trace on A.*

Proof.

Necessity. Let K be a state space of A with the Banach ball property. For atoms $u_0, \vartheta_0 \in U$ denote $h = u_0 \vee \vartheta_0, h = He$, $H \in \mathcal{P}$. Then, according to [3], im $H = H(A)$ is a homogeneous factor of type I_2 with the state space $K_0 = imH^* \cap K$. Let $\hat{u}_0, \hat{\vartheta}_0$ be the extreme points K corresponding to u_0 and ϑ_0 respectively. Since K has the Banach ball property, then the face FACE($\{\hat{u}_0, \hat{\vartheta}_0\}$) $= K_0$ is isomorphic to the smooth and strictly convex unit ball of some Banach space and there exists a trace τ on $H(A)$ ([4],lemma 3). It follows from [4] that $\tau(u) = \tau(\vartheta)$ for all atoms u and ϑ from A.

Let $e = \sum_{i=1}^{n} u_i, \ u_i \perp u_j \ (i \neq j)$, is the orthogonal decomposition of the unit element e into atoms. We denote by $\hat{u}_i$ the extreme point of K corresponding to u_i. Let us also set $\tau = \frac{1}{n} \sum_{i=1}^{n} \hat{u}_i$. Then we have that $\tau \in K, \ \tau$ is a trace and it does not depend on choice of decomposition of e.

Sufficiency. Let a trace exist on A . Then on every subfactor $A_1 \subset A$ such that $A_1 = R(A), \quad R \in \mathcal{P}$, there exist a trace. Let ρ, σ be a pair of extreme points of K. We denote by u and ϑ corresponding minimal projective units (atoms) for ρ and σ respectively [5]. Let us also set $He = h = uv\vartheta$. Then from [4], we have that $(H(A), h)$ is a generalized spin factor with the state space FACE($\{\rho, \sigma\}$).Q.E.D.

It appears that, we can pick out generalized spin factor among homogeneous factor of type I_2.

Corollary. Let A be a homogeneous factor of type I_2. A is the generalized spin factor of and only if A has a trace.

Let A be on order unit space, τ be a trace on A. For every $a \in A$ we denote $\| a \|_1 = \tau(|a|)$, where $|a|$ is the module of the element a [1]. The map $\| \cdot \|_1 \colon A \to \mathbf{R}$ is a norm on A if the trace is a faithful trace. Finally, the completion of A with respect to L_1-norm we denote by $L_1(A)$.

Theorem 2 *Let A be a homogeneous factor of type I_n and $V^* = A, \quad \tau$ is a trace on A. Spaces $L_1(A)$ and V are order and isometrically isomorphic if and only if A is a JBW - factor of type I_n.*

Proof. In the case when A is a JBW-factor with a trace order and isometric isomorphism of spaces $L_1(A)$ and V is proved in [6].

Conversely, let $L_1(A) \cong V$ and u, v be some pair atoms in A. We denote $He = h = uv\vartheta$. Then $(H(A), h)$ is a homogeneous factor of type I_2 with the state space $K_0 = K \cap imH^* = FACE(\{\hat{u}, \hat{\vartheta}\})$. It is proved in [4] that in this case, if $L_1(H(A))$ order and isometrically isomorphic to $V_0 = H^*(V)$, then K_0 is the Hilbert ball. Since the atoms $u, \vartheta \in U$ (extreme points of K) are arbitrary, we have that K possesses the Hilbert ball property. From theorem 7.2 in paper [1], we have that A is a JBW-factor. Q.E.D.

References

[1] Alfsen E.M., Shultz F.W. State spaces of Jordan algebras. *Acta Math.*,140:3-4(1978), 155-190.

[2] Alfsen E.M., Shultz F.W. On non commutative spectral theory and Jordan algebras. *Proc.Lomdon Math. Soc.*, 38(1979), 497-516.

[3] Berdikulov M.A., Odilov S.T. Generalized spin factor. *Uzb.Math. Journal.* 2(1994), 15-20.(Russian)

[4] Berdikulov M.A. Homogeneous order unit space of type I_2. *Izv. Acad.Nauk. UzSSR. ser.phys.-math.nauk.*, 4(1990), 8-14.(Russian)

[5] Ayupov Sh.A., Yadgorov N.Dz. Geometry of the state space of modular Jordan algebras. *Izv.Acad.Nauk Russian*, 57:6, 199-211.(Russian)

[6] Ayupov Sh.A. Integration on Jordan algebras. *Izv. Acad. Nauk SSSR ser.math.*, 47(1983), 3-25.(Russian)

E-Mail : Avazbek@saturn.silk.org

ON MARKOV RANDOM FIELDS ON UHF ALGEBRAS

N.N.GANIKHODZHAEV, F.M.MUKHAMEDOV

Institute of Mathematics,
Uzbekistan Academy of Sciences
F. Hodjaev str., 29, 700143 Tashkent, Uzbekistan.

Markov random fields on Z^n were introduced by R.A.Dobrushin[1]. Futher Markov random fields on infinite trees were studied by F.Spitzer[2]. In his work, in the set of all Markov random fields on the Cayley tree, a class of Markov chains was selected.

Naturally a problem of describing of Markov random fields and Markov chains in the case of quantum lattice systems arises. In the present paper Markov random fields for quantum lattice systems are defined, which have been called quantized Markov random fields. As in [2] the class of quantized Markov chains are selected and the same problems considered in [2] are solved. In this article the case when the Cayley tree is lattice Z is considered.

Consider the C^* -algebra $M = \otimes_Z Mat_2(C)$, where $Mat_2(C)$ matrix algebra $2x2$ on the field of complex numbers C. We denote by $e_1, \quad e_2, \quad e_3, \quad e_4$ the matrix units of $Mat_2(C)$, i.e.

$$e_1 = e_{11}, \quad e_2 = e_{22}, \quad e_3 = e_{12}, \quad e_4 = e_{21}.$$

We introduce the following notations:

$$\omega(a_{i_k}^{(k)} \mid a_{i_1}^{(1)}, ..., a_{i_{k-1}}^{(k-1)}, a_{i_{k+1}}^{(k+1)}, ..., a_{i_n}^{(n)}) =$$

$$= \frac{\omega(a_{i_1}^{(1)} \otimes ... \otimes a_{i_{k-1}}^{(k-1)} \otimes a_{i_k}^{(k)} \otimes a_{i_{k+1}}^{(k+1)} \otimes ... \otimes a_{i_n}^{(n)})}{\omega(a_{i_1}^{(1)} \otimes ... \otimes a_{i_{k-1}}^{(k-1)} \otimes \mathbf{1} \otimes \mathbf{a}_{\mathbf{i}_{\mathbf{k}+1}}^{(\mathbf{k}+1)} \otimes ... \otimes \mathbf{a}_{\mathbf{i}_{\mathbf{n}}}^{(\mathbf{n})})},$$

where ω is a state on M and $\mathbf{1}, \mathbf{a}_{\mathbf{i}_{\mathbf{p}}}^{(\mathbf{p})} \in \mathbf{Mat}_{\mathbf{2}}(\mathbf{C}), \quad \mathbf{p} = \mathbf{1}, \mathbf{2}, ..., \mathbf{n}.$

Y. Khakimdjanov et al (eds.), Algebra and Operator Theory, 187-192.

Definition 1. A state ω on M is called *quantized Markov random field* (q.m.r.f.), if

$$\omega(e_{i_1}^{(1)} \otimes ... \otimes e_{i_n}^{(n)}) \neq 0,$$

for any $i_k \in \{1,2\}$, $k = 1,2,...,n$, $n \in N$, in addition, if for $j \in Z$

$$\omega(e_{i_1}^{(j+1)} \otimes ... \otimes e_{i_{n-1}}^{(j+n-1)} \otimes \mathbf{1} \otimes \mathbf{e}_{\mathbf{i_{n+1}}}^{(\mathbf{j+n+1})}) \neq \mathbf{0}$$

$$\omega(e_{i_{n-1}}^{(j+n-1)} \otimes \mathbf{1} \otimes \mathbf{e}_{\mathbf{i_{n+1}}}^{(\mathbf{j+n+1})}) \neq \mathbf{0}$$

for any $n \in N \cup \{0\}, i_p \in \{1,2,3,4\}, p = 1,...,n+1$ then the following equation is hold:

$$\omega(e_{i_n}^{(j+n)} \mid e_{i_1}^{(j+1)},...,e_{i_{n-1}}^{(j+n-1)}, e_{i_{n+1}}^{(j+n+1)}) =$$

$$= \omega(e_{i_n}^{(j+n)} \mid e_{i_{n-1}}^{(j+n-1)}, e_{i_{n+1}}^{(j+n+1)}) \tag{1}$$

and conditional states (1) are invariant over shifts of Z .

If

$$\omega(e_{i_1}^{(j+1)} \otimes ... \otimes e_{i_{n-1}}^{(j+n-1)} \otimes \mathbf{1} \otimes \mathbf{e}_{\mathbf{i_{n+1}}}^{(\mathbf{j+n+1})}) = \mathbf{0},$$

$$\omega(e_{i_{n-1}}^{(j+n-1)} \otimes \mathbf{1} \otimes \mathbf{e}_{\mathbf{i_{n+1}}}^{(\mathbf{j+n+1})}) = \mathbf{0},$$

$$\omega(e_{i_1}^{(j+1)} \otimes ... \otimes e_{i_{n-1}}^{(j+n-1)} \otimes e_{i_n}^{(j+n)} \otimes e_{i_{n+1}}^{(j+n+1)}) = 0,$$

$$\omega(e_{i_{n-1}}^{(j+n-1)} \otimes e_{i_n}^{(j+n)} \otimes e_{i_{n+1}}^{(j+n+1)}) = 0,$$

then we put

$$\omega(e_{i_n}^{(j+n)} \mid e_{i_1}^{(j+1)},...,e_{i_{n-1}}^{(j+n-1)}, e_{i_{n+1}}^{(j+n+1)})$$

$$= \omega(e_{i_n}^{(j+n)} \mid e_{i_{n-1}}^{(j+n-1)}, e_{i_{n+1}}^{(j+n+1)}) = 1.$$

Definition 2. Let ω_1 and ω_2 be two states on M. We say that ω_1 is *subequivalent* to ω_2 and denote by $\omega_1 \sim^s \omega_2$ if the following equations are hold:

$$\omega_1(e_{i_n}^{(j+n)} \mid e_{i_{n-1}}^{(j+n-1)}, e_{i_{n+1}}^{(j+n+1)}) = \omega_2(e_{i_n}^{(j+n)} \mid e_{i_{n-1}}^{(j+n-1)}, e_{i_{n+1}}^{(j+n+1)}).$$

Similarly to [2] we select a class of quantized Markov chains.

Let $P = (p_{ij})$ be a stochastic matrix such that $p_{ij} > 0$ and let $\Omega = \{1,2\}^Z$ and Σ be σ - algebra which generated by cylindric subsets of Ω. On the measurable space (Ω, Σ) Markov measure is defined as the following one:

$$\mu(i_1, i_2, ..., i_{n-1}, i_n) = p_{i_1} p_{i_1 i_2} ... p_{i_{n-1} i_n},$$

where $\pi = (p_1, p_2)$ is the unique invariant measure for the matrix P, i.e. $\pi P = \pi$.

Let $diagMat_2(C)$ be the algebra generated by elements e_1 and e_2 and let $MC = \otimes_Z diagMat_2(C)$ be C^*- algebra. Let ε be the conditional exspectation from M on MC. Existence and uniquness of such operator were proved in [3].

We define a functional $\omega_p(x) = \mu(\varepsilon(x)), x \in M$, where we identify i with e_i ($i = 1,2$). Then ω_p is a state.

Proposition. ω_p is q.m.r.f..

Proof. Let $i_k \in \{1,2\}$, $k = 1,...,n+1$, $j \in Z$ .Then we have

$$\frac{\omega_p(e_{i_1}^{(j+1)} \otimes ... \otimes e_{i_{n-1}}^{(j+n-1)} \otimes e_{i_n}^{(j+n)} \otimes e_{i_{n+1}}^{(j+n+1)})}{\omega_p(e_{i_1}^{(j+1)} \otimes ... \otimes e_{i_{n-1}}^{(j+n-1)} \otimes \mathbf{1} \otimes \mathbf{e}_{\mathbf{i}_{\mathbf{n}+1}}^{(\mathbf{j}+\mathbf{n}+\mathbf{1})})} =$$

$$= \frac{p_{i_1} p_{i_1 i_2} \cdots p_{i_{n-1} i_n}}{p_{i_1} p_{i_1 i_2} \cdots p_{i_{n-1} i_{n+1}}^{(2)}} =$$

$$= \frac{p_{i_{n-1} i_n} p_{i_n i_{n+1}}}{p_{i_{n-1} i_{n+1}}^{(2)}} = \frac{p_{i_{n-1}} p_{i_{n-1} i_n} p_{i_n i_{n+1}}}{p_{i_{n-1}} p_{i_{n-1} i_{n+1}}^{(2)}} =$$

$$\frac{\omega_p(e_{i_{n-1}}^{(j+n-1)} \otimes e_{i_n}^{(j+n)} \otimes e_{i_{n+1}}^{(j+n+1)})}{\omega_p(e_{i_{n-1}}^{(j+n-1)} \otimes \mathbf{1} \otimes \mathbf{e}_{\mathbf{i}_{\mathbf{n}+1}}^{(\mathbf{j}+\mathbf{n}+\mathbf{1})})} =$$

$$\omega_p(e_{i_n}^{(j+n)} \mid e_{i_{n-1}}^{(j+n-1)}, e_{i_{n+1}}^{(j+n+1)}).$$

If there exists $k_0 \in \{n-1, n, n+1\}$ for which $i_{k_0} = 3$ or 4 then it is easy to see, that the condition of the definition 1 is fulfiled and, therefore (1) is true. Then it follows that ω_p is q.m.r.f.. The propositon is proved.

Defined q.m.r.f. ω_p is called quantized Markov chain. We denote by Q^J the set of all q.m.r.f. on M. We put

$$Q_p^J = \{\omega \in Q^J \mid \omega_p \sim^s \omega\},$$

S^J - set of all quantized Markov chains.

Our aim is to find the cardinality of the set $Q_p^J \cap S^J$.

Theorem.

$$Card(Q_p^J \cap S^J) = 1,$$

where Card(.) - cardinality of a set.

Proof. Let $\omega \in Q_p^J \cap S^J$. We denote by μ^ω restricton of ω on MC. Then μ^ω is a measure on (Ω, Σ). It follows from the definition 1, that μ^ω is a Markov random field [2]. In the notations of [2] from the definition 2 it follows that $\mu^\omega \in J_\mu$, where J_μ is the set of all Markov random fields discribed in [2].

It follows from[2] that $\mu^\omega = \mu$. Thus, we obtain that $\omega = \omega_p$ on MC. Since $\omega \in Q_p^J \cap S^J$ then from the definition 2 we get

$$\omega(e_i^{(j)}) = \omega_p(e_i^{(j)}), \tag{2}$$

$i = 1, 2, 3, 4, \quad j \in Z$.

Since $\omega_p(e_i^{(j)}) = 0$ for $i = 3, 4$ then

$$\omega(e_i^{(j)}) = 0 \tag{3}$$

for $i = 3, 4, \quad j \in Z$.

Turning again into the definition 2 we also get

$$\frac{\omega(e_{i_1}^{(n-1)} \otimes e_{i_2}^{(n)})}{\omega(e_{i_1}^{(n-1)} \otimes \mathbf{1})} = \frac{\omega_p(e_{i_1}^{(n-1)} \otimes e_{i_2}^{(n)})}{\omega_p(e_{i_1}^{(n-1)} \otimes \mathbf{1})},$$

$$\frac{\omega(e_{i_1}^{(n-1)} \otimes e_{i_2}^{(n)})}{\omega(e_{i_1}^{(n-1)})} = \frac{\omega_p(e_{i_1}^{(n-1)} \otimes e_{i_2}^{(n)})}{\omega_p(e_{i_1}^{(n-1)})}.$$

If $i_1 \neq 3, 4$ then keeping in mind $\omega = \omega_p$ on MC we obtain

$$\omega(e_{i_1}^{(n-1)} \otimes e_{i_2}^{(n)}) = \omega_p(e_{i_1}^{(n-1)} \otimes e_{i_2}^{(n)}) \tag{4}$$

for any $i_2 \in \{1, 2, 3, 4\}, n \in Z$.

If $i_1 = 3$ or 4 then

$$\omega_p(e_{i_1}^{(n-1)} \otimes e_{i_2}^{(n)}) = \omega_p(e_{i_1}^{(n-1)})) = 0$$

and then from the definition 1 we have

$$\frac{\omega_p(e_{i_1}^{(n-1)} \otimes e_{i_2}^{(n)})}{\omega_p(e_{i_1}^{(n-1)})} = 1..$$

From (3) we obtain

$$\omega(e_{i_1}^{(n-1)} \otimes e_{i_2}^{(n)}) = 0, \tag{5}$$

for any $i_2 \in \{1, 2, 3, 4\}, \quad n \in Z$.

So from (4), (5) we get

$$\omega(e_{i_1}^{(n-1)} \otimes e_{i_2}^{(n)}) = 0, \tag{6}$$

where at least one of i_1, i_2 equals 3 or 4.

Using again the definition 2 we have

$$\frac{\omega(e_{i_1}^{(n-1)} \otimes e_{i_2}^{(n)} \otimes e_{i_3}^{(n+1)})}{\omega(e_{i_1}^{(n-1)} \otimes e_1^{(n)} \otimes e_{i_3}^{(n+1)}) + \omega(e_{i_1}^{(n-1)} \otimes e_2^{(n)} \otimes e_{i_3}^{(n+1)})} =$$

$$= \frac{\omega_p(e_{i_1}^{(n-1)} \otimes e_{i_2}^{(n)} \otimes e_{i_3}^{(n+1)})}{\omega_p(e_{i_1}^{(n-1)} \otimes e_1^{(n)} \otimes e_{i_3}^{(n+1)}) + \omega_p(e_{i_1}^{(n-1)} \otimes e_2^{(n)} \otimes e_{i_3}^{(n+1)})}.$$

If $i_3 = 3$ or 4 then

$$\omega_p(e_{i_1}^{(n-1)} \otimes e_{i_2}^{(n)} \otimes e_{i_3}^{(n+1)}) = 0,$$

for any $i_1, i_2 \in \{1,2,3,4\}$.

Consequently, it follows from the definition 1

$$\omega(e_{i_1}^{(n-1)} \otimes e_{i_2}^{(n)} \otimes e_{i_3}^{(n+1)}) = \omega(e_{i_1}^{(n-1)} \otimes e_1^{(n)} \otimes e_{i_3}^{(n+1)}) + \omega(e_{i_1}^{(n-1)} \otimes e_2^{(n)} \otimes e_{i_3}^{(n+1)}). \quad (7)$$

If $i_2 = 1$ or 2 we have

$$\omega(e_{i_1}^{(n-1)} \otimes e_{i_2}^{(n)} \otimes e_{i_3}^{(n+1)}) = 0, \quad (8)$$

for any $i_1 \in \{1,2,3,4\}$.

If $i_2 = 3$ or 4 then it follows from (7),(8)

$$\omega(e_{i_1}^{(n-1)} \otimes e_{i_2}^{(n)} \otimes e_{i_3}^{(n+1)}) = 0, \quad (9)$$

for any $i_1 \in \{1,2,3,4\}$.

Analogously if $i_1 = 3$ or 4 we get the following equation

$$\omega(e_{i_1}^{(n-1)} \otimes e_{i_2}^{(n)} \otimes e_{i_3}^{(n+1)}) = 0, \quad (10)$$

for any $i_1, i_2 \in \{1,2,3,4\}, n \in Z$.

From (8),(9),(10) we conclude that if there exists $i_{k_0} = 3$ or 4 , $k_0 \in \{1,2,3\}$ then

$$\omega(e_{i_1}^{(n-1)} \otimes e_{i_2}^{(n)} \otimes e_{i_3}^{(n+1)}) = 0, \quad (11)$$

for any $n \in Z$.

Finally, using the definition 1 and (11) with analogously reasoning we get

$$\omega(e_{i_1}^{(j+1)} \otimes \ldots \otimes e_{i_n}^{(j+n)}) = 0,$$

if there exists $i_{k_0} = 3$ or 4 , $0 \leq r \leq n$. Thus we get $\omega = \omega_p$ on M. **Q.E.D.**

References

[1] Dobrushin R.L. *Teor. ver. i primen.* 2(1968), 201-229.

[2] Spitzer F. it The Ann. Prob., 3:3(1975), 387-398.

[3] Stratila S. Modular theory in Operator Algebras, Bucuresti: Editura Academici, 1981, 492 p.

E-Mail : N.N. Ganikhodzhaev : nasyr@uwd.freenet.uz , F.M. Mukhamedov : root@im.tashkent.su

INJECTIVITY, AMENABILITY, SEMIDISCRETENESS AND HYPERFINITENESS IN REAL W^*-ALGEBRAS

A.RAKHIMOV, B.BOYKABILOV

Institute of Mathematics,
Uzbekistan Academy of Sciences
F. Hodjaev str., 29, 700143 Tashkent, Uzbekistan.

Abstract In this paper we shall define semidiscreteness and amenability for real W^*-algebra and consider their connection with the injectivity and hyperfiniteness of real W^*-algebras. It is well known that for W^*-algebras the properties of injectivity, amenability, semidiscreteness and hyperfiniteness coincide (see [W], [C], [H], [EL]). Here we shall consider that for real W^*-algebra these properties are not equivalent in general, namely the hyperfiniteness and the injectivity are not equivalent.

Let H be a complex Hilbert space, $B(H)$ denote the algebra of all bounded linear operators on H. W^* - a l g e b r a - is a weakly closed complex $*$-algebra of operators on a Hilbert space H containing the identity operator $\mathbf{1}$. Recall that W^*-algebras are also called *von Neumann algebra.* A real *-algebra $\mathfrak{R}$ in $B(H)$ is called a *real W^*-algebra* if it is closed in the weak operator topology and satisfies the conditions $\mathfrak{R}\bigcap i\mathfrak{R} = \{0\}$, $\mathbf{1} \in \mathfrak{R}$.

Let M be a complex or real W^*-algebra. A linear mapping $\alpha : M \longrightarrow M$ is called a **-automorphism* (respectively a *-antiautomorphism) if

1) $\alpha(x^*) = \alpha(x)^*$;
2) $\alpha(xy) = \alpha(x)\alpha(y)$ (respectively $\alpha(xy) = \alpha(y)\alpha(x)$),

for all $x, y \in M$. We shall denote by $Aut(M)$ the group of all *-automorphisms , by $Ant(M)$ the group of all *-antiautomorphisms of M. Two *-automorphisms or *-antiautomorphisms α and β are said to be *conjugate*,

Y. Khakimdjanov et al (eds.), Algebra and Operator Theory, 193-201.

if $\alpha = \theta \cdot \beta \cdot \theta^{-1}$ for some *-automorphism θ.

Let $\mathfrak{R}$ and S be W^*-algebras, $\Phi : \mathfrak{R} \to S$ a linear map with $\Phi(\mathbf{1}) = \mathbf{1}$. Denote by $M_n(\mathfrak{R})$ the algebra of $n \times n$ matrix in $\mathfrak{R}$. Suppose that $\Phi_n : M_n(\mathfrak{R}) \to M_n(S)$ is the map defined as

$$\Phi_n\Big([a_{ij}]\Big) = \Big[\Phi(a_{ij})\Big].$$

Similarly to complex W^*-algebras (see [EL]), we consider the following definitions: a maps Φ is called n-**positive**, if the map Φ_n is positive; Φ is **completyly positive**, if it is n-positive for all n; Φ is **morphism**, if it is completely positive and $\Phi(\mathbf{1}) = \mathbf{1}$.

Definition 1. A real W^*-algebra $\mathfrak{R}$ is called *semidiscrete* if the identity map on $\mathfrak{R}$ can be approximated in the topology of simple-weak* convergence by completely positive normal linear maps of finite rank which preserve the identity.

Proposition 1 *A real W^*-algebra $\mathfrak{R}$ is semidescrete if and only if the enveloping W^*-algebra $\mathcal{U}(\mathfrak{R})$ is semidiscrete.*

Proof. Let $\mathfrak{R}$ be semidescrete, i.e. there is a sequence of normal morphisms of finite rank $\{\Phi_m\}$, such that $\{\Phi_m\}$ converges to the identity map $I_{\mathfrak{R}}$ *-weakly on $\mathfrak{R}$. Let $I_{\mathcal{U}}$ be the identity map on $\mathcal{U}$. It is clear that $I_{\mathcal{U}} = I_{\mathfrak{R}} + iI_{\mathfrak{R}}$. Then it is obvious that $\{\overline{\Phi}_m = \Phi_m + i\Phi_m\}$ is a sequence of normal morphisms of finite rank and $\overline{\Phi}_m$ *-weakly converges to the identity map $I_{\mathcal{U}}$. Therefore, $\mathcal{U}$ is semidiscrete.

Conversely, let $\mathcal{U}$ be semidiscrete and suppose that $E : \mathcal{U} \to \mathfrak{R}$ is the canonical conditional expectation (see [HS]). It is obvious that (see Lemma 1.3 (iii) from [EL]) the projection E is a normal morphism from $\mathcal{U}$ to $\mathfrak{R}$. Let $\{\overline{\Phi}_m\}$ be the sequence of normal morphisms of finite rank which converges to the identity map $I_{\mathcal{U}}$ *-weakly. Put $\Phi_m = E \circ \overline{\Phi}_m$. It is clear that $\{\Phi_m\}$ is a sequence of normal morphisms of finite rank and $\{\Phi_m\}$ converges to the identity map $I_{\mathfrak{R}}$ *-weakly, i.e. $\mathfrak{R}$ is semidiscrete. ∎

Let $\mathfrak{R}$ and S be real W^*-algebras, and let $\mathfrak{R} \otimes S$ denote their algebraic tensor product, that is each element of $\mathfrak{R} \otimes S$ is an expression of the form $x = \sum_i a_i \otimes b_i$, where $a_i \in \mathfrak{R}$, $b_i \in S$. Let $\mathcal{S}(\mathfrak{R} \otimes S)$ denote the set of all linear functionals f on the vector space $\mathfrak{R} \otimes S$ which satisfy the conditions

$$f(xx^*) \geq 0, \quad f(\mathbf{1}) = \mathbf{1}, \quad \text{for all } x \in \mathfrak{R} \otimes S.$$

For each $f \in \mathcal{S}(\Re \otimes S)$, let $\{\pi_f, \xi_f, H_f\}$ be GNS-representation constructed by f, where ξ_f is a cyclic separating unit vector, and π_f is a homomorphism from $\Re \otimes S$ into the Hilbert space H_f, such that $f(x) =< \pi_f(x)\xi_f, \xi_f >$, $x \in \Re \otimes S$. For each $f \in \mathcal{S}(\Re \otimes S)$, define the seminorm on $\Re \otimes S$ as $\rho_f(x) = \|\pi_f(x)\|$, $x \in \Re \otimes S$. For $\Gamma \subset \mathcal{S}(\Re \otimes S)$ put $\rho_\Gamma = \sup\{\rho_f : f \in \Gamma\}$. If ρ_Γ is a norm, then we call Γ a **separating** subset of $\mathcal{S}(\Re \otimes S)$, and denote by $\mathcal{S}(\Re \otimes_\Gamma S)$ the real W^*-algebra obtained by completing $\Re \otimes S$ with respect to ρ_Γ. Let $\Re^* \otimes S^*$ denote the vector space tensor product of $\Re^*$ and S^*, considered as a space of linear functions on $\Re \otimes S$. We define the following sets in $\mathcal{S}(\Re \otimes S)$:

$$\begin{aligned}
&Min = (\Re^* \otimes S^*) \bigcap \mathcal{S}(\Re \otimes S), \\
&Max = \mathcal{S}(\Re \otimes S), \\
&Bin = \{f \in Max : \text{ the map } (r, s) \to f(r \otimes s) \\
&\text{is separately weak}^* \text{ continuous }\}.
\end{aligned}$$

We denote by ρ_{Min}, ρ_{Max}, and ρ_{Bin} the norms on $\Re \otimes S$ generated by the sets Min, Max, Bin respectively. The completions of $\Re \otimes S$ by those norms we denote respectively by $\Re \otimes_{Min} S$, $\Re \otimes_{Max} S$, and $\Re \otimes_{Bin} S$. It is clear that $Min \subseteq Bin \subseteq Max$, therefore, we have: $\rho_{Min} \leq \rho_{Bin} \leq \rho_{Max}$.

We repeat step by step the scheme of proof of Theorem 4.1 [EL], to obtain the following Theorem:

Theorem 1 *Let $\Re$ be a real W^*-algebra. Then $\Re$ is semidiscrete if and only if for any real W^*-algebra S,*

$$\Re \otimes_{Min} S \;=\; \Re \otimes_{Bin} S.$$

Definition 2. Let $\Re$ be a real W^*-algebra, and $\eta : \Re \otimes \Re' \;\to\; B(H)$ a *-homomorphism defined by:

$$\eta(x \otimes x') \;=\; x \cdot x', \quad x \in \Re, x' \in \Re'.$$

$\Re$ is called *amenable*, if η has a bounded extension to $\Re \otimes_{Min} \Re'$.

Theorem 2. *Let $\Re$ be a real W^*-algebra. Then $\Re$ is amenable if and only if $\Re$ is semidiscrete.*

Proof. Let $\Re$ be semidiscrete. Suppose that $\eta : \Re \otimes \Re' \to B(H)$ is a *-homomorphism defined by $\eta(x \otimes x') = xx'$, $x \in \Re, x' \in \Re'$. Let ω_ξ be the normal state on $B(H)$, corresponding to the unit vector $\xi \in H$. Since $f = \omega_\xi \cdot \eta$ is a linear functional with $f(\mathbf{1}) = 1$, $f(x^*x) \geq 0$ and the map $(x, x') \to f(x \otimes x')$ is separately weak* continuous (therefore, $f \in Bin$), we have

$$\|\eta(x)\| = \sup_{\|\xi\|=1} [(\omega_\xi \cdot \eta)(x^*x)]^{1/2} \leq \|x\|_{bin}$$

By Theorem 1 we have $\|\eta(x)\| \leq \|x\|_{min}$, therefore, there exists a *-homomorphism $\overline{\eta}$, extending η to $\Re \otimes \Re'$, i.e. $\Re$ is amenable.

Conversely, let $\Re$ be amenable. Then the *-homomorphism $\eta : \Re \otimes \Re' \to B(H)$ defined as $\eta(x \otimes x') = xx'$, $x \in \Re, x' \in \Re'$, has an extension to $\Re \otimes_{Min} \Re'$, which we denote again by η. Let ξ be the unit vector in H such that the normal state ω_ξ on $B(H)$, corresponding to ξ, has the restriction to $\Re$ and which it is $\alpha_\Re$ - invariant (where $\alpha_\Re$ is the involutive *-antiautomorphism of $\mathcal{U}(\Re)$, generating $\Re$). Then, it is easy to see that the linear functional $f = \omega_\xi \cdot \eta$ is in $Min \cap Bin$. Fix ξ. Denoting by $e' \in \Re'$ the projection with range $\overline{\Re\xi}$, we may spatially identify $\pi_\rho(\Re)$ with $\Re e'$. Similarly to the complex case (see Lemmas 4.2 - 4.4 from [EL]) we can construct a completely positive map $T : \Re' \to \Re_*$, corresponding to the state ω_ξ, such that $T(\mathbf{1}) = \rho$ is a state, and also construct the morphism: $\theta_\rho^{-1} \cdot T : \Re' \to e'\Re'e'$. For $x' \in \Re'$ we have:

$$y = \theta_\rho^{-1}(T(x')) \in e'\Re'e'$$

and $T(x')(x) = < xy'\xi, \xi >$, for $x \in \Re$. On the other hand we have:

$$T(x')(x) = < xx'\xi, \xi > = < xe'x'e'\xi, \xi >,$$

and hence $\theta_\rho^{-1} \cdot T(x') = e'r'e'$. Since $Min \cap Bin$ is *-weakly dense in $S_{Min}(\Re \otimes \Re')$, T is a weakly * limit of the net of weak* continuous completely positive maps $T_\gamma : \Re' \to \Re_*$ with finite rank. From $T(\Re') \subset \Re_*$ it follows that we can assume, that $T_\gamma(\mathbf{1}) = \rho$, therefore, $\{\theta_\rho^{-1} \cdot T_\gamma\}$ is the net of morphisms with finite rank which weak* converges to the morphism $\theta_\rho^{-1} \cdot T$. Now let σ be a normal state on $e'\Re'e'$. We define the net of normal morphisms

$\Phi_\gamma : e'\Re' e' \to e'\Re' e'$ as:

$$\Phi_\gamma(x) = \{\theta_\rho^{-1} \cdot T_\gamma(x + \sigma(x)(\mathbf{1} - e'))\}.$$

It is easy to see that the morphism Φ_γ has finite rank and the net $\{\Phi_\gamma\}$ converges to the identity map $I_{e'\Re' e'}$ in the simple weak* topology, i.e. the subalgebra $e'\Re' e'$ is semidiscrete, therefore, $\Re e'$ is semidiscrete. Then $\Re$ is semidiscrete too. ■

Theorem 3 *For real W^*-algebras the notions of amenability, semidiscreteness and injectivity are equivalent.*

Proof Theorem 2 implies equivalence of amenability and semidiscreteness. Since the notions of amenability, semidiscreteness and injectivity are equivalent in the enveloping W^*-algebra (see [C], [H]), then by Proposition 1 the proof of Theorem 3 is complete. ■

Recall that a real W^*-algebra $\Re$ is said to be *hyperfinite* if there exists an increasing sequence $\{R_n\}$ of finite dimensional real W^*-subalgebras $\{R_n\}$ of $\Re$ with $\mathbf{1} \in R_n$ such that $\bigcup_{n\geq 1} R_n$ is weakly dense in $\Re$.

Proposition 2. *If a real W^*-algebra $\Re$ is hyperfinite, then the enveloping W^*-algebra $\mathcal{U}(\Re)$ is hyperfinite.*

Proof Let $\Re$ be hyperfinite, i.e. there exists an increasing sequence of finite dimensional real W^*-subalgebras $\{R_n\}$ of $\Re$ with $\mathbf{1} \in R_n$ such that $\bigcup_{n\geq 1} R_n$ is weakly dense in $\Re$. Then it is clear that the sequence $\{\mathcal{U}(R_n)\}$ is increasing and $\bigcup_{n\geq 1} \mathcal{U}(R_n)$ is weakly dense in $\mathcal{U}$, i.e. $\mathcal{U}$ is hyperfinite. ■

Recall the following two results from [S].

Lemma 1 [S, Lemma 2.2]. *Let $\Re$ be an irreducible real W^*-algebra acting on the n-dimensional Hilbert space $\mathbf{C}^n$, $n < \infty$. Then there exists a conjugate linear isometry J on $\mathbf{C}^n$ such that $JxJ = x$, for all $x \in \Re$, and either $J^2 = \mathbf{1}$, or $J^2 = -\mathbf{1}$. Furthermore if $J^2 = \mathbf{1}$, then $\Re$ is isomorphic to the real $n \times n$ matrices, and if $J^2 = -\mathbf{1}$, then n is even and $\Re$ is isomorphic to the $n/2 \times n/2$ matrices over the quaternions.*

Lemma 2 [S, Corollary 2.3]. *Let $\Re_1$ and $\Re_2$ be irreducible real W^*-*

algebras acting on the finite dimensional Hilbert spaces $\mathbf{C}^{n_1}$ *and* $\mathbf{C}^{n_2}$ *respectively. Let* J_k *be the conjugate linear isometry on* $\mathbf{C}^{n_k}$ *associated with* $\mathfrak{R}_k$, *as in Lemma 1,* $k = 1, 2$. *Let* $\mathfrak{R}_1 \otimes \mathfrak{R}_2$ *be the tensor product of* $\mathfrak{R}_1$ *and* $\mathfrak{R}_2$ *acting on* $\mathbf{C}^{n_1} \otimes \mathbf{C}^{n_2}$. *Then the conjugate linear isometry associated with* $\mathfrak{R}_1 \otimes \mathfrak{R}_2$ *is* $J_1 \otimes J_2$, *hence we have:*

(i) if $J_k^2 = \mathbf{1}$ *or* $J_k^2 = -\mathbf{1}$, *for both* $k = 1, 2$, *then* $\mathfrak{R}_1 \otimes \mathfrak{R}_2$ *is isomorphic to the real* $n_1 n_2 \times n_1 n_2$ *matrices;*

(ii) if $J_1^2 = \mathbf{1}$ and $J_2^2 = -\mathbf{1}$, or $J_1^2 = -\mathbf{1}$ and $J_2^2 = \mathbf{1}$, then $\mathfrak{R}_1 \otimes \mathfrak{R}_2$ is isomorphic to the $n_1 n_2/2 \times n_1 n_2/2$ matrices over the quaternions.

Proposition 3. *Let* $\mathfrak{R}$ *be a properly infinite real factor. Then the following conditions are equivalent:*

(1) $\mathfrak{R}$ *is hyperfinite ;*

(2) $\mathfrak{R}$ *is countably generated, and given elements* $x_1, x_2, ..., x_n$ *in* $\mathfrak{R}$ *and any normal state* ψ *of* $\mathfrak{R}$, *and* $\varepsilon > 0$, *there exist a finite dimensional real* W^**-subalgebra* B *of* $\mathfrak{R}$ *and* $y_1, y_2, ..., y_n$ *elements in* B *such that* $\|y_k - x_k\|_\psi^\# \leq \varepsilon$, $k = 1, 2, ..., n$;

(3) $\mathfrak{R}$ *is the weak closure of the union of an increasing sequence* $\{R_n\}$ *of real factors with* $\mathbf{1} \in R_n$, *such that* R_n *is isomorphic to the real* $2^n \times 2^n$ *matrices.*

Proofs of (1) $\Rightarrow$ (2), and of (3) $\Rightarrow$ (1) are trivial. Proof of (2) $\Rightarrow$ (3) is carried out easily, similarly to the proof of Theorem 3 in [EW]. We shall show that a finite subfactor B, and therefore, subfactors $\{R_n\}$ can be choosed such that they are isomorphic to the real matrices. Indeed, it is clear that a real subfactor B of $\mathfrak{R}$ is isomorphic to an irreducible real factor acting on the 2^m-dimensional complex Hilbert space for some m having the required approximative properties. If B is isomorphic to the real $2^n \times 2^n$ matrices we are through. If not, we can apply Lemma 1.2 in [S] twice to the relative commutant $B' = B' \bigcap \mathfrak{R}$ to find a real subfactor R_4 of B' isomorphic to the real 4×4 matrices, i.e. $R_4 \cong M_4(\mathbf{R})$. By Lemma 2 we obtain $R_4 = \mathbb{Q} \otimes \mathbb{Q}$, where $\mathbb{Q}$ denotes the quaternions represented as complex 2×2 matrices. If we identify $\mathcal{B} = B \otimes Q \subset \mathfrak{R}$ with a real subfactor of $\mathfrak{R}$, then by Lemma 2 $\mathcal{B}$ is isomorphic to the real $2^k \times 2^k$ matrices for some k.

Using Proposition 3, similarly to Corollary 2.10 from [S] we can prove the following theorem.

Theorem 4 *Let M be a properly infinite factor. Then the conjugacy class of involutive *-antiautomorphism generating hyperfinite real factors, is unique.*

Proof. Let α be an involutive *-antiautomorphism of M and let $\mathfrak{R} = \{x \in M : \alpha(x) = x^*\}$ be the real factor generated by α which is hyperfinite by condition of Theorem. By Proposition 3 there exists an increasing sequence $\{R_n\}$ of subalgebras of $\mathfrak{R}$, which are isomorphic to the algebra of real $2^n \times 2^n$ matrices, and such that $\mathbf{1} \in R_n$ and $\bigcup\limits_{n\geq 1} R_n$ is weakly dense in $\mathfrak{R}$. Let $M_n = R_n + iR_n$ be the factor of type I_{2^n}. Denote by t_n the transpose map on M_n such that $R_n = \{x \in M_n : t_n(x) = x^*\}$. Since R_n is the algebra of real matrices and the sequence $\{R_n\}$ is weakly approximating $\mathfrak{R}$, then $\alpha(x) = t_n(x)$, for all $x \in M_n$. Thus, any *-antiautomorphism α is the inductive limit of the transpose maps t_n. Since in each matrix algebra all transpose maps are conjugate by a *-automorphism of $\bigcup\limits_{n\geq 1} M_n$ and since the conjugating *-automorphisms weakly converge to unique *-automorphism of M, then every two involutive *-antiautomorphisms of M generating hyperfinite real factors will be conjugate by a *-automorphism of M. ∎

It is well-known there exists two classes of non-isomorphic injective real factors of type III_λ, $0 < \lambda < 1$ generating (as the complex enveloping W^*-algebras) the same injective factor of type III_λ. The first of these factors is generated by an involutive *-antiautomorphism with $mod(\alpha) = 1$, and the second factor is generated by an involutive *-antiautomorphism with $mod(\alpha) = \sqrt{\lambda}$, where $mod(\cdot)$ is fundamental homomorphism for $Aut \bigcup Ant$, defined in classical works A.Connes. We shall construct a hyperfinite real factor $\mathfrak{R}$ of type III_λ, $0 < \lambda < 1$, with $mod(\alpha_{\mathfrak{R}}) = 1$ (which we are called *the real Powers factor*).

Let $M_2(\mathrm{R})$ be the algebra of real 2×2 matrices and let φ_λ be the state on $M_2(\mathrm{R})$ defined as

$$\varphi_\lambda \begin{pmatrix} x_{11} & x_{12} \\ x_{21} & x_{22} \end{pmatrix} = \frac{1}{1+\lambda}(\lambda x_{11} + x_{22}).$$

Similarly to the construction of complex Powers factor it is easy to see that if $\lambda = 1$, the algebra $\mathfrak{R}_\lambda = \otimes_1^\infty (M_2(\mathrm{R}), \varphi_\lambda)$ is a real factor of type II_1, and if $\lambda = 0$ then the algebra $\mathfrak{R}_\lambda$ is a real factor of type I_∞, and when $\lambda \in (0,1)$,

the algebra $\Re_\lambda$ is a real factor of type III_λ. Since the state $\omega = \otimes_1^\infty \varphi_\lambda$ on $\Re_\lambda$ satisfies $\sigma_T^{\overline{\omega}} = id$, then we have: $mod(\alpha_{\Re_\lambda}) = 1$.
Now Theorem 4 implies the following

Theorem 5 *A real injective W^*-factor R of type III_λ, $0 < \lambda < 1$, is hyperfinite if and only if $mod(\alpha_\Re) = 1$.*

Note that a real injective factor S of type III_λ, $0 < \lambda < 1$ with $mod(\alpha_S) = \sqrt{\lambda}$ can not be approximated by increasing sequence of real or quaternionian matrix subalgebras of S, although its enveloping (complex) factor is approximated by an increasing sequence of complex matrix subalgebras.

We summarize the above results in the following

Theorem 6 *Let $\Re$ be a real factor and let α be the involutive *-antiautomorphism of $\mathcal{U}(\Re)$ generating $\Re$. Then for $\Re$ the following implications are valid:*

$$\begin{array}{ccccc} \textit{Injectivity} & \Leftrightarrow & \textit{Amenability} & \Leftrightarrow & \textit{Semidiscreteness} \\ \Uparrow & & & & \Downarrow \textit{ iff } mod(\alpha_\Re) = 1 \\ & & \textit{Hyperfiniteness} & & \end{array}$$

Proof Theorem 3 implies the equivalence of properties of injectivity, amenability and semidiscreteness for real factors. By Proposition 2 the hyperfiniteness implies the injectivity. Theorem 5 implies that if a real factor $\Re$ is injective then it satisfies the condition $mod(\alpha_\Re) = 1$, where $\alpha_\Re$ is the involutive *-antiautomorphism of $\mathcal{U}(\Re)$ generating $\Re$, if and only if $\Re$ is hyperfinite. ■

References

[BR] Bratteli O., Robinson D. Operator algebras and quantum statistical mechanics. Springer-Verlag, New York Heidelber, Berlin 1979.

[C] Connes A. Classification of injective factors. *Ann. Math.*, 104:1(1976), 73-115.

[EL] Effros E., Lance E. Tensor products of operator algebras. *Advances in Math.*, 25(1977), 34 p.

[EW] Elliott G.A., Woods E.J. The equivalence of various definition for a properly infinite von Neumann algebra to be approximately finite dimensional. *Proc. Amer. Math. Soc.*, 60(1976), 175-179.

[H] Haagerup U. A new proof of the equivalence of injectivity and hyperfiniteness for factors on a separable Hilbert space. *Journal of Functional Analysis.*, 62(1985), 160-201.

[HS] Haagerup U., Stormer E. Positive projections of von Neumann algebras onto JW-algebras. *J. Reports on Math. Physics.*, 36(1995), 317-330.

[S] Stormer E. Real structure in the hyperfinite factor. *Duke Math. J.* 47:1(1980), 145-153.

[W] Wasserman S. Injective W^*-algebras. *Math. Proc. Cambridge Phil. Soc.*, 82(1977), 39-47.

E-Mail : Abdugafur Rakhimov : aratex @ hal.freenet.uz

CONTRACTIVE PROJECTIONS ON FACIALLY SYMMETRIC SPACES

N.J.YADGOROV, M.M.IBRAGIMOV

Institute of Mathematics,
Uzbekistan Academy of Sciences
F. Hodjaev str., 29, 700143 Tashkent, Uzbekistan.

Facially Symmetric spaces were introduced and studied in details in work [1,2]. Examples of such spaces are Hilbert spaces, predual of a von Neumann algebra or a JBW^*- algebra, and more general-predual spaces of JBW^*- triples. In this work we shall consider contractive projections on facially symmetric spaces. Terminology and notation are from [1 - 4].

Let Z be a real or complex normed space. Elements $f, g \in Z$ are said to be orthogonal and denoted $f \diamond g$, if $\|f+g\| = \|f-g\| = \|f\|+\|g\|$. Subsets $S, T \subset Z$ are said to be orthogonal, if $f \diamond g$ for all $(f,g) \in S \times T$. For a subset S of Z, $S^\diamond$ denotes $\{f \in Z : f \diamond g$ for all $g \in S\}$ and is called orthogonal complement of S. A norm exposed face of unit ball Z_1 of Z is a nonempty set (necessarily $\neq Z_1$) of the form $F_x = \{f \in Z : f(x) = 1\}$, where $x \in Z^*, \|x\| = 1$. An element $u \in Z^*$ is called a projective unit if $\|u\| = 1$ and $\langle u, F_u^\diamond \rangle = 0$. $\mathcal{F}$ and U denote the collections of norm exposed faces of Z_1 and projective units in Z^*, respectively.

We define a symmetric face to be norm exposed face F in Z with the following property: there is a linear isometry S_F of Z onto Z with $S_F^2 = I$, such that the fixed point set of S_F is $(\overline{sp}F) \oplus F^\diamond$ (topological direct sum). A generalized tripotent is a projective unit $u \in U$ with the property that $F = F_u$ is a symmetric face and $S_F^* u = u$ for some choice of symmetry S_F, corresponding to F. GU and $S\mathcal{F}$ denote the collections of generalized tripotents and symmetric faces respectively.

Definition. A real or complex normed space Z is said to be *weakly facially symmetric* (WFS), if every norm exposed face in ∂Z_1 is symmetric.

Definition. A WFS space is said to be a *strongly facially symmetric* space (abbreviation SFS space), if for every norm exposed face F in Z_1 and every

Y. Khakimdjanov et al (eds.). Algebra and Operator Theory, 203-206.

$y \in Z^*$ with $\|y\| = 1$ and $F \subset F_y$ we have $S_F^* y = y$, where S_F denotes a symmetry associated with F.

For each symmetric face F we define contractive projections $P_k(F)$ for $k = 0, 1, 2$ on Z as follows: First $P_1(F) = \frac{1}{2}(I - S_F)$ is the projection of Z onto $\{f \in Z : S_F f = -f\}$. Next we define $P_2(F)$ and $P_0(F)$ as the projections of Z onto $\overline{sp} F$ and $F^\diamond$ respectively.

These projections are called generalized Peirce projections.

A contractive projection Q on a normed space Z is said to be neutral if for each $\varphi \in Z$, $\|Q\varphi\| = \|\varphi\|$ implies $Q\varphi = \varphi$. A normed space Z is neutral if for every symmetric face F, the projection $P_2(F)$ corresponding to some choice of symmetry S_F, is neutral.

Proposition. *Let Z be a WFS− space and $Q : Z \to Z$ arbitrary neutral contractive projection. If $x \in Q^*(Z^*)$ and $\|x\| = 1$, then $F_x \subset Q(Z)$; conversely, if $x \in GU$ and $F_x \subset Q(Z)$, then $x \in Q^*(Z^*)$.*

Proof. Let $x \in Q^*(Z^*)$ and $\|x\| = 1$. Then $Q^*x = x$. Consequently for each $\rho \in F_x$ we have:

$$\langle Q\rho, x\rangle = \langle \rho, Q^*x\rangle = \langle \rho, x\rangle = 1. \tag{1}$$

As far as Q contractive projection,

$$\|Q\rho\| \leq \|\rho\|, \qquad i.e. \qquad Q\rho \in Z_1 \tag{2}$$

From (1) and (2) we have that $Q\rho \in F_x$. Then $\|Q\rho\| = \|\rho\| = 1$. Since Q neutral projection, $Q\rho = \rho$. In other words Q displays each element F_x in itself, and so

$$Q(F_x) = F_x. \tag{3}$$

As far as $Q(F_x) \subset Q(Z)$, from (3) is seen that $F_x \subset Q(Z)$.

Now we will prove the converse. Let $x \in GU$ and $F_x \subset Q(Z)$. Then $Qf = f$ for any $f \in F_x$. Consequently we have:

$$\langle f, Q^*x\rangle = \langle Qf, x\rangle = \langle f, x\rangle = 1 \tag{4}$$

and as far as Q^* contractive,

$$\|Q^*x\| = \sup |\langle g, Q^*x\rangle| \leq \|x\| = 1 \tag{5}$$

From (4) and (5) it follows that $\|Q^*x\| = \|x\| = 1$. Since Q^* is neutral, $Q^*x = x$. This means that $x \in Q^*(Z^*)$. Proposition is proved.

Remark 1. This result is proved in the theorem 2.3. from [2] for neutral weakly facially symmetrical spaces for generalized Peirce projections $P_k(F)$, where $k \in \{0, 2\}$.

If Y is a closed subspace of a normed space Z, the collections of norm exposed faces and symmetric faces in ∂Y_1 will be denoted by $\mathcal{F}_Y$ and $S\mathcal{F}_Y$ respectively. For any $K \in \mathcal{F}_Y$, by the Hahn-Banach theorem, $K = F_x \cap Y$ for some $x \in Z^*$.

Lemma 1. *Let Z be a $WFS-$ space and $Q : Z \to Z$ arbitrary contractive projection. If $K \in \mathcal{F}_{Q(Z)}$ and F_y is the smallest norm exposed face, for which $K = F_y \cap Q(Z)$, then $Q(F_y) = K$.*

Proof. Since Q is a contractive projection, $\|Q^*y\| = sup|\langle g, Q^*y\rangle| \leq \|y\| = 1$. But on the other hand for any $f \in K$ we have: $f \in Z_1$, $\langle f, y\rangle = 1$ and $Qf = f$. Therefore, $\langle f, Q^*y\rangle = \langle Qf, y\rangle = \langle f, y\rangle = 1$. Consequently, $\|Q^*y\| = \langle f, Q^*y\rangle = 1$ and $K \subset F_{Q*y}$. As far as F_y- the smallest norm exposed faces, contained in K, then $F_y \subset F_{Q^*y}$. Hence for arbitrary $g \in F_y$ we have:

$$\langle Qg, y\rangle = \langle g, Q^*y\rangle = 1 \tag{6}$$

Since $Qg \in Z_1$, from (56) follows that $QF_y \subset F_y$. Then $QF_y \subset F_y \cap Q(Z)$, i.e. $QF_y \subset K$.

Now we will prove the converse. As far as $K = F_y \cap Q(Z)$, $K \subset F_y$, and so $Q(K) \subset Q(F_y)$. But $Q(K) = K$, consequently $K \subset Q(F_y)$. Thereby $K = Q(F_y)$. Lemma is proved.

Lemma 2. *Let Z be a $WFS-$ space and $Q : Z \to Z$ an arbitrary neutral contractive projection. If $K \in \mathcal{F}_{Q(Z)}$, then $K \in \mathcal{F}_Z$.*

Proof. Let $K \in \mathcal{F}_{Q(Z)}$. Then, as it was confirmed above, there is a norm exposed faces $F \in \mathcal{F}_Z$ such, that $K = F \cap Q(Z)$. We choose the least from these faces. Let this be F_y. Then on the strength of lemma 2, $Q(F_y) = K$, consequently $Q(F_y) \subset F_y$. Thus, if $f \in F_y$, then $Qf \in F_y$. Then $\|Qf\| = \|f\|$. From the neutrality of Q it follows that $Qf = f$. So $F_y = Q(F_y)$. Therefore, $K = F_y$, i.e. $K \in \mathcal{F}_Z$. Lemma is proved.

Theorem 3. *Let Z be a $WFS-$ space and $Q : Z \to Z$ an arbitrary neutral contractive projection. Then $Q(Z)$ is a WFS space.*

Proof. Let $K \in \mathcal{F}_{Q(Z)}$. Then, by the lemma 2 $K \in \mathcal{F}_Z$. As far as Z is a WFS space, $K \in S\mathcal{F}_Z$. So there exists a symmetry S_K on Z corresponding to K. The set of fixed point of this symmetry is $\overline{sp}K \oplus K^\diamond$. Then $S_K|_{Q(Z)}$ leaves invariant $\overline{sp}K \oplus (K^\diamond \cap Q(Z))$. Now we will prove that S_K is invariant on $Q(Z)$. Let $f \in Q(Z)$. Then $f = f_0 + f_1 + f_2$, where $f_0 \in P_0(K)(Z) \cap Q(Z)$, $f_1 \in P_1(K)(Z)$, $f_2 \in P_2(K)(Z)$.

It is clear, that $f_0, f_2 \in Q(Z)$. As far as $Qf = f$, i.e. $Q(f_0 + f_1 + f_2) = f_0 + f_1 + f_2$, $f_0 + Qf_1 + f_2 = f_0 + f_1 + f_2$. Thus $Qf_1 = f_1$, i.e. $f_1 \in Q(Z)$.

Therefore $S_K f = S_K(f_0 + f_1 + f_2) = f_0 - f_1 + f_2 \in Q(Z)$, i.e. S_K is invariant on $Q(Z)$. This means that $S_K|_{Q(Z)}$ is a symmetry on $Q(Z)$ corresponding to K. Then $K \in S\mathcal{F}_{Q(Z)}$. So $Q(Z)$ is a WFS space. Theorem is proved.

Corollary. Let Z be a $SFS-$ space and $Q : Z \to Z$ an arbitrary neutral contractive projection. Then $Q(Z)$ is SFS space.

Remark 2. Theorem 3 gives an affirmative answer to the question, suggested in work [2,p.. 119] in the case of neutrality of contractive projection.

References

[1] Friedman Y, Russo B. *Quart, J.Math. Oxford*, 37:2(1986), 263-277.

[2] Friedman Y, Russo B. *Math. Proc. Camb. Phil. Soc.*, 106(1988), 107-124.

[3] Friedman Y, Russo B. *Proc. London. Math. Soc.*, 65(1990), 142-174.

[4] Yadgorov N.J *Dokl. AN RUz.*, 5(1996), 6-8.

THE PROPERTY (tα) FOR LOCALLY COMPACT CONNECTED GROUPS

K.K.MUMINOV

Departement of Mathematics,
Tashkent State University,
700095, Tashkent, Uzbekistan

Introduction

In the work [5] various classes of topological groups $t\alpha, t\beta$ - *groups)* have been introduced, the representations of which have one or another properties connected with existence of invariant vectors or invariant functionals. In the same paper it was noticed that any locally compact $t\beta$ - *groups* is compact. It is natural to expect that the analogous result remains true for $t\alpha$ - *groups*, too, in spite of the fact that a $t\alpha$ -representation is not, generally, a $t\beta$ -representation. In the present work the positive solution of this problem is given in the class of connected locally compact groups.

Let (E,t) be a separable topological vector space over the field C of complex numbers, let GL (E) be the group of all continuous linear transformations of E into itself having continuous inverse.

Let G be a group. A homomorphism $\rho : G \rightarrow GL(E)$ is called a linear representation (action) of the group G in E, and E is called a G-module.

A linear representation $\rho : G \rightarrow GL(E)$ of a topological group (G, τ) in E is said to be continuous, if the map $((G, \tau), (E, t)) \rightarrow (E, t)$ defined by the formula $(g, x) \rightarrow \rho(g)x$ is continuous with respect to each argument. Everywhere below we consider only continuous representations.

Denote by E^G the set $\{x \in E : \rho(g)x = x \text{ for all } g \in G\}$. It is clear that if $\rho : G \rightarrow GL(E)$ is a continuous representation of the group G in E, then E^G is a closed subspace in E. Denote by E' the space of all continuous linear functionals on E.

We call a linear representation $\rho : G \rightarrow GL(E)$ a $t\alpha$ - representation if for

Y. Khakimdjanov et al (eds.), Algebra and Operator Theory, 207-211.

any $x \in E^G, x \neq 0$, there exists a continuous G-invariant functional $f \in E'$ such that $f(x) \neq 0$. In the case when $E^G = \{0\}$ the representation ρ is also considered a $t\alpha$- representation.

Recall that a closed subspace V of (E,t) is said to be t-complemented if there exists a closed subspace W of E such that $V \cap W = \{0\}$ and E is a topological direct sum of the subspaces V and W. In addition, it is considered that there exist continuous projections from E onto V and W. In this case the subspace W is called a t-complement of V and the notation E=V⊕W is used.

A continuous linear representation $\rho : G \to GL(E)$ of a group (G, τ) in (E,t) is called a $t\gamma$ representation if for every t-complemented G-invariant subspace in E there exists a G-invariant t-complement.

A continuous linear representation $\rho : G \to GL(E)$ of a group (G, τ) in (E,t) is called a $t\beta$ - representation if for every non-zero continuous G-invariant linear functional f on E there exists G-invariant element $x \in E$ such that $f(x) \neq 0$.

A group is (G, τ) called a $t\alpha$ -group (resp., $t\beta$ group,$t\gamma$ -group) if each its continuous representation is a $t\alpha$ - (resp., $t\beta$-, $t\gamma$-) representation.

A topological group G is said to be of class $ft\alpha$ (resp., $ft\beta$-, $ft\gamma$-) if each its finite-dimensional continuous representation is a $t\alpha$-(resp., $t\beta$-, $t\gamma$-) representation.

Our main result is the following.

Theorem 1 *Every locally compact connected $t\alpha$- group (G, τ) is a compact group.*

The proof of the theorem consists of several steps.

Proposition 1 *Let (G, τ) be a $t\alpha$- group, let N be a closed normal divisor, let t be the factor-topology in G/N. Then (G/N, t) is a $t\alpha$- group.*

Proposition 2 *The group $G = G_0 \times L$ with the normal divisor $L \cong R^n$, $n > 0$, is not a $ft\alpha$- group.*

Lemma 1 If (G, τ) is a semisimple connected Lie group which is a $t\alpha$ -group, then G is a compact group.

Proof Since G is a semisimple Lie group, the correspondent Lie algebra U=Lie G has trivial center [1]. Let G_0 be the center of G, then U_0 =Lie G_0 is the center of G, i.e. U_0 is a trivial algebra, in addition, the Lie algebras Lie (G/G_0) and $U/U_0 = U$ are isomorphic, hence the groups G and G/G_0 are locally isomorphic ([2], p.109).

If we show that the group G/G_0 is compact, then it will follow from ([1], p.409,462) that the 1-connected Lie group G_1 with the Lie algebra Lie G_1 =Lie $(G/G_0) = U$ is compact, too. But then $G = G_1/N$ is also a compact group, where N is some discrete normal divisor from the center of G_1 ([2], p.110). Thus it sufficies to show that connected semisimple Lie group with trivial center being a $t\alpha$- group is a compact group.

Let G=NAK be the Ivisava decomposition for G ([2], p.116), let M be the centralizer of A in K and let P=NAM be the minimal parabolic subgroup in G.

Let us fix $\lambda \in (LieA)^*$ and consider the character $x_\lambda(a)$, setting

$$x_\lambda(expx) = expi\lambda(x), \quad x \; LieA$$

The character x_λ can be extended a representation of the group P which is trivial on N and on M.

We consider

$$E^\infty(\lambda) = \{f \in C^\infty(G) : f(ax) = x_\lambda(a)f(x)\}$$

It is clear that $E^\infty(\lambda)$ is G-module with respect to the right regular representation of G. Denote by f_k the restriction of a function $f \in E^\infty(\lambda)$ onto K. The map $E^\infty(\lambda) \ni f \to f_k$ is injective, hence the sesquilinear form $(f, g) = \int_k f(x)\bar{g}(x)dx$ defines a scalar product on $E^\infty(\lambda)$. Let be the Hilbert space which is obtained by completing $E^\infty(\lambda)$ with respect to the norm $\|f\| = \sqrt{(f,f)}$. $H(\lambda)$ are G-modules and have the following properties [2]:

1) $H(\lambda)' = H(\lambda + 2\rho)$, where ρ is the half-sum of bounded positive roots from $(LieA)^*$;

2) $C_0 \hookrightarrow H(\lambda) \Leftrightarrow \lambda = 0$, where C_0 is the trivial G-module with the action $(g - e)C_0 = 0$ for all $g \in G$.

Suppose G is not a compact group. Then the group A is not trivial, $\rho \neq 0$ and, since $H(O)' = H(2\rho)$ does not contain C_0, G is not a $t\alpha$- group. The obtained contradiction implies that G is a compact group.

Lemma 2 *Let G be a connected Lie group and tα- group. Then G is a compact group.*

Proof It follows from the structure theorem for connected Lie groups that G=SR (the Levi-Maltsev decomposition), where S is the maximal connected semisimple subgroup in G, R is the maximal connected resolvable normal divisor in G (i.e.radical in G), in addition, $S \cap R$ is a discrete subgroup ([2], p.114).

Since G is a $t\alpha$- group, G/R is a $t\alpha$- group, too (see Proposition 1). We have $G/R \approx S/S \cap R$, hence $S/S \cap R$ is a $t\alpha$- group. The group $S \cap R$ is discrete, therefore its Lie algebra Lie $(S \cap R)$ is trivial, which imples the equalities

$$Lie(S/S \cap R) = Lie(S)/Lie(S \cap R) = LieS$$

Therefore the Lie algebra Lie (S/S∩R) is semisimple and so the Lie group S/S∩R is semisimple. It remains to note that S/S∩R is a connected Lie group. Thus, by lemma 1, $G/R \approx S/S \cap R$ is a compact group.

We show now that the group R is compact. Let R' be the derived group in R. Then R' is a normal divisor in R and in G ([3], p.28) in addition, R/R' is abelian group ([4], p.317).

We have G=SR and $G/R' = S(R/R')$. Since the group R/R' is abelian and connected, we have $R/R' = T^k \times R^n$, where T^k is k-dimensional torus and the maximal compact subgroup in R/R'.

We show that $adg(T^k) \subset T^k$ for all $g \in S$. Consider $B_g = \{g^{-1}ag : a \in T^k\}, g \in S$. Since R is a normal divisor in G, R/R' is a normal divisor in G/R', so $B_g \subset R/R'$.

Since $a \to g^{-1}ag = adg(a)$ is continuous, $B_g = adg(T^k)$ is a compact subgroup of R/R' hence $B_g \subset T^k$. Therefore, T^k is a normal divisor in G/R'. So we can consider the group

$$(G/R')T^k = (G/R(R/R'))/T^k = G/R \times ((T^k \times R^n)/T^k) = (G/R)R^n.$$

It follows from Proposition 1 that $(G/R')/T^k$ is a $t\alpha$- group, hence $(G/R)R^n$ is a $t\alpha$- group, too. By Proposition 2, we have n=0. Thus, the group R/R' is compact.

Let R'' be the second derivative of the group R. We consider

$$G/R'' = G/R \cdot R/R'' = G/R' \cdot R'/R'' = (G/R \cdot R/R') \cdot R'/R''$$

The group R'/R'' is abelian and connected. Repeating the above arguments we get that the group R'/R'' is compact, too. Repeating this process m times, where m is the degree of resolvability of R, we get that $R^{(m-1)}/R^{(m)}$ is a compact group, in addition, $R^{(m)} = \{e\}$ is compact. Then (see [3], p.118) the groups $R^{(m-1)}, R^{(m-2)}, ..., R^1$ and R are compact, too. Since G/R and R are compact, G is a compact group, too.

Proof of theorem 1 Since (G, τ) is a locally compact group, for any neighborhood U of the unit element e there exists closed normal divisor $N \subset U$ such that G/N is Lie group with respect to the factor-topology t ([2], p.112).

Since G is locally compact, the neighborhood U can be regarded as a compact one, so N is a compact subgroup of G.

Since G is a $t\alpha$- group, by Proposition 1, (G/N,t) is also a $t\alpha$- group. Thus, (G/N,t) is a connected Lie group which is a $t\alpha$- group. It follows from Lemma 2 that (G/N,t) is a compact group. Hence the group (G, τ) is compact, too (see [3], p.118).

Theorem 1 and ([5], Proposition 2, Theorem 1) imply the following.

Corollary *Let G be a locally compact connected group. The following conditions are equivalent:*

1) G *is a* $t\alpha$ *-group;*
2) G *is a* $t\beta$ *- group;*
3) G *is a* $t\gamma$ *- group;*
4) G *is a compact group.*

Proposition 3 *Let G be a* $ft\alpha$*- group, N be a normal divisor of G. Then G/N is a* $ft\alpha$*- group, too.*

Further, we will use the following variant of Lemma 2.

Lemma 3 *Let G be a connected* $ft\alpha$*-Lie group. Then there exists a compact normal divisor R of G such that G/R is a semisimple Lie group.*

Proof By the Levi-Maltsev decomposition, we have G=SR, where S is a connected semisimple subgroup of G, R is the maximal connected resolvable normal divisor of G. As it was mentioned in the proof of Lemma 2, G/R=S/S∩R is a semisimple connected Lie group.

Using Proposition 3 and repeating the proof of Lemma 2 we get $R/R' = T^k * R^n$ and $(G/R)R^n = (G/R')/T^k$ is a $ft\alpha$- group, where R' is the derivative group of R. It is clear that the assertion of Proposition 2 holds true for $ft\alpha$-groups,too, i.e. $(G/R)R^n$ is not a $ft\alpha$-group for $n \neq 0$. Thus, n=0 which implies compactness of the group R/R'. Repeating the end of the proof of Lemma 2 we obtain that the group R is compact, too.

Theorem 2 *Let G be a locally compact connected $ft\alpha$-group. Then there exists a compact normal divisor N of G such that G/N is a semisimple connected Lie $ft\alpha$- group.*

Proof It follows the proof of Theorem 1 that there exists a compact normal divisor N_1 of G such that G/N_1 is a connected $ft\alpha$-Lie group. By Lemma 3, there exists a compact normal divisor R of G/N_1 such that $(G/N_1)/R$ is a semisimple connected $ft\alpha$-group. Let

$$N = \{g \in G : gN_1 \in R\}$$

Then N is a normal subgroup of G. Since the groups $N/N_1 = R$ and N_1 are compact, N is also a compact subgroup of G. In addition, $G/N = (G/N_1)/R$ is a semisimple connected $ft\alpha$-Lie group.

References

[1] Zelobenko D.P. Compact Lie groups and their representations.

[2] Zelobenko D.P., Stern A.I. Representations of Lie groups. Moscow: Nauka, 1983.

[3] Pontryagin L.S. Continuous groups. Moscow: Nauka, 1984.

[4] Kostrikin A.I. Introdiution to algebra. Moscow: Nauka, 1977.

[5] Muminov K.K. Various topological variants of Linear reduction and their interconnections. *Dokl.Acad.Nauk UzSSR*, 3(1991), 5-6.

E-mail: mumin@saturn.silk.org.

GRUPOS CUÁNTICOS

Ch. KASSEL

Institut de Recherche Mathématique Avancée,
Université Louis Pasteur - C.N.R.S.,
7 rue René Descartes, 67084 Strasbourg Cedex, France.

Abstract. These notes provide a concise introduction to the theory of quantum groups from the point of view of braided tensor categories.

Estas son las notas del curso que dí en Agosto de 1994 en el XI° Coloquio Latinoamericano de Álgebra en Mendoza, Argentina y en el Coloquio Franco-Ouzbeko. El objetivo de éste fue dar una introducción a la teoría de los grupos cuánticos desde el punto de vista de las categorías.

Los grupos cuánticos de los que hablamos son las llamadas "álgebras envolventes cuánticas" de Drinfeld-Jimbo. Estas son deformaciones formales de las álgebras envolventes de las álgebras de Lie semisimples. La definición de estas álgebras es bastante complicada; lo haremos solo en el caso más simple, cuando el álgebra de Lie es $\mathfrak{sl}(2)$. Las álgebras envolventes cuánticas tienen una propiedad común muy importante: sus categorías de módulos son categorías trenzadas. Este último concepto es el más estudiado en este curso. También consideraremos otras categorías trenzadas interesantes, que no son categorías de módulos, como las construídas con objetos 1-dimensionales en el plano.

El contenido de estas notas es el siguiente. Empezamos definiendo categorías trenzadas, ilustrando este concepto con dos tipos de ejemplos:

(i) el primer ejemplo serán categorías de módulos sobre ciertas álgebras, llamadas biálgebras trenzadas. Las álgebras envolventes cuánticas de Drinfeld-Jimbo son de este tipo.

(ii) construiremos la categoría trenzada $\mathcal{B}$ utilizando ciertos objetos 1-dimensionales en el plano, llamados trenzas. La categoría $\mathcal{B}$ es universal entre todas

Y. Khakimdjanov et al (eds.), Algebra and Operator Theory, 213-236.

las categorías trenzadas.

Después haremos una construcción categórica del doble cuántico de Drinfeld. El método mostrado aquí es más conceptual y más fácil de entender que la construcción original de Drinfeld. Esto nos permitirá obtener la R-matriz universal para el álgebra envolvente cuántica de $\mathfrak{sl}(2)$.

En la Lección IV damos una versión infinitesimal del concepto de categoría trenzada, esto es, el concepto de una categoría infinitesimal simétrica. La categoría de módulos sobre un álgebra de Lie semisimple es un ejemplo de estas categorías. Existe una categoría simétrica infinitesimal universal $\mathcal{A}$, que también construiremos en estas notas.

Terminamos este curso mostrando cómo construir una categoría trenzada dada cualquier categoría infinitesimal simétrica. Por un importante teorema de Drinfeld, si aplicamos este método a la categoría de módulos sobre un álgebra de Lie semisimple, recuperamos la categoría de módulos sobre la correspondiente álgebra envolvente cuántica. Por otro lado, considerando la categoría $\mathcal{A}$ de la Lección IV, tenemos una representación universal del grupo de trenzas.

Mi cálido agradecimiento a Silvia Cécere, María Julia Redondo y Andrea Solotar quienes me ayudaron a escribir estas notas y las tradujeron al español.

El plan de este curso es el siguiente.

Lección I. Categorías tensoriales

I.1. Definición. Recordamos una definición clásica (ver [Mac71], [Kas95], Capítulo XI).

Definición 1. *Una categoría tensorial $\mathcal{C}$ es una categoría provista de un producto tensorial $\otimes : \mathcal{C} \times \mathcal{C} \longrightarrow \mathcal{C}$, esto es, dados objetos V, W de la categoría, $V \otimes W$ es un objeto de $\mathcal{C}$, y dados morfismos $f : V \longrightarrow V'$ y $g : W \longrightarrow W'$, se le asocia un morfismo $f \otimes g : V \otimes W \longrightarrow V' \otimes W'$ de $\mathcal{C}$ que verifica:*

(i) $\otimes$ es asociativo en los objetos y en los morfismos,

(ii) $\mathrm{id}_{V\otimes W} = \mathrm{id}_V \otimes \mathrm{id}_W$, V, W de $\mathcal{C}$,

(iii) Si V, V', V'', W, W', W'' son objetos de $\mathcal{C}$, y $f : V \longrightarrow V'$, $g : W \longrightarrow W'$, $f' : V' \longrightarrow V''$, $g' : W' \longrightarrow W''$ son morfismos de $\mathcal{C}$, entonces

$$(f' \otimes g') \circ (f \otimes g) = (f' \circ f) \otimes (g' \circ g),$$

(iv) Existe un objeto I de $\mathcal{C}$ tal que $I \otimes V = V \otimes I = V$, para todo objeto V, y $\mathrm{id}_V \otimes f = f \otimes \mathrm{id}_I = f$ para todo morfismo f en $\mathcal{C}$.

I.2. La categoría de los espacios vectoriales. Si k es un cuerpo, $\mathcal{C}$ la categoría de k-espacios vectoriales, $\otimes$ el producto tensorial usual, entonces $\mathcal{C}$ es una categoría tensorial con $I = k$.

Sea G un grupo. Es claro que la subcategoría de espacios vectoriales con una acción a izquierda de G es también tensorial. El producto tensorial es el mismo que en el caso de los espacios vectoriales. Además, si V y W son G-módulos, entonces $V \otimes_k W$ es un G-módulo con la acción

$$g(v \otimes w) = gv \otimes gw \tag{1.1}$$

para todo $g \in G$, $v \in V$ y $w \in W$. El grupo actúa trivialmente en el espacio vectorial unidimensional k, es decir, $gv = v$ para todo $g \in G$ y $v \in k$.

I.3. Módulos sobre una biálgebra. La categoría de G-módulos es un caso particular de la categoría A-mod de módulos a izquierda sobre una k-álgebra A, asociativa y con unidad, tomando en este caso $A = k[G]$ el álgebra de grupo.

Ahora buscamos condiciones sobre la k-álgebra A que impliquen que A-mod es una subcategoría tensorial de la categoría de los espacios vectoriales. Si V, W son A-módulos, entonces $V \otimes_k W$ es un $A \otimes A$-módulo. Si existe un morfismo de k-álgebras $\Delta : A \to A \otimes A$, entonces $V \otimes_k W$ resulta un A-módulo definiendo para $a \in A$, $v \in V$ y $w \in W$

$$a(v \otimes w) = \Delta(a)(v \otimes w) = \sum_{(a)} a'v \otimes a''w \tag{1.2}$$

si $\Delta(a) = \sum_{(a)} a' \otimes a''$. Es fácil ver que para que se cumplan (i–iii) de la Definición 1, el morfismo Δ debe ser coasociativo, es decir que

$$(\Delta \otimes \mathrm{id}_A)\Delta = (\mathrm{id}_A \otimes \Delta)\Delta. \tag{1.3}$$

Como objeto I hay que tomar al cuerpo k. Para que k sea un A-módulo, necesitamos un morfismo de k-álgebras $\varepsilon : A \to k$. Entonces k resulta un A-módulo definiendo para $a \in A$, $v \in k$

$$av = \varepsilon(a)v. \tag{1.4}$$

Tenemos tambien las relaciones

$$(\varepsilon \otimes \mathrm{id}_A)\Delta = (\mathrm{id}_A \otimes \varepsilon)\Delta = \mathrm{id}_A. \tag{1.5}$$

Una terna (A, Δ, ε) satisfaciendo Condiciones (1.3) y (1.5) se llama una *biálgebra.* Los morfismos Δ y ε son la *comultiplicación* y la *counidad* de la biálgebra A.

El álgebra de grupo $k[G]$ de I.2 es una biálgebra con comultiplicación y counidad dadas por

$$\Delta(g) = g \otimes g \quad \text{y} \quad \varepsilon(g) = 1 \tag{1.6}$$

para todo $g \in G$.

El álgebra envolvente $U(\mathfrak{g})$ del álgebra de Lie $\mathfrak{g}$ es también una biálgebra, con comultiplicación y counidad dadas por

$$\Delta(x) = x \otimes 1 + 1 \otimes x \quad \text{y} \quad \varepsilon(x) = 0 \tag{1.7}$$

para todo $x \in \mathfrak{g}$. Por lo tanto, la categoría $U(\mathfrak{g})$-mod de $\mathfrak{g}$-módulos es también un subcategoría tensorial de la categoría de los espacios vectoriales.

En estos ejemplos se verifica que $\Delta = \Delta^{\mathrm{op}}$, donde $\Delta^{\mathrm{op}} = \sigma \circ \Delta$, y σ es la trasposición (12). En este caso, (A, Δ, ε) se llama una k-biálgebra *coconmutativa.* Veremos ahora un ejemplo de una biálgebra que no es conmutativa ni coconmutativa.

I.4. El álgebra envolvente cuántica de $\mathfrak{sl}(2)$. Los primeros ejemplos de biálgebras eran conmutativos o coconmutativos. Drinfeld y Jimbo [Dri87], [Jim85] construyeron ejemplos naturales de biálgebra no conmutativa ni coconmutativa, a partir de las álgebras de Lie semisimples $\mathfrak{g}$, que vamos a describir para el caso $\mathfrak{g} = \mathfrak{sl}(2)$.

Recordemos que

$$\mathfrak{sl}(2) = \left\{ \begin{pmatrix} a & b \\ c & d \end{pmatrix} \in M_2(\mathbf{C}) \mid a + d = 0 \right\}$$

y está generada como álgebra por H, X, Y, donde

$$H = \begin{pmatrix} 1 & 0 \\ 0 & -1 \end{pmatrix}, \quad X = \begin{pmatrix} 0 & 1 \\ 0 & 0 \end{pmatrix} \quad \text{y} \quad Y = \begin{pmatrix} 0 & 0 \\ 1 & 0 \end{pmatrix}$$

con las relaciones

$$[H, X] = 2X, \quad [H, Y] = -2Y, \quad [X, Y] = H. \tag{1.8}$$

Llamamos $U_h(\mathfrak{sl}(2))$ a la $\mathbf{C}[[h]]$-álgebra generada por X, Y, H con las relaciones

$$[H, X] = 2X, \quad [H, Y] = -2Y \tag{1.9}$$

y

$$[X, Y] = \frac{\exp(hH/2) - \exp(-hH/2)}{\exp(h/2) - \exp(-h/2)} = \frac{\sinh(hH/2)}{\sinh(h/2)}. \tag{1.10}$$

La estructura de biálgebra está definida por la comultiplicación Δ y la counidad ε:

$$\Delta(H) = 1 \otimes H + H \otimes 1, \tag{1.11}$$

$$\Delta(X) = X \otimes \exp(hH/4) + \exp(-hH/4) \otimes X, \tag{1.12}$$

$$\Delta(Y) = Y \otimes \exp(hH/4) + \exp(-hH/4) \otimes Y, \tag{1.13}$$

$$\varepsilon(H) = \varepsilon(X) = \varepsilon(Y) = 0. \tag{1.14}$$

Es evidente que $U_h(\mathfrak{sl}(2))$ no es conmutativa ni coconmutativa. Además $[X, Y] \equiv H$ módulo h. Más generalmente, para $h = 0$ recuperamos el caso clásico, esto es,

$$U_h(\mathfrak{sl}(2))/hU_h(\mathfrak{sl}(2)) \cong U(\mathfrak{sl}(2)) \tag{1.15}$$

como biálgebras.

I.5. Una categoría de dibujos. En los ejemplos de categorías que hemos visto, nunca especificamos los morfismos porque era claro a quienes nos referíamos. Ahora describiremos una categoría $\mathcal{D}$ de naturaleza completamente diferente, para la cual la información esencial está en los morfismos, no en los objetos.

Los objetos de la categoría $\mathcal{D}$ son los números naturales. Un morfismo en $\mathcal{D}$ del objeto m en el objeto n es cualquier figura que uno quiera dibujar con las siguientes propiedades: la figura está en la banda horizontal $\mathbf{R} \times [0, 1]$; tiene extremos, que son intervalos topológicos, que unen m puntos distintos de la recta $\mathbf{R} \times \{0\}$ con n puntos distintos de la recta $\mathbf{R} \times \{1\}$. Consideramos el conjunto de las figuras a menos de homeomorfismos, con los extremos fijos.

La composición en $\mathcal{D}$ está dada por: sean $f : m \longrightarrow n$ y $g : n \longrightarrow p$ morfimos en $\mathcal{D}$ representados por figuras D y D' respectivamente. Por definición, D tiene n extremos superiores y D' tiene la misma cantidad de extremos inferiores. La composición $g \circ f$ es la clase de homeomorfismo de la figura que se obtiene poniendo D' arriba de D, pegando los correspondientes extremos, y comprimiendo la figura resultante en la banda $\mathbf{R} \times [0,1]$. La identidad de un objeto $n > 0$ esta representada por la unión $\{1, \ldots, n\} \times [0,1]$ de intervalos. La identidad del objeto 0 es la figura vacía.

Afirmamos que $\mathcal{D}$ es una categoría tensorial. En los objetos, el producto tensorial está dado por la adición: $m \otimes n = m + n$. En los morfismos, se define así: sean $f : m \longrightarrow n$ y $g : p \longrightarrow q$ morfismos de $\mathcal{D}$ representados por figuras D y D' respectivamente. Definimos $f \otimes g$ como la clase de homeomorfismo de la figura obtenida poniendo D' a la derecha de D, de tal manera que ambas figuras sean disjuntas. El lector puede verificar que todos los axiomas de la Definición 1 se satisfacen. Es claro que 0 es el objeto unidad. Observar que su conjunto de endomorfimos es muy grande: es el conjunto de clases de homeomorfismo de todas las figuras en la banda abierta $\mathbf{R} \times]0,1[$.

La categoría $\mathcal{D}$ es anecdótica. Fue desarrollada para este curso, solo para considerar un ejemplo sencillo muy diferente a los usuales. De todos modos, en lo que sigue encontraremos otras dos categorías de figuras construídas en esta forma. Ambas serán muy importantes en la teoría de grupos cuánticos.

Lección II. Categorías trenzadas

II.1. Definición. El siguiente concepto es fundamental para este curso.

Definición 2. *Dada una categoría tensorial $\mathcal{C}$, un trenzado de $\mathcal{C}$ es una familia de isomorfismos*

$$c_{V,W} : V \otimes W \longrightarrow W \otimes V$$

para todos los objetos V, W de $\mathcal{C}$ tales que:

(i) Si $f : V \longrightarrow V'$, $g : W \longrightarrow W'$,

$$c_{V',W'} \circ (f \otimes g) = (g \otimes f) \circ c_{V,W}. \tag{2.1}$$

(ii) Los siguientes triángulos conmutan:

$$\begin{array}{ccc} U \otimes V \otimes W & \xrightarrow{c_{U \otimes V, W}} & W \otimes U \otimes V \\ {\scriptstyle \mathrm{id}_U \otimes c_{V,W}} \searrow & & \nearrow {\scriptstyle c_{U,W} \otimes \mathrm{id}_V} \\ & U \otimes W \otimes V & \end{array} \tag{2.2}$$

y

$$\begin{array}{ccc} U \otimes V \otimes W & \xrightarrow{c_{U, V \otimes W}} & V \otimes W \otimes U \\ {\scriptstyle c_{U,V} \otimes \mathrm{id}_W} \searrow & & \nearrow {\scriptstyle \mathrm{id}_V \otimes c_{U,W}} \\ & V \otimes U \otimes W & \end{array} \tag{2.3}$$

para todos los objetos U, V, W. Una categoría tensorial $\mathcal{C}$ provista de un trenzado se dice una categoría trenzada.

Si $\mathcal{C}$ es la categoría de k-espacios vectoriales, y $c_{V,W} : V \otimes W \to W \otimes V$ es la trasposición $\sigma_{V,W}(v \otimes w) = w \otimes v$, la categoría $\mathcal{C}$ resulta una categoría trenzada.

II.2. La ecuación de Yang-Baxter.

Proposición 1. *En una categoría trenzada, el siguiente hexágono conmuta para todo objeto U, V, W:*

$$\begin{array}{ccc}
U \otimes V \otimes W & \xrightarrow{c_{U,V} \otimes \mathrm{id}_W} & V \otimes U \otimes W \\
{\scriptstyle \mathrm{id}_U \otimes c_{V,W}} \downarrow & & \downarrow {\scriptstyle \mathrm{id}_V \otimes c_{U,W}} \\
U \otimes W \otimes V & & V \otimes W \otimes U \\
{\scriptstyle c_{U,W} \otimes \mathrm{id}_V} \downarrow & & \downarrow {\scriptstyle c_{V,W} \otimes \mathrm{id}_U} \\
W \otimes U \otimes V & \xrightarrow{\mathrm{id}_W \otimes c_{U,V}} & W \otimes V \otimes U
\end{array}$$

DEMOSTRACIÓN.— Observar que si completamos el hexágono con las flechas $c_{U,V\otimes W}$ y $c_{U,W\otimes V}$, entonces los triángulos conmutan. Ahora basta ver que el rectángulo interior conmuta, y esto es inmediato a partir de la relación (2.1). *QED*

Corolario 1. *Para todo objeto V en una categoría trenzada, el automorfismo $c = c_{V,V}$ de $V \otimes V$ satisface la ecuación de Yang-Baxter*

$$(c \otimes \mathrm{id})(\mathrm{id} \otimes c)(c \otimes \mathrm{id}) = (\mathrm{id} \otimes c)(c \otimes \mathrm{id})(\mathrm{id} \otimes c) \tag{2.4}$$

Los grupos cuánticos fueron construídos originariamente para encontrar soluciones a la ecuación de Yang-Baxter (ver [Dri87]).

II.3. Biálgebras trenzadas. Veamos cuándo resulta trenzada la categoría tensorial de A-módulos.

Proposición 2. *Sea (A, Δ, ε) una biálgebra y sea $\mathcal{C}$ la categoría tensorial de A-módulos. La categoría $\mathcal{C}$ es trenzada sí y sólo si existe $R \in A \otimes A$ tal que:*

(a) R es inversible en el álgebra $A \otimes A$,

(b) para todo $a \in A$

$$\Delta^{op}(a) = R\Delta(a)R^{-1}, \tag{2.5}$$

(c)

$$(\Delta \otimes \mathrm{id})(R) = R_{13}R_{23} \quad y \quad (\mathrm{id} \otimes \Delta)(R) = R_{13}R_{12} \tag{2.6}$$

donde $R_{12} = R \otimes 1$, $R_{23} = 1 \otimes R$ y $R_{13} = \sum_{(R)} R^{(1)} \otimes 1 \otimes R^{(2)}$ donde $R = \sum_{(R)} R^{(1)} \otimes R^{(2)}$.

Si A satisface las condiciones de la proposición 2, A se llamará una *biálgebra trenzada*. Drinfeld [Dri87], [Dri89] usaba la terminología "álgebra cuasitriangular". El elemento R se llama la *R-matriz universal* de A. Si A es *coconmutativa*, por ejemplo, $A = k[G]$ ó $A = U(\mathfrak{g})$, puede tomarse $R = 1 \otimes 1$ y el trenzado la trasposición σ.

DEMOSTRACIÓN.— $\Leftarrow$) Se define $c_{V,W} : V \otimes W \to W \otimes V$ a partir de la R-matriz universal por:

$$c_{V,W}(v \otimes w) = \sigma\big(R(v \otimes w)\big) = [R(v \otimes w)]_{21} \tag{2.7}$$

Es claro que $c_{V,W}$ es un isomorfismo que verifica (2.1), y para ver que los triángulos (2.2–2.3) conmutan, es suficiente usar la condición (2.6) de la proposición.

$\Rightarrow$) Sea $R = [c_{A,A}(1 \otimes 1)]_{21} \in A \otimes A$. Como $c_{A,A}$ es un isomorfismo, R es inversible. Observemos también que por la funtorialidad de c, dados $v \in V$ y $w \in W$, el siguiente diagrama conmuta:

$$\begin{array}{ccc} A \otimes A & \xrightarrow{c_{A,A}} & A \otimes A \\ {\scriptstyle \bar{v} \otimes \bar{w}} \downarrow & & \downarrow {\scriptstyle \bar{w} \otimes \bar{v}} \\ V \otimes W & \xrightarrow{c_{V,W}} & W \otimes V \end{array}$$

donde $\bar{v}$ y $\bar{w}$ son las aplicaciones A-lineales $A \to V$ y $A \to W$ definidas por $\bar{v}(1) = v$ y $\bar{w}(1) = w$. Luego

$$\begin{aligned} c_{V,W}(v \otimes w) &= c_{V,W}\big((\bar{v} \otimes \bar{w})(1 \otimes 1)\big) \\ &= (\bar{w} \otimes \bar{v})\big(c_{A,A}(1 \otimes 1)\big) \\ &= R_{21}(w \otimes v) \\ &= [R(v \otimes w)]_{21}. \end{aligned}$$

Por la A-linealidad de $c_{V,W}$, sigue fácilmente que $\Delta^{op}(a)R = R\Delta(a)$ para todo $a \in A$. La condición (c) se verifica usando la conmutatividad de (2.2–2.3). *QED*

II.4. El caso $U_h(\mathfrak{sl}(2))$. El objetivo de este párrafo es describir una subcategoría de $U_h(\mathfrak{sl}(2))$-módulos que es una categoría trenzada. Recordemos que $U(\mathfrak{sl}(2))$ es semisimple, es decir, todo $\mathfrak{sl}(2)$-módulo V de dimensión finita es suma directa de submódulos simples V_n con $\dim V_n = n+1$ y tales que la acción ρ de $\mathfrak{sl}(2)$ sobre V_n está dada en una base adecuada por

$$\rho(H) = \begin{pmatrix} n & & & & \\ & n-2 & & & \\ & & \ddots & & \\ & & & -n+2 & \\ & & & & -n \end{pmatrix},$$

$$\rho(X) = \begin{pmatrix} 0 & n & & & \\ 0 & 0 & n-1 & & \\ & & & \ddots & \\ & & & & 1 \\ & & & & 0 \end{pmatrix},$$

$$\rho(Y) = \begin{pmatrix} 0 & & & & \\ 1 & 0 & & & \\ 0 & 2 & 0 & & \\ & & \ddots & \ddots & \\ & & & n & 0 \end{pmatrix}.$$

Cada $\mathfrak{sl}(2)$-módulo V_n puede ser deformado a un $U_h(\mathfrak{sl}(2))$-módulo $\widetilde{V}_n$ libre de rango $n+1$ sobre $\mathbf{C}[[h]]$ con la acción

$$\rho_h(H) = \begin{pmatrix} n & & & & \\ & n-2 & & & \\ & & \ddots & & \\ & & & -n+2 & \\ & & & & -n \end{pmatrix},$$

$$\rho_h(X) = \begin{pmatrix} 0 & [n] & & & \\ 0 & 0 & [n-1] & & \\ & & & \ddots & \\ & & & & [1] \\ & & & & 0 \end{pmatrix},$$

$$\rho_h(Y) = \begin{pmatrix} 0 & & & & \\ [1] & 0 & & & \\ 0 & [2] & 0 & & \\ & & \ddots & \ddots & \\ & & & [n] & 0 \end{pmatrix},$$

donde $[n] = \sinh(nh/2)/\sinh(h/2)$. Observar que $[n]$ tiende a n cuando h tiende a 0.

Consideremos la categoría $U_h(\mathfrak{sl}(2))$-mod$_f$ cuyos objetos son las sumas directas finitas de los módulos $\widetilde{V}_n$. Dado cualquier $\mathfrak{sl}(2)$-mod V de dimensión finita, notamos $\widetilde{V}$ al único objeto de $U_h(\mathfrak{sl}(2))$-mod$_f$ tal que $\widetilde{V}/h\widetilde{V} = V$ como $\mathfrak{sl}(2)$-módulo. La categoría $U_h(\mathfrak{sl}(2))$-mod$_f$ es tensorial, con producto tensorial dado por $\widetilde{V} \otimes \widetilde{W} = \widetilde{(V \otimes_k W)}$.

Teorema 1. *La categoría $U_h(\mathfrak{sl}(2))$-mod$_f$ está trenzada, con trenzado inducido por la R-matriz universal*

$$R_h = \sum_{\ell \geq 0} \frac{(q-q^{-1})^\ell}{[\ell]!} q^{-\ell(\ell+1)/2} \exp\left[\frac{h}{2}\left(\frac{H \otimes H}{2} + \frac{1}{2}(\ell H \otimes 1 - 1 \otimes \ell H)\right)\right] \left(X^\ell \otimes Y^\ell\right)$$

donde $q = \exp(h/2)$, $[\ell] = (q^\ell - q^{-\ell})/(q - q^{-1})$ y $[\ell]! = [1][2]\dots[\ell]$.

El lector puede intentar probar directamente que R_h satisface las condiciones de la Definición 2. Pero esto es difícil. En III.3 encontraremos otra expresión de R_h usando la construcción del doble cuántico.

II.5. Grupos de trenzas: definición algebraica. Ahora explicaremos por qué las categorías consideradas en esta lección se llaman trenzadas. Sea n un entero > 1.

Definición 3. *El grupo B_n es el grupo generado por $\sigma_1, \ldots, \sigma_{n-1}$, con las relaciones*

$$\sigma_i\sigma_j = \sigma_j\sigma_i \quad si \mid i-j \mid \geq 2 \quad y \quad \sigma_i\sigma_{i+1}\sigma_i = \sigma_{i+1}\sigma_i\sigma_{i+1}$$

para todos i, j. Un elemento de B_n es llamado una trenza con n hebras.

Agregando la relación $\sigma_i^2 = 1$ para todo i, se obtiene una presentación del grupo simétrico $\mathfrak{S}_n$.

Ahora describiremos una relación que existe entre los grupos de trenzas y las categorías trenzadas. Si V es un objeto cualquiera de una categoría trenzada, entonces el grupo B_n actúa naturalmente sobre $V^{\otimes n}$, es decir, existe un morfismo de grupos

$$\rho_n : B_n \longrightarrow \mathrm{Aut}(V^{\otimes n})$$

que en los generadores $\sigma_1, \ldots, \sigma_{n-1}$ está dado por

$$\rho_n(\sigma_i) = \mathrm{id} \otimes \ldots \otimes \underbrace{c_{V,V}}_{i,i+1} \otimes \ldots \otimes \mathrm{id}. \tag{2.8}$$

Del Corolario 1 se deduce que esta fórmula define un morfismo de grupos en B_n.

II.6. Descripción topológica de las trenzas. Nosotros representamos una trenza con n hebras como la unión de n intervalos inmersos en $\mathbf{R} \times [0,1]$ tales que

(i) los extremos de los intervalos son los puntos $(1,0), \ldots, (n,0)$ y $(1,1), \ldots, (n,1)$ del plano,

(ii) para todo $t \in [0,1]$ salvo un número finito, la intersección de la recta $\mathbf{R} \times \{t\}$ con la trenza consiste en n puntos distintos. Más aún, suponemos que todas las singularidades de la inmersión son puntos dobles ordinarios. En cada punto doble, existe un orden de las hebras que pasan por allí, que significa que una de ellas pasa por encima de la otra, y esto lo representamos interrumpiendo el dibujo de la hebra que pasa por debajo en un entorno del punto doble.

Consideramos las trenzas a menos de homeomorfismos. Esto significa un homeomorfismo de $\mathbf{R} \times [0,1]$ que fija los extremos, que manda puntos dobles en puntos dobles, y que preserva el orden de las hebras en los puntos dobles.

El producto TT' de dos trenzas T y T' con n hebras se define como en la categoría $\mathcal{D}$ de la Lección I.5, esto es, poniendo T arriba de T', pegando los extremos y comprimiendo la trenza resultante dentro de $\mathbf{R}\times[0,1]$. Este producto tiene la unidad representada por la unión $\{1, \ldots, n\} \times [0,1]$ de intervalos.

Representamos el generador σ_k por la unión de los intervalos $\{i\} \times [0,1]$ para $1 \leq i \leq n$ y $i \neq k, k+1$ y los intervalos $[(k,0),(k+1,1)]$ y $[(k+1,0),(k,1)]$, donde el penúltimo intervalo pasa por encima del último intervalo.

II.7. La categoría trenzada universal. Consideremos la categoría $\mathcal{B}$ cuyos objetos son los números naturales, y tal que

$$\mathrm{Hom}_{\mathcal{B}}(n,m) = \begin{cases} \emptyset & \text{si } n \neq m, \\ B_n & \text{si } n = m, \end{cases}$$

donde B_n es el grupo de la Definición 3 y $B_0 = B_1 = \{1\}$. La composición la definimos por el producto en los grupos trenzados o, equivalentemente, como en la categoría $\mathcal{D}$ de I.5. La identidad id_n del objeto n es la unidad del grupo B_n. La categoría $\mathcal{B}$ es tensorial con producto tensorial definido en los objetos y en los morfismos como en el caso de $\mathcal{D}$.

Nosotros afirmamos que $\mathcal{B}$ es trenzada. Para todo par de objetos (n,m), definimos un morfismo $c_{n,m} : n \otimes m \to m \otimes n$ como el elemento de B_{n+m} definido con los generadores $\sigma_1, \ldots, \sigma_{n+m}$ por $c_{n,0} = c_{0,n} = \mathrm{id}_n$ y

$$c_{n,m} = (\sigma_m \sigma_{m-1} \ldots \sigma_1)(\sigma_{m+1}\sigma_m \ldots \sigma_2) \ldots (\sigma_{m+n-1}\sigma_{m+n-2} \ldots \sigma_n). \tag{2.9}$$

Proposición 3. *La familia de morfismos $(c_{n,m})_{n,m \geq 0}$ es un trenzado para la categoría tensorial $\mathcal{B}$.*

La demostración se deja al lector. La categoría $\mathcal{B}$ es la categoría trenzada universal, más precisamente:

Teorema 2. *Sea $\mathcal{C}$ una categoría trenzada y V un objeto de $\mathcal{C}$. Entonces existe un único funtor $F_V : \mathcal{B} \to \mathcal{C}$ tal que $F_V(0) = I$, $F_V(1) = V$,*

$$F_V(n \otimes m) = F_V(n) \otimes F_V(m), \quad F_V(T \otimes T') = F_V(T) \otimes F_V(T')$$

para todos los objetos n, m y todos los morfismos T, T' y

$$F_V(\sigma_1) = c_{V,V} \in \mathrm{Aut}(V \otimes V).$$

Demostración.— En los objetos, el funtor está definido por $F_V(n) = V^{\otimes n}$. En los morfismos T de B_n, está definido por $F_V(T) = \rho_n(T)$, donde ρ_n es la representación construída en II.5. *QED*

Lección III. El doble cuántico

¿ Cómo obtener biálgebras trenzadas, es decir, biálgebras A tales que A-mod sea una categoría trenzada ? Existe un método creado por Drinfeld que asocia a toda biálgebra A otra biálgebra $D(A)$ que es trenzada. Esta construcción no es sencilla. Nosotros veremos una construcción categórica simple, que a cada categoría tensorial $\mathcal{C}$ le asocia una categoría trenzada $\mathcal{D}(\mathcal{C})$, de forma tal que

$\mathcal{D}(A\text{-mod}) = D(A)\text{-mod}$. Luego aplicaremos este método para "re"-encontrar la expresión de R_h en $U_h(\mathfrak{sl}(2))$.

III.1. Definición. Sea $(\mathcal{C}, \otimes, I)$ una categoría tensorial. Definimos una nueva categoría $\mathcal{D}(\mathcal{C})$, llamada el *doble cuántico* de $\mathcal{C}$, como sigue. Los objetos de $\mathcal{D}(\mathcal{C})$ son pares $(V, c_{X,V})$, donde V es un objeto de $\mathcal{C}$, y $c_{X,V} : X \otimes V \to V \otimes X$ es una familia de isomorfismos tales que

$$c_{Y,V}(f \otimes \mathrm{id}_V) = (\mathrm{id} \otimes f)\, c_{X,V} \tag{3.1}$$

y

$$c_{X\otimes Y,V} = (c_{X,V} \otimes \mathrm{id}_Y)(\mathrm{id}_X \otimes c_{Y,V}) \tag{3.2}$$

para todo par (X, Y) de objetos y todo morfismo $f : X \to Y$ de $\mathcal{C}$.

Un morfismo de $(V, c_{X,V})$ en $(W, c_{X,W})$ en $\mathcal{D}(\mathcal{C})$ es un morfismo $f : V \to W$ de $\mathcal{C}$ tal que

$$(f \otimes \mathrm{id}_X)\, c_{X,V} = c_{X,W}\, (\mathrm{id}_X \otimes f) \tag{3.3}$$

para todo objeto X en $\mathcal{C}$.

Se tiene entonces el siguiente:

Teorema 3. *La categoría $\mathcal{D}(\mathcal{C})$ es una categoría trenzada tal que*

(i) el producto tensorial de dos objetos está dado por

$$(V, c_{X,V}) \otimes (W, c_{X,W}) = (V \otimes W, c_{X,V\otimes W})$$

donde $c_{X,V\otimes W} = (\mathrm{id}_V \otimes c_{X,W})(c_{X,V} \otimes \mathrm{id}_W)$,

(ii) la unidad es (I, id),

(iii) el trenzado $(V, c_{X,V}) \otimes (W, c_{X,W}) \to (W, c_{X,W}) \otimes (V, c_{X,V})$ está dado por $c_{V,W}$.

DEMOSTRACIÓN.— Hace falta ver que (i) $(V, c_{X,V}) \otimes (W, c_{X,W})$ es un objeto de la categoría, (ii) si f y g son morfismos en la categoría, entonces $f \otimes g$ también, (iii) $c_{V,W}$ es un morfismo en $\mathcal{D}(\mathcal{C})$, (iv) $c_{V,W}$ es un trenzado. Todo esto se verifica fácilmente. *QED*

III.2. Aplicación a la categoría de A-módulos. Dada una biálgebra (A, Δ, ε), se quiere describir la categoría $\mathcal{D}(A\text{-mod})$ en términos de A. Para hacer esto, necesitamos la siguiente definición.

Definición 4. *Dada una biálgebra (A, Δ, ε), un A-bimódulo cruzado es un A-módulo a izquierda V con una aplicación lineal $\Delta_V : V \to V \otimes A$ tal que*

(i) se tiene

$$(\Delta_V \otimes \mathrm{id}_A) \circ \Delta_V = (\mathrm{id}_V \otimes \Delta) \circ \Delta_V, \tag{3.4}$$

$$(\mathrm{id}_V \otimes \varepsilon) \circ \Delta_V = \mathrm{id}_V, \tag{3.5}$$

(ii) para todo $a \in A$ y $v \in V$ tenemos

$$\sum_{(a)} \sum_{(v)} a' v_V \otimes a'' v_A = \sum_{(a)} \sum_{(v)} (a''v)_V \otimes (a''v)_A\, a' \tag{3.6}$$

donde $\Delta(a) = \sum_{(a)} a \otimes a$ *y* $\Delta_V(v) = \sum_{(v)} v_V \otimes v_A$.

Teorema 4. *La categoría* $\mathcal{D}(A\text{-mod})$ *es equivalente a la categoría de los A-bimódulos cruzados.*

DEMOSTRACIÓN.— De todo A-bimódulo cruzado V obtenemos un elemento $(V, c_{X,V})$ de $\mathcal{D}(A\text{-mod})$ definido por

$$c_{X,V}(x \otimes v) = \Delta_V(1 \otimes x) = \sum_{(v)} v_V \otimes v_A\, x. \tag{3.7}$$

Cada uno puede verificar que $c_{X,V}$ satisface los axiomas que definen un objeto de $\mathcal{D}(A\text{-mod})$.

Recíprocamente, sea $(V, c_{X,V})$ un objeto de $\mathcal{D}(A\text{-mod})$. Definimos una aplicación lineal $\Delta_V : V \to V \otimes A$ por

$$\Delta_V(v) = c_{A,V}(1 \otimes v)$$

para todo $v \in V$. La naturalidad (3.1) de $c_{X,V}$ en X implica que se puede recuperar de Δ_V como en la ecuación (3.7). La relación (3.2) para $X = Y = A$ implica (3.4). La A-linealidad de $c_{A,V}$ implica (3.6). *QED*

De los Teoremas 3–4 se deduce que la categoría de los A-bimódulos cruzados es trenzada. El trenzado entre dos A-bimódulos cruzados está dado por (3.7).

III.3. Bimódulos cruzados en el caso $\mathfrak{sl}(2)$. Apliquemos ahora el Teorema 4 a una sub-biálgebra de $U_h(\mathfrak{sl}(2))$. La idea de Drinfeld para encontrar la R-matriz universal R_h, fue construir $D(A)$ que es una biálgebra trenzada, y por lo tanto, tiene una R-matriz universal $\widehat{R}$, que se aplica en R_h mediante una suryección.

Recordemos que $U_h(\mathfrak{sl}(2))$ es el álgebra generada por X, H, Y con las relaciones (1.9–1.10). Sea B_h la subálgebra de $U_h(\mathfrak{sl}(2))$ generada por H, X. Sea $E = Xe^{hH/4}$. Tenemos que $[H, E] = 2E$. De (1.11–1.14) se deduce que B_h es una sub-biálgebra de $U_h(\mathfrak{sl}(2))$. A ella le queremos aplicar el Teorema 4.

Sea V un B_h-módulo. Determinemos cuándo es un bimódulo cruzado. Si lo es, existe una aplicación lineal Δ_V que satisface las condiciones de la Definición 4. Como $\{H^m E^n\}_{m,n \geq 0}$ es una base topológica de B_h, la aplicación Δ_V es de la forma

$$\Delta_V = \sum_{m,n \geq 0} \Delta^{m,n} \otimes H^m E^n \tag{3.8}$$

donde $(\Delta^{m,n})_{m,n \geq 0}$ es una familia de endomorfismos de V.

La relación (3.4) nos permite expresar el producto de dos elementos de esta familia como una combinación lineal de los elementos de la familia. Una inducción sencilla muestra que

$$\Delta^{m,n} = \frac{q^{n(n-1)/2}}{m! [n]!} \Delta_1^m \Delta_2^n \tag{3.9}$$

para todo m, n, donde $\Delta_1 = \Delta^{1,0}$ y $\Delta_2 = \Delta^{0,1}$. La relación (3.5) implica que $\Delta^{0,0} = \mathrm{id}_V$.

La relación (3.6) implica relaciones entre la familia $(\Delta^{m,n})_{m,n}$ y los operadores H y E en V. Más precisamente, uno puede probar que V es un B_h-módulo cruzado sí y sólo si tiene cuatro endomorfismos H, E, Δ_1, Δ_2 actuando sobre él, tales que sus conmutadores están dados por

$$[H,E] = 2E, \qquad [\Delta_1,\Delta_2] = -\frac{h}{2}\Delta_2, \tag{3.10}$$

$$[H,\Delta_1] = 0, \qquad [H,\Delta_2] = -2\Delta_2, \tag{3.11}$$

$$[E,\Delta_1] = -\frac{h}{2}E, \qquad [E,\Delta_2] = \exp(hH/2) - \exp(2\Delta_1). \tag{3.12}$$

En otras palabras, un B_h-bimódulo cruzado es lo mismo que un módulo sobre el álgebra D_h generado por cuatro generadores H, E, Δ_1, Δ_2 y las relaciones (3.10–3.12).

Como la categoría de B_h-bimódulos cruzados es trenzada (Teoremas 3–4), el álgebra D_h debe ser trenzada (Proposición 2). Por (2.7), (3.7) y (3.8) su R-matriz universal tiene que ser

$$\widehat{R} = \sum_{m,n\geq 0} H^m E^n \otimes \Delta^{m,n}.$$

Combinando esto con (3.9), tenemos que

$$\widehat{R} = \sum_{m,n\geq 0} \frac{q^{n(n-1)/2}}{m![n]!} H^m E^n \otimes \Delta_1^m \Delta_2^n. \tag{3.13}$$

El próximo lema relaciona a la biálgebra trenzada D_h con el álgebra envolvente cuántica $U_h(\mathfrak{sl}(2))$.

Lema 1. *Existe un morfismo sobreyectivo de biálgebras $\chi : D_h \longrightarrow U_h(\mathfrak{sl}(2))$ dado por*

$$\chi(E) = X\exp(hH/4), \quad \chi(H) = H,$$

$$\chi(\Delta_1) = \frac{h}{4}H, \quad \chi(\Delta_2) = (\exp(h/4) - \exp(-3h/4))\,Y.$$

Esto implica que, como D_h es trenzada con la R-matriz universal $\widehat{R}$, entonces $U_h(\mathfrak{sl}(2))$ es trenzada con la R-matriz universal $(\chi\otimes\chi)(\widehat{R})$. Tenemos

$$(\chi\otimes\chi)(\widehat{R}) = \sum_{m,n\geq 0} \frac{q^{n(n-1)/2}}{m![n]!} \chi(H)^m \chi(E)^n \otimes \chi(\Delta_1)^m \chi(\Delta_2^n). \tag{3.14}$$

Un cálculo rápido muestra que el lado derecho de (3.14) es igual a la expresión dada por R_h en el Teorema 1, y esto prueba lo último.

Lección IV. Trenzados infinitesimales

IV.1. Definición. Como hemos visto en el caso de $\mathfrak{sl}(2)$, la categoría de módulos sobre un álgebra envolvente cuántica depende de un parámetro h. Por ejemplo, el trenzado $c_{V,W}$ se puede extender a una serie formal en h:

$$c_{V,W} = \sigma_{V,W}\left(\mathrm{id}_{V\otimes W} + h\, r_{V,W} + \cdots\right) \tag{4.1}$$

donde $\sigma_{V,W}$ es la trasposición y $r_{V,W}$ es un endomorfismo de $V\otimes W$. Estudiemos el término lineal $r_{V,W}$. Como $c_{V,W}$ satisface los axiomas (2.1–2.3), éstos se traducen en condiciones sobre $r_{V,W}$.

Sea $\mathcal{S}$ una *categoría simétrica*, es decir, una categoría trenzada cuyo trenzado σ satisface la condición adicional

$$\sigma_{W,V} \circ \sigma_{V,W} = \mathrm{id}_{V\otimes W} \tag{4.2}$$

para todo objeto V, W de $\mathcal{S}$. Estos trenzados se llamarán *simetrías*. Además suponemos que $\mathcal{S}$ es $\mathbf{C}$-lineal, que significa que todos sus conjuntos de morfismos son espacios vectoriales complejos, y la composición y el producto tensorial de morfismos es $\mathbf{C}$-bilineal.

Definición 5. *Sea $\mathcal{S}$ una categoría simétrica y $\mathbf{C}$-lineal. Un trenzado infinitesimal de $\mathcal{S}$ es una familia de endomorfismos $t_{V,W} \in \mathrm{End}(V\otimes W)$ definida para todo objeto V, W tal que:*

$$(f\otimes g)\, t_{V,W} = t_{V',W'}\,(f\otimes g) \tag{4.3}$$

$$\sigma_{V,W} \circ t_{V,W} = t_{W,V} \circ \sigma_{V,W}, \tag{4.4}$$

$$t_{U,V\otimes W} = t_{U,V}\otimes \mathrm{id}_W + (\sigma_{U,V}\otimes \mathrm{id}_W)^{-1}(\mathrm{id}_V \otimes t_{U,W})(\sigma_{U,V}\otimes \mathrm{id}_W) \tag{4.5}$$

para todo objeto U, V, W y todo morfismo $f \in \mathrm{Hom}_{\mathcal{S}}(V,V')$ y $g \in \mathrm{Hom}_{\mathcal{S}}(W,W')$.

Si $\mathcal{S}$ tiene un trenzado infinitesimal se llamará una categoría simétrica infinitesimal.

Si $t_{V,W}$ se define como en (4.1) por

$$t_{V,W} = r_{V,W} + \sigma_{V,W}^{-1}\, r_{W,V}\, \sigma_{V,W}$$

entonces $t_{V,W}$ satisface (4.3–4.5).

IV.2. Trenzados infinitesimales sobre categorías de módulos. Como en II.3, buscamos álgebras A cuya categoría de módulos sea simétrica infinitesimal. Sea (A,Δ,ε) una biálgebra *coconmutativa*. Sabemos por II.3 que la categoría de A-módulos es trenzada, con trenzado dado por la trasposición, que obviamente, es una simetría. El próximo resultado clasifica todos los trenzados infinitesimales en esta categoría.

Proposición 4. *Sea (A, Δ, ε) una biálgebra coconmutativa, y sea $\mathcal{S}$ la categoría de A-módulos con σ la trasposición. Entonces $\mathcal{S}$ es una categoría simétrica infinitesimal sí y sólo si existe $t \in A \otimes A$ tal que:*

(i) $[\Delta(a), t] = 0$ *para todo* $a \in A$,

(ii) t *es un tensor simétrico y,*

(iii) t *pertenece a* $\mathrm{Prim}(A) \otimes \mathrm{Prim}(A)$ *donde* $\mathrm{Prim}(A)$ *es el conjunto de elementos* $a \in A$ *tales que* $\Delta(a) = 1 \otimes a + a \otimes 1$.

DEMOSTRACIÓN.— $\Leftarrow$) Si $v, w \in \mathcal{S}$, sea $t_{V,W} \in \mathrm{End}(V \otimes W)$ definido por:

$$t_{V,W}(v \otimes w) = t(v \otimes w) = \sum_i x_i v \otimes y_i w \tag{4.6}$$

donde $t = \sum_i x_i \otimes y_i$. Es fácil verificar que las condiciones (i–iii) implican (4.3–4.5).

$\Rightarrow$) Sea $t = \sum_i x_i \otimes y_i \in A \otimes A$ definido por $t = t_{A,A}(1 \otimes 1)$. Veamos que se verifican (i–iii). La condición (i) se verifica porque $t_{A,A}$ es A-lineal. La condición (ii) es clara.

Probemos la condición (iii). Si aplicamos ambos lados de (4.5) al elemento $1 \otimes 1 \otimes 1$, con $U = V = W = A$, tenemos

$$\sum_i x_i \otimes \Delta(y_i) = \sum_i x_i \otimes (y_i \otimes 1 + 1 \otimes y_i).$$

Podemos suponer que los elementos x_i son linealmente independientes. Entonces tenemos $\Delta(y_i) = y_i \otimes 1 + 1 \otimes y_i$, es decir, todos los elementos y_i pertenecen a $\mathrm{Prim}(A)$. Por lo tanto, t pertenece a $A \otimes \mathrm{Prim}(A)$. Terminamos la demostración usando la simetría de t. *QED*

Podemos aplicar la Proposición 4 al caso en que A sea el álgebra envolvente de un álgebra de Lie $\mathfrak{g}$ sobre un cuerpo de característica cero. Entonces los trenzados infinitesimales de la categoría simétrica de $\mathfrak{g}$-módulos están en biyección con 2-tensores $\in \mathfrak{g} \otimes \mathfrak{g}$ que son simétricos e invariantes bajo la representación adjunta.

Consideremos el caso especial de un álgebra de Lie compleja semisimple $\mathfrak{g}$. Sea $\{x_i\}$ una base ortonormal de $\mathfrak{g}$ con respecto a la forma de Killing $< x, y > = \mathrm{Tr}(\mathrm{ad}(x)\mathrm{ad}(y))$. Es bien sabido que el 2-tensor $t = \sum x_i \otimes x_i$ es simétrico e invariante. Entonces la categoría de $\mathfrak{g}$-módulos es una categoría simétrica infinitesimal con la trasposición como simetría y con trenzado infinitesimal inducido por la acción de t.

IV.3. Un ejemplo gráfico. El resto de esta Lección está dedicado a desarrollar un ejemplo gráfico de una categoría simétrica infinitesimal. Empezamos con algunas notaciones. Para todo entero $n \geq 2$, sea $\mathfrak{p}_n$ el álgebra de Lie compleja generada por los generadores $(t^{ij})_{1 \leq i \neq j \leq n}$ y las relaciones

$$t^{ij} = t^{ji}, \tag{4.7}$$

$$[t^{ij}, t^{ik} + t^{jk}] = 0, \tag{4.8}$$

$$[t^{ij}, t^{k\ell}] = 0 \tag{4.9}$$

para índices distintos i, j, k, ℓ. Llamamos $U(\mathfrak{p}_n)$ a su álgebra envolvente. El grupo simétrico $\mathfrak{S}_n$ actúa sobre $U(\mathfrak{p}_n)$ por la regla $\sigma \cdot t^{ij} = t^{\sigma^{-1}(i)\sigma^{-1}(j)}$ donde $\sigma \in \mathfrak{S}_n$. Podemos considerar el producto cruzado A_n de $\mathfrak{S}_n$ por $U(\mathfrak{p}_n)$. Recordemos que A_n es un álgebra cuyo espacio vectorial subyacente es el mismo que el del álgebra de grupo $U(\mathfrak{p}_n)[\mathfrak{S}_n]$, pero la multiplicación esta determinada por la regla

$$\sigma\, t^{ij} = (\sigma \cdot t^{ij})\, \sigma \tag{4.10}$$

para todo $\sigma \in \mathfrak{S}_n$. Además $A_0 = A_1 = \mathbf{C}$.

Ahora consideremos la categoría $\mathcal{A}$ cuyos objetos son los números naturales, y tal que

$$\mathrm{Hom}_{\mathcal{A}}(n,m) = \begin{cases} \emptyset & \text{si } n \neq m, \\ A_n & \text{si } n = m. \end{cases}$$

La composición está definida por el producto en el álgebra A_n. La identidad id_n es la unidad de A_n.

Para darle a la categoría $\mathcal{A}$ una estructura de categoría infinitesimal simétrica, daremos primero una descripción gráfica de sus elementos. Como en el caso de las trenzas, estarán representados por figuras consideradas a menos de homeomorfismos en la banda $\mathbf{R} \times [0,1]$. Estas figuras estarán llamadas *diagramas de cuerdas.* El diagrama de cuerdas del producto de dos elementos se obtiene, como en el caso de las trenzas, pegando el diagrama de cuerdas de un elemento arriba del diagrama de cuerdas del otro elemento.

Para representar gráficamente los elementos de A_n es suficiente representar a sus generadores. El álgebra A_n tiene dos tipos de generadores: los generadores t^{ij} de $\mathfrak{p}_n$ y las trasposiciones $s_k = (k, k+1)$ del grupo simétrico $\mathfrak{S}_n$.

Representemos al generador t^{ij} por n intervalos verticales paralelos con una línea punteada horizontal entre el i-ésimo y el j-ésimo intervalo. Una línea punteada horizontal se llama una *cuerda.* Representemos la trasposición s_k por la unión de los intervalos $\{i\} \times [0,1]$ para $1 \leq i \leq n$ y $i \neq k, k+1$ y los intervalos $[(k,0),(k+1,1)]$ y $[(k+1,0),(k,1)]$ de $\mathbf{R} \times [0,1]$. En el grupo simétrico σ_k es igual a su inverso. Esto significa que, contrariamente a lo que sucede con las trenzas, aquí no hay diferencia entre cruzar por adelante o por atrás. Observar que s_k está representada por un diagrama de cuerdas sin cuerdas. Dejamos al lector que agregue las relaciones que resultan de las Relaciones (4.8–4.10). La relación entre los diagramas de cuerdas correspondientes a (4.8) es conocida en la teoría de invariantes de grado finito de nudos bajo la terminología "relación de cuatro términos" (para más información, ver [BN92] y [Kas95], Capítulo XX).

Con esta representación gráfica, es fácil dar una estructura tensorial a la categoría $\mathcal{A}$. Esto está hecho, así como hicimos para las categorías $\mathcal{D}$ en I.5 y $\mathcal{B}$ en II.7. La fórmula (2.9) define una simetría en $\mathcal{A}$.

Ahora construiremos un trenzado infinitesimal. Para todo objeto n, $m \geq 1$, definimos un endomorfismo $t_{n,m}$ del objeto $n \otimes m$ como el elemento

$$t_{n,m} = \sum_{i=1}^{n} \sum_{j=1}^{m} t^{i,j+n} \in A_{n+m}. \tag{4.11}$$

Si n ó $m = 0$, tenemos $t_{n,m} = 0$. Dejamos como ejercicio la demostración del próximo resultado (sugerencia: la relación (4.8) se usa para probar la naturalidad (4.3) de $t_{n,m}$).

Proposición 5. *La familia de endomorfismos $(t_{n,m})_{n,m\geq 0}$ es un trenzado infinitesimal de la categoría $\mathcal{A}$.*

La categoría $\mathcal{A}$ tiene la siguiente propiedad universal.

Teorema 5. *Sea $\mathcal{A}$ una categoría simétrica infinitesimal $\mathbf{C}$-lineal con simetría σ y trenzado infinitesimal t. Dado un objeto V de $\mathcal{S}$, existe un único funtor $G_V : \mathcal{A} \to \mathcal{S}$ tal que $G_V(0) = I$, $G_V(1) = V$,*

$$G_V(n \otimes m) = G_V(n) \otimes G_V(m), \quad G_V(D \otimes D') = G_V(D) \otimes G_V(D')$$

para todos objetos n, m y todos morfismos D, D', y

$$G_V(s_1) = \sigma_{V,V} \in \mathrm{Aut}(V \otimes V) \quad y \quad G_V(t^{12}) = t_{V,V} \in \mathrm{End}(V \otimes V).$$

Lección V. Integración de categorías simétricas infinitesimales

A partir de una categoría simétrica infinitesimal $\mathcal{S}$, construímos una categoría trenzada $\mathcal{S}[[h]]$. Como veremos más adelante, esto tiene consequencias muy importantes, gracias a un resultado fundamental de Drinfeld.

V.1. Categorías trenzadas revisadas. Antes de comenzar a construir $\mathcal{S}[[h]]$, necesitamos extender el concepto de categoría trenzada. Empezamos ampliando la condición (i) de la Definición 1 suponiendo la existencia de una *isomorfismo de asociatividad* natural

$$a : (U \otimes V) \otimes W \to U \otimes (V \otimes W)$$

definido para toda terna U, V, W de objetos de la categoría $\mathcal{C}$ (cuando a es la identidad, recuperamos la Definición 1). En esta nueva situación, tenemos muchas elecciones de colocar los paréntesis en el producto tensorial de n objetos.

Todas ellas son isomorfas. Por ejemplo, tomemos cuatro objetos U, V, W, Z. Hay cinco formas de colocar paréntesis en $U \otimes V \otimes W \otimes Z$. El siguiente diagrama, a menudo llamado el pentágono de Mac Lane, muestra dos formas de pasar de $((U\otimes V)\otimes W)\otimes Z$ a $U\otimes(V\otimes(W\otimes Z))$ usando el isomorfismo de asociatividad a.

$$\begin{array}{ccc} ((U\otimes V)\otimes W)\otimes Z & \xrightarrow{a_{U\otimes V,W,Z}} & (U\otimes V)\otimes(W\otimes Z) \\ {\scriptstyle a_{U,V,W}\otimes \mathrm{id}_Z}\downarrow & & \downarrow{\scriptstyle a_{U,V,W\otimes Z}} \\ (U\otimes(V\otimes W))\otimes Z & & U\otimes(V\otimes(W\otimes Z)) \\ {\scriptstyle a_{U,V\otimes W,Z}}\searrow & & \nearrow{\scriptstyle \mathrm{id}_U\otimes a_{V,W,Z}} \\ & U\otimes((V\otimes W)\otimes Z) & \end{array} \tag{5.1}$$

Nosotros supondremos que dados cuatro objetos cualesquiera, el pentágono de Mac Lane conmuta. Esto significa que entre dos formas de colocar los paréntesis en el producto tensorial de cuatro objetos, existe un único isomorfismo natural expresado con el isomorfismo de asociatividad. Mac Lane probó que la conmutatividad del pentágono también implica que existe un único isomorfismo natural, expresado con el isomorfismo de asociatividad, entre dos formas de colocar los paréntesis en el producto tensorial de n objetos, donde n es un entero positivo arbitrario (ver [Mac71]).

También tenemos que tener en cuenta al isomorfismo de asociatividad a en la definición del trenzado c. Esto significa que debemos reemplazar los dos triángulos conmutativos (2.2–2.3) de la Definición 2 por dos hexágonos conmutativos. El primer hexágono es

$$\begin{array}{ccc} (U\otimes V)\otimes W & \xrightarrow{c_{U\otimes V,W}} & W\otimes(U\otimes V) \\ {\scriptstyle a_{U,V,W}}\downarrow & & \uparrow{\scriptstyle a_{W,U,V}} \\ U\otimes(V\otimes W) & & (W\otimes U)\otimes V \\ {\scriptstyle \mathrm{id}_U\otimes c_{V,W}}\downarrow & & \uparrow{\scriptstyle c_{U,W}\otimes \mathrm{id}_V} \\ U\otimes(W\otimes V) & \xrightarrow{a^{-1}_{U,W,V}} & (U\otimes W)\otimes V \end{array} \tag{5.2}$$

El segundo hexágono se obtiene a partir del primero, reemplazando el isomorfismo c por su inverso c^{-1}.

V.2. Series de Drinfeld. Una serie formal $\Phi(A,B)$ en dos variables A, B que no conmutan se llama una *serie de Drinfeld* si $\Phi(A,B)$ es solución del siguiente sistema de ecuaciones:

$$\Phi(0,0) = 1,$$

$$\Phi(t_{12}, t_{23}+t_{24})\,\Phi(t_{13}+t_{23}, t_{34}) = \Phi(t_{23}, t_{34})\,\Phi(t_{12}+t_{13}, t_{24}+t_{34})\,\Phi(t_{12}, t_{23}),$$

$$\exp(\tfrac{t_{13}+t_{23}}{2}) = \Phi(t_{13}, t_{12})\,\exp(\tfrac{t_{13}}{2})\,\Phi(t_{13}, t_{23})^{-1}\,\exp(\tfrac{t_{23}}{2})\,\Phi(t_{12}, t_{23}),$$

$$\exp(\tfrac{t_{13}+t_{12}}{2}) = \Phi(t_{23}, t_{13})^{-1}\,\exp(\tfrac{t_{13}}{2})\,\Phi(t_{12}, t_{13})\,\exp(\tfrac{t_{12}}{2})\,\Phi(t_{12}, t_{23})^{-1},$$

donde $(t_{ij})_{1\leq i\neq j\leq 4}$ verifica las relaciones (4.7–4.9). Observar que $\Phi(A,B)=1$ no es una serie de Drinfeld. En realidad,

$$\exp(\frac{t_{13}+t_{23}}{2}) \neq \exp(\frac{t_{13}}{2})\exp(\frac{t_{23}}{2})$$

porque t_{13} y t_{23} no conmutan. De todos modos, tenemos el siguiente resultado.

Teorema 6. *Existen series de Drinfeld.*

Drinfeld [Dri89], [Dri90] probó este teorema exhibiendo una solución particular Φ_{KZ} de la siguiente manera. Consideremos la ecuación diferencial lineal

$$\frac{dG(z)}{dz} = \frac{1}{2i\pi}\left(\frac{A}{z}+\frac{B}{z-1}\right)G(z) \tag{5.3}$$

definida para $z \in \mathbf{C}\setminus\{0,1\}$, donde A, B son operadores que no conmutan. La ecuación (5.3) tiene singularidades regulares en 0, 1, ∞. Sea $G(z)$ alguna solución. No está definida en 0, 1, pero su comportamiento asintótico en estos puntos es bien conocido. Escribamos

$$G(z) = z^{hA/2i\pi}(1-z)^{hB/2i\pi}V(z).$$

La función $V(z)$ está definida en un entorno de $[0,1]$. Drinfeld probó que $\Phi_{\mathrm{KZ}} = V(1)V(0)^{-1}$ es una serie de Drinfeld. En realidad, es la única solución conocida hasta hoy. Podemos resolver (5.3) usando el método de aproximación de Picard. Por ejemplo, podemos calcular los primeros términos de Φ_{KZ}:

$$\Phi_{\mathrm{KZ}}(A,B) = 1+\frac{1}{24}[A,B]+\frac{\zeta(3)}{(2i\pi)^3}([[A,B],B]-[A,[A,B]])+\cdots. \tag{5.4}$$

Aquí $\zeta(3)$ es el valor de la función zeta de Riemann en el entero 3. La serie de Drinfeld Φ_{KZ} fue encontrada por Drinfeld en relación con las ecuaciones de Knizhnik-Zamolodchikov (para más información, ver [Dri89], [Dri90], [Kas95], Capítulo XIX).

V.3. La construcción de Drinfeld-Cartier. Ahora estamos listos para definir $\mathcal{S}[[h]]$ cuando $\mathcal{S}$ es una categoría simétrica infinitesimal $\mathbf{C}$-lineal. Seguiremos [Dri90] y [Car93].

Los objetos de $\mathcal{S}[[h]]$ son los mismos que los objetos de $\mathcal{S}$. Un morfismo en $\mathcal{S}[[h]]$ de V en W es una serie formal $\sum_{n\geq 0} f_n h^n$, donde $f_n \in \mathrm{Hom}_{\mathcal{S}}(V,W)$ para todo n. La composición está definida usando la composición en $\mathcal{S}$ y la multiplicación usual de las series formales. La identidad de un objeto V en $\mathcal{S}[[h]]$ es la serie formal constante $\sum_{n\geq 0} f_n h^n$ donde $f_0 = \mathrm{id}_V$ y $f_n = 0$ cuando $n>0$.

La categoría $\mathcal{S}[[h]]$ tiene un producto tensorial obtenido extendiendo $\mathbf{C}[[h]]$-linealmente al producto tensorial de $\mathcal{S}$.

Ahora le daremos a $\mathcal{S}[[h]]$ una estructura de categoría trenzada en el sentido ampliado de V.1. Para hacer esto, definimos un isomorfismo de asociatividad y un trenzado. Ambos estarán definidos como series formales. Para el isomorfismo de asociatividad, necesitamos elegir una serie de Drinfeld Φ como está definida en V.2. Sean

$$a_{U,V,W} = \Phi(h\, t_{U,V} \otimes \mathrm{id}_W, h\, \mathrm{id}_U \otimes t_{V,W}) \tag{5.5}$$

donde $t_{U,V}$, $t_{V,W}$ son los trenzados infinitesimales de la categoría $\mathcal{S}$. El trenzado está definido por

$$c_{V,W} = \sigma_{V,W} \circ \exp(h\, t_{V,W}/2) \tag{5.6}$$

donde $\sigma_{V,W}$ es la simetría en $\mathcal{S}$.

Es fácil verificar que las ecuaciones que definen una serie de Drinfeld implican la conmutatividad del pentágono (5.1) y de los hexágonos (5.2). Como la categoría trenzada que acabamos de construir depende de una serie de Drinfeld Φ, hacemos notar esto llamando a la categoría trenzada $\mathcal{S}_\Phi[[h]]$.

V.4. El teorema fundamental de Drinfeld. Sea $\mathfrak{g}$ un álgebra de Lie semisimple compleja. Como mencionamos anteriormente, Drinfeld y Jimbo construyeron una $\mathbf{C}[[h]]$-biálgebra trenzada $U_h(\mathfrak{g})$ tal que

$$U_h(\mathfrak{g})/hU_h(\mathfrak{g}) \cong U(\mathfrak{g})$$

como biálgebras. Podemos considerar la categoría $U_h(\mathfrak{g})$-mod$_f$ de $U_h(\mathfrak{g})$-módulos que son libres de rango finito como $\mathbf{C}[[h]]$-módulos. Esta categoría es trenzada.

Por otro lado, por IV.2 sabemos que la categoría $U(\mathfrak{g})$-mod$_f$ de $\mathfrak{g}$-módulos de dimensión finita es simétrica infinitesimal cuando la equipamos con la trasposición como simetría y con el trenzado infinitesimal inducido por el 2-tensor $t = \sum_i x_i \otimes x_i$ que es dual a la forma de Killing. Podemos aplicar el procedimiento de integración de V.3 a $U(\mathfrak{g})$-mod$_f$.

Drinfeld ([Dri89], [Dri90]) probó el siguiente resultado, que muestra la importancia del procedimiento de integración del párrafo anterior. Esta es una interpretación categórica de la demostración dada por Drinfeld de un teorema de Kohno sobre la monodromía de ecuaciones de Knizhnik-Zamolodchikov.

Teorema 7. *Para toda serie de Drinfeld Φ, las categorías trenzadas $U_h(\mathfrak{g})$-mod$_f$ y $(U(\mathfrak{g})\text{-mod}_f)_\Phi\,[[h]]$ son equivalentes.*

V.5. Una representación universal del grupo de trenzas. Volvamos a la categoría simétrica infinitesimal universal $\mathcal{A}$ construída en IV.3. Fijemos una serie de Drinfeld Φ. Por V.3 conseguimos una categoría trenzada $\mathcal{A}_\Phi[[h]]$. Consideremos el objeto 1 de esta categoría. Por el Teorema 2, existe un único funtor $Z = F_1 : \mathcal{B} \to \mathcal{A}_\Phi[[h]]$ que preserva el producto tensorial y el trenzado tal que $Z(1) = 1$. Este funtor da una representación de B_n por un grupo de series formales cuyos coeficientes son diagramas de cuerdas con n hebras: para cualquier trenza T con n hebras tenemos

$$Z(T) = \sum_{k\geq 0} Z_k(T)\, h^k \tag{5.7}$$

donde $Z_k(T)$ es un elemento del álgebra A_n que es una combinación lineal de diagramas de cuerdas con k cuerdas. Por ejemplo, si σ_1 es el generador de B_2,

$$Z(\sigma_1) = s_1 \circ \exp(ht^{12}/2) = \sum_{k \geq 0} \frac{h^k}{2^k k!} s_1 \circ (t^{12})^k. \qquad (5.8)$$

Tomando "trazas", obtenemos un invariante de lazos a valores en diagramas de cuerdas en el círculo. Este es el invariante universal de Kontsevich (ver [BN92], [Kas95], Capítulo XX).

El funtor Z es universal para todos los funtores F de la categoría $\mathcal{B}$ en cualquier categoría trenzada de la forma $\mathcal{S}_\Phi[[h]]$, que preserva todas las estructuras. En realidad, sea V el objeto $F(1)$ de $\mathcal{S}$. Por el Teorema 5 existe un único funtor $G : \mathcal{A} \to \mathcal{S}$ que preserva todas las estructuras y tal que $G(1) = V$. Este funtor se extiende a un funtor $G_\Phi[[h]] : \mathcal{A}_\Phi[[h]] \to \mathcal{S}_\Phi[[h]]$. Consideremos la composición $G_\Phi[[h]] \circ Z : \mathcal{B} \to \mathcal{S}_\Phi[[h]]$. Ella manda el objeto 1 en el objeto V. Por la unicidad establecida en el Teorema 2, tenemos

$$F = G_\Phi[[h]] \circ Z, \qquad (5.9)$$

que prueba que F se factoriza a través de Z.
El funtor Z se puede usar para probar el siguiente isomorfismo de Kohno [Koh85] entre completados del álgebra de grupo del grupo de trenzas B_n y del álgebra A_n definida en IV.3:

$$\varprojlim_k \mathbf{C}[B_n]/I^k \cong \varprojlim_k A_n/J^k, \qquad (5.10)$$

donde I es el ideal bilátero de $\mathbf{C}[B_n]$ generado por los elementos $\sigma_1 - \sigma_1^{-1}, \ldots, \sigma_{n-1} - \sigma_{n-1}^{-1}$, y J es el ideal de A_n generado por los elementos $\{t^{ij}\}_{1 \leq i \neq j \leq n}$. La demostración que usa el funtor Z está en [KT95], Apéndice C.

Bibliografía

El contenido de este curso fue tomado esencialmente de [Kas95]. En esta monografía el lector interesado encontrará más detalles. También puede ser muy útil el artículo [Kas93].

La definición original de álgebras envolventes cuánticas es de Drinfeld [Dri87] y Jimbo [Jim85]. Es muy recomendable leer el artículo de Drinfeld [Dri87], pues aquí se explica la motivación de la aparición de los grupos cuánticos a partir de la física.

Las categorías trenzadas fueron inventadas por Joyal y Street [JS93]. En este tema, le aconsejo al lector mirar también [FY89], [JS91].

La construcción del doble se debe a Drinfeld (ver [Dri87]). La construcción del doble de una biálgebra está explicado en [Kas95], Capítulo IX. El contenido de la Lección III se puede encontrar en [KT92].

Con respecto a las Lecciones IV–V, se puede ver [Car93], [Dri89], [Dri90] y [Kas95], Capítulos XIX–XX. En [RT90] y [Tur94] hay aplicaciones de álgebras envolventes cuánticas a nudos y 3-variedades. El invariante de Kontsevich y los diagramas de cuerdas se pueden ver también en [BN92].

Finalmente, hay varios libros sobre grupos cuánticos y temas relacionados, por ejemplo, [CP94], [Gui95], [Jos95], [KRT97], [Lus93], [Maj95], [SS94], [Tur94].

[BN92] D. BAR NATAN, *On the Vassiliev knot invariants*, Topology 34 (1995), 423–472.

[Car93] P. CARTIER, *Construction combinatoire des invariants de Vassiliev-Kontsevich des nœuds*, Comptes-Rendus Acad. Sc. Paris 316 (1993), 1205–1210.

[CP94] V. CHARI, A. PRESSLEY, *A guide to quantum groups*, Cambridge University Press, Cambridge, 1994.

[Dri87] V. G. DRINFELD, *Quantum groups*, Proc. I.C.M. Berkeley 1986, Amer. Math. Soc., Providence, RI, vol. 1 (1987), 798–820.

[Dri89] V. G. DRINFELD, *Quasi-Hopf algebras*, Algebra i Analiz 1:6 (1989), 114–148 (= Leningrad Math. J. 1 (1990), 1419–1457).

[Dri90] V. G. DRINFELD, *On quasitriangular quasi-Hopf algebras and a group closely connected with* $\mathrm{Gal}(\bar{\mathbf{Q}}/\mathbf{Q})$, Algebra i Analiz 2:4 (1990), 149–181 (= Leningrad Math. J. 2 (1991), 829–860)

[FY89] P. J. FREYD, D. YETTER, *Braided compact closed categories with applications to low-dimensional topology*, Adv. Math. 77 (1989), 156–182.

[Gui95] A. GUICHARDET, *Groupes quantiques. Introduction au point de vue formel*, InterEditions and CNRS Editions, Paris, 1995.

[Jim85] M. JIMBO, *A q-difference analogue of* $U(\mathfrak{g})$ *and the Yang-Baxter equation*, Lett. Math. Phys. 10 (1985), 63–69.

[Jos95] A. JOSEPH, *Quantum groups and their primitive ideals*, Ergebnisse der Math., vol. 29, Springer-Verlag, Heidelberg, 1995.

[JS91] A. JOYAL, R. STREET, *The geometry of tensor calculus, I*, Adv. Math. 88 (1991), 55–112.

[JS93] A. JOYAL, R. STREET, *Braided tensor categories*, Adv. Math. 102 (1993), 20–78.

[Kas93] C. KASSEL, *Invariants des nœuds, catégories tensorielles et groupes quantiques*, Gazette des Mathématiciens, 56 (1993), 63–80, Société Mathématique de France, Paris.

[Kas95] C. KASSEL, *Quantum groups*, Graduate Texts in Math., vol. 155, Springer-Verlag, New York-Heidelberg-Berlin, 1995. ISBN 0-387-94370-6.

[KRT97] C. KASSEL, M. ROSSO, V. TURAEV, *Quantum groups and knot invariants*, Panoramas et Synthèses, vol. 5, Soc. Math. France, Paris, 1997. ISBN 2-

85629-055-8, ISSN 1272-3835.

[KT92] C. KASSEL, V. TURAEV, *Double construction for monoidal categories*, Acta Mathematica 175 (1995), 1–48.

[KT95] C. KASSEL, V. TURAEV, *Chord diagram invariants of framed tangles and graphs*, aparecerá en Duke Math. J.

[Koh85] T. KOHNO, *Série de Poincaré-Koszul associée aux groupes de tresses pures*, Invent. Math. 82 (1985), 57–75.

[Lus93] G. LUSZTIG, *Introduction to quantum groups,* Progr. Math., vol. 110, Birkhäuser, Boston 1993.

[Mac71] S. MAC LANE, *Categories for the working mathematician,* Springer-Verlag, New York-Heidelberg-Berlin 1971.

[Maj95] S. MAJID, *Foundations of quantum group theory,* Cambridge University Press, Cambridge 1995.

[RT90] N. YU. RESHETIKHIN, V. G. TURAEV, *Ribbon graphs and their invariants derived from quantum groups*, Commun. Math. Phys. 127 (1990), 1–26.

[SS94] S. STERNBERG, S. SHNIDER, *Quantum groups — from coalgebras to Drinfeld algebras —*, International Press, Cambridge, Mass., 1994.

[Tur94] V. G. TURAEV, *Quantum invariants of knots and 3-manifolds*, W. de Gruyter, Berlin 1994.

MATHEMATICS SUBJECT CLASSIFICATION (1991):
16W30, 17B37, 18D10, 20F36, 81R50

E-mail: kassel@math.u-strasbg.fr

ON THE GROUP OF WEAK AUTOMORPHISMS OF A FAMILY OF EQUIVALENCE RELATIONS

Th.GSCHWEND, O.H.KEGEL

Mathematisches Institut
Albert-Ludwigs-Universität
Eckertstr.1
D-79104 FREIBURG i. Br.
Germany

Abstract Let E be a set of equivalence relations on the set Ω. We call a permutation π of Ω an *e-permutation*, $e \in E$, if it leaves every e-class of Ω invariant. We consider the set $S(E) := \bigcup_{e \in E} S(e)$ of all *E-permutations* of Ω or *weak automorphisms* of E. If E is directed by the natural partial order the set $S(E)$ is a group - the *E-symmetric group* - our *group of weak automorphisms* of E. If the family E is well-behaved (*homogeneous*) and Ω is countable then the class of all E-symmetric groups can be characterised up to permutation-isomorphism by the *Steinitz type*. Some normal subgroups of $S(E)$ are pointed out.

1 Introduction

Let Ω be an infinite set and e an equivalence relation on Ω. The permutation π of Ω is an *e-permutation* if it leaves every e-class of Ω invariant. The set $S(e)$ of all e-permutations is a subgroup of $S(\Omega)$, the symmetric group on Ω; it is isomorphic to the Cartesian product of the symmetric groups on the e-classes of Ω. If E is a set of equivalence relations on Ω, we shall consider the set $S(E) := \bigcup_{e \in E} S(e)$ of all *E-permutations* of Ω or *weak E-automorphisms*.

Y. Khakimdjanov et al (eds.), Algebra and Operator Theory, 237-248.

If we endow the family E with some smoothness properties (*homogeneity*) and if Ω is countable such sets of equivalence relations are classified in a *three-class-society*. Further, the class of all E-symmetric groups will be characterised up to permutation-isomorphism by the *Steinitz type*. Similar applications of *Steinitz arguments* are presented by Kroshko/Sushchansky in [5] for the complete classification of limit groups of countable direct systems of finite symmetric groups with strictly diagonal embeddings.

As subgroups of the symmetric group $S(\Omega)$ on Ω the groups $S(E)$ contain $S_0(\Omega)$ and $Alt(\Omega)$, the only non-trivial normal subgroups of $S(\Omega)$ if Ω is countable. In this case we point out a few further normal subgroups and their relative position in $S(E)$. Unfortunately, we are far from describing all the normal subgroups of $S(E)$.

2 Foundations

Consider the set $S(E)$ of all *E-permutations* on an infinite set Ω as mentioned in the section above. On the set E of equivalence relations on Ω there is a natural partial order. For equivalence relations e and e' we shall say: e' is *coarser* than e ($e' \geq e$) if every e'-class of Ω is the union of e-classes. If for the relations e, e' one has $e' \geq e$ then one has $S(e') \supseteq S(e)$. The following four propositions are immediate:

Proposition 2.1 *If the set E of equivalence relations on Ω is directed, i.e. if to any pair $e', e'' \in E$ there is an $e \in E$ with $e \geq e'$ and $e \geq e''$, then the set $S(E)$ is a subgroup of $S(\Omega)$.*

The set E of equivalence relations on Ω is *transitive* if for any two elements $\omega, \omega' \in \Omega$ there is an $e \in E$ such that ω and ω' are in the same e-class of Ω ($\omega \sim_e \omega'$).

Proposition 2.2 *If the set E of equivalence relations on Ω is directed and transitive then, for every $n \in \mathbb{N}$, the group $S(E)$ acts n-transitively on Ω; the group $S_0(\Omega)$ of all finitary permutations of Ω is a normal subgroup of $S(E)$.*

Two sets E and E' of equivalence relations on Ω are *of the same type* if for every $e \in E$ there is an $e' \in E'$ with $e' \geq e$ and if for every $e' \in E'$ there is an $e \in E$ with $e \geq e'$.

Proposition 2.3 *If the sets E and E' of equivalence relations on Ω are of the same type then the sets $S(E)$ and $S(E')$ of permutations of Ω coincide.*

For the set E of equivalence relations on Ω and $e \in E$ define the subset $E_e := \{e' \in E | e' \geq e\}$.

Proposition 2.4 *If the set E of equivalence relations on Ω is directed then, for every $e \in E$, the subset E_e of E is directed and of the same type as E.*

The equivalence relation e on Ω is *finite (bounded)* if every e-class of Ω has only finitely (boundedly) many elements and the set E is *finite (bounded)* if every $e \in E$ is finite (bounded). If there is an infinite e-class of Ω then e as well as E is called *infinite.* If the equivalence relation e' on Ω is coarser than the equivalence relation e, then e' is called a *finite (bounded) coarsening* of e if every e'-class of Ω is the union of finitely (boundedly) many e-classes. If there is an e'-class that is union of infinitely many e-classes, we shall call e' an *infinite coarsening* of e.

If the set E of equivalence relations on Ω is directed one can characterise some properties of E by properties of the group $S(E)$.

Theorem 2.5 *If the set E of equivalence relations on Ω is directed one has*

(1) The group $S(E)$ is torsion if and only if every $e \in E$ is bounded. In this case $S(E)$ is a locally finite group.

(2) The group $S(E)$ is locally residually finite if and only if every $e \in E$ is finite.

(3) There is an infinite $e \in E$ if and only if $S(E)$ contains an infinite, finitely generated subgroup which is not residually finite. In this case the subset E_e, which is of the same type as E, consists of infinite equivalence relations on Ω only.

Proof: (1) If $S(E)$ is a torsion group then so is every $S(e)$, $e \in E$. Assuming that the e-classes are not of bounded size, one obtains cycles of unbounded size and support in distinct e-classes. In the Cartesian product $S(e)$ these cycles yield an element of infinite order. Contradiction. If, on the other hand the equivalence relation $e \in E$ is bounded, $S(e)$ has finite exponent. As Cartesian product of finite groups of bounded order $S(e)$ is locally finite, see [6].

(2) Let G be a finitely generated subgroup of $S(E)$, and assume that every $e \in E$ is finite. Then there is an $e \in E$ with $G \subseteq S(e)$. Thus, as a subgroup of the Cartesian product $S(e)$ of finite symmetric groups, G is residually finite.

(3) Let $e \in E$ have an infinite class $[\omega]_e$ in Ω. Since there are infinite, finitely generated groups which are not residually finite, see [2], one can embed them (by the right regular representation) into the symmetric group $S([\omega]_e)$, hence into $S(e) \subseteq S(E)$. With e every $e' \geq e$ is infinite.

Considering sets E of equivalence relations on Ω, it is useful to observe that they come in two types:

Proposition 2.6 *A given set E of equivalence relations on Ω is either an infinite coarsening if for every $e \in E$ there is an infinite coarsening $e' \in E$ of e, or E is a finite coarsening if there is an element $e \in E$ such that every coarsening $e' \in E$ of e is a finite coarsening of e.*

This dichotomy is particularly useful if E is directed; we are going to refine it in the next section.

Up to the end of this section we suppose that E has a strictly ascending sequence $T = (t_i)_{i\in\mathbb{N}}$ of the same type as E; in this case T is said to be an *approximating tower* of E. Thus we may restrict attention to a suitable approximating tower.

If two e-classes on Ω have the same cardinality, we say that they have the same *class-type*. Recall that if for $e' \geq e$ every e'-class is union of e-classes of in general different class-types. If we group together every t_i-class contained in $[\omega]_{t_{i+1}}$, seperately for every (t_i-) class-type appearing, the different cardinalities of the resulting unions form the *level-type* of $[\omega]_{t_{i+1}}$; another description of $|[\omega]_{t_{i+1}}|$ which refers to the partition of $[\omega]_{t_{i+1}}$ in its t_i-classes of different class-types. If we have for E, respectively E', two approximating towers $T = (t_i)_{i\in\mathbb{N}}$, respectively $T' = (t'_i)_{i\in\mathbb{N}}$, we are interested in the question whether a t_{i+1}-class and a t'_{i+1}-class have the *same level-type*, i.e. the distinct unions of t_i- respectively t'_i-classes of each class-type appearing, have the same cardinality in each case. We call $T = (t_i)_{i\in\mathbb{N}}$ and $T' = (t'_i)_{i\in\mathbb{N}}$ *type-isomorphic* if for every $i \geq 1$ there are equivalence classes of t_i and t'_i of the same level-type. From the following theorem we obtain necessary conditions for two sets of equivalence relations to be *permutation-equivalent*, i.e. they are transformed into one another by a permutation of $S(\Omega)$.

Theorem 2.7 *Let E and E' be sets of equivalence relations with approximating towers $T = (t_i)_{i\in\mathbb{N}}$ and $T' = (t'_i)_{i\in\mathbb{N}}$, defined on two underlying sets Ω and Ω', respectively. If T and T' are type-isomorphic then:*

$$(S(E), \Omega) \simeq (S(E'), \Omega').$$

Proof: First construct a bijection β, step by step, of the underlying sets with respect to the different partitions in every step. Since T and T' are type-isomorphic there exist a t_1-class and a t'_1-class having the same level-type. Choose arbitrary elements $\omega \in \Omega$, $\omega' \in \Omega'$ of a t_0- respectively t'_0-class of the same class-type and fix them. Define the bijection β_0 by $[\omega]_{t_0} \longrightarrow [\omega']_{t'_0}$. Assuming the bijection β_i has been already chosen, adjust β_{i+1} to an extension of the bijection β_i on $[\omega]_{t_{i+1}}$ in the following way: β_{i+1} maps every t_i-class contained in $[\omega]_{t_{i+1}}$ onto a t'_i-class of the same class-type in $[\omega']_{t'_{i+1}}$. By assumption $[\omega]_{t_{i+1}}$ and $[\omega']_{t'_{i+1}}$ have the same level-type, so β_{i+1} is an one-to-one correspondence between t_i-classes and t'_i-classes of the same class-type. Hence β_{i+1} is a bijection which preserves every t_j-class (and every t'_j-class) for $j \leq i$ in $[\omega]_{t_{i+1}}$, respectively $[\omega']_{t'_{i+1}}$. Iterating this procedure, one gets β as a limit of expanding bijections β_i and therefore β is a bijection.

In the second part of the proof we have to construct a suitable group-isomorphism $\alpha : S(E) \longrightarrow S(E')$ so that the following property holds for all $\omega \in \Omega$ and

$\pi \in S(E)$: $\omega^{\pi\beta} = \omega^{\beta\pi^{\alpha}}$. For every $\pi \in S(E)$ there exists an $t_i \in T \subseteq E$ so that $\pi \in \prod_{\omega\in\Omega} S([\omega]_{t_i})$. By the action of β_i from the first part π induces a permutation $\pi^{\alpha_i} \in \prod_{\omega'\in\Omega'} S([\omega']_{t'_i})$ because any two elements of the same t'_i-class have their images in a t_i-class under the permutation $\pi^{\alpha_i} = \beta_i^{-1}\pi\beta_i$. Hence we get:

$$\prod\nolimits_{\omega\in\Omega} S([\omega]_{t_i}) \overset{\alpha_i}{\cong} \prod\nolimits_{\omega'\in\Omega'} S([\omega']_{t'_i}).$$

Finally, if we define α as the limit of the expanding group-isomorphisms α_i, $i \in I\!N$, this yields the isomorphism since

$$S(E) = \bigcup_{i\in I\!N} \prod_{\omega\in\Omega} S([\omega]_{t_i}) \overset{\alpha}{\cong} \bigcup_{i\in I\!N} \prod_{\omega'\in\Omega'} S([\omega']_{t'_i}) = S(E')$$

and because E, T respectively E', T' are of the same type.

3 Characterisations

Theorem 2.7 suggests a programme for describing isomorphism types of all E-symmetric groups, i.e. to distinguish different approximating towers. For answering the underlying question what kind of influence has a set E of equivalence relations defined on Ω on the structure of the affiliated group of E-permutations the latest theorem instructs us to classify the sets E respectively the approximating towers T which are of the same kind. If we consider from now on sets of equivalence relations defined on Ω having some smoothness properties we get a remarkable *three-class-society* describing such sets of equivalence relations.

First it is useful to describe the underlying set Ω via ascending families of e-classes. If the set E is directed we write $\Omega = \bigcup_{e\in E}[\omega]_e$, for all $\omega \in \Omega$; call such an ascending family $\{[\omega]_e\}_{e\in E}$ an *approximating family* of Ω. From now on we suppose that E has an approximating tower T. Hence Ω is covered by an *approximating sequence* $\{[\omega]_{t_i}\}_{i\in I\!N}$, for $t_i \in T$. the equivalence relation e is said to be *balanced* if all e-classes have the same cardinality, and further we call E ($\aleph_0$-)*balanced*, if every $e \in E$ is balanced (and all e-classes are countable at most).

Now let E be $\aleph_0$-balanced and have an approximating tower, then Ω is countable. For $e' \geq e$ we define the *index* $|e' : e|$ *of* e *in* e': In case every e'-class of Ω is the union of exactly n e-classes of Ω for some $n \in I\!N$ then $|e' : e| = n$, otherwise $|e' : e| = \aleph_0$. For short, we call the $\aleph_0$-balanced set E of equivalence relations having an approximating tower *homogeneous* if the index can be defined for any two $e, e' \in E$ with $e' \geq e$.

If E is homogeneous every approximating family of the underlying set Ω belongs to one of the following classes of a *three-class-society*:

1. For every $e \in E$ there exists a coarser relation $e' \in E$ with $|e' : e| = \aleph_0$.
2. In E there exists a relation e so that for every relation $e' \in E$ coarser than e one has: $|e' : e| < \aleph_0$ and $|[\omega]_e| = \aleph_0$.
3. For any two relations e and e' of E with $e' \geq e$ one has: $|e' : e| < \aleph_0$ and $|[\omega]_e|$ is finite.

Now we may translate the necessary condition of theorem 2.7 to the case of homogeneous sets of equivalence relations. For every $e \in E$ there is only one class-type hence the approximating towers T and T' are type-isomorphic if and only if there are equipotent equivalence classes $[\omega]_{t_0}$ and $[\omega']_{t'_0}$, respectively, and every t_i and t'_i has the same index in any coarser equivalence relation t_j, respectively t'_j, for $t_i, t_j \in T$ and $t'_i, t'_j \in T'$.

Corollary 3.1 *Let E and E' be sets of homogeneous equivalence relations with approximating towers $T = (t_i)_{i \in \mathbb{N}}$ and $T' = (t'_i)_{i \in \mathbb{N}}$, defined on two sets Ω, respectively Ω'; for both underlying sets there are approximating sequences $\{[\omega]_{t_i}\}_{i \in \mathbb{N}}$ and $\{[\omega']_{t'_i}\}_{i \in \mathbb{N}}$, for $t \in T$, $t' \in T'$ and $\omega \in \Omega$, $\omega' \in \Omega'$ with at least two equipotent equivalence classes $[\omega]_{t_0}$, respectively $[\omega']_{t'_0}$. If these towers have the same index sequences $\{|t_{i+1} : t_i|\}_{i \in \mathbb{N}}$, respectively $\{|t'_{i+1} : t'_i|\}_{i \in \mathbb{N}}$, then*

$$(S(E), \Omega) \simeq (S(E'), \Omega').$$

Hence we get immediately a complete description for the Upper Class of the above society, since for every $t_i \in T$ there exists a coarser relation $t_j \in T$ with $|t_j : t_i| = \aleph_0$.

Corollary 3.2 *For any two sets E, E' of the Upper Class the two groups $S(E)$ and $S(E')$ with their actions on Ω and Ω', respectively, are permutation-equivalent.*

What can be said about the group $S(E)$ of all E-permutations if E does not belong to the Upper Class? To examine this question further we want to construct a group in which a given approximating tower T can be realised.

It will be convenient to consider as the underlying set Ω a countable abelian group A which is not finitely generated. Clearly A is union of its finitely generated subgroups A_e, for $e \in \mathbb{N}$. For every index e we define an equivalence relation in the following way: $x \sim_e y$ holds if and only if $x = ya$ for some $a \in A_e$. Note, that in this case E is homogeneous and has an approximating tower $T = (t_i)_{i \in \mathbb{N}}$ such that $A = \bigcup_{i \in \mathbb{N}} A_{t_i}$. For a fixed approximating tower T we construct a corresponding abelian group $A(T) =: A$ as follows:

Denote for short by $A_i := A_{t_i}$ and let $A_1 := \mathbb{Z}_{|t_1 : t_0|}$, the cyclic group of order $|t_1 : t_0|$. Let $A_0 = 1$, the identity subgroup; if A_i is already chosen, then inductively define A_{i+1} as $A_{i+1} := A_i \oplus \mathbb{Z}_{|t_{i+1} : t_i|}$. But, put $\mathbb{Z}_{|t_{i+1} : t_i|} := \mathbb{Z}$, if and only if $|t_{i+1} : t_i| = |A_{i+1} : A_i|$ is infinite. Put $A := \bigcup_{i\in\mathbb{N}} A_i$. This yields a *approximating sequence of subgroups* $\{A_i\}_{i\in\mathbb{N}}$ of A which meet all requirements, because A is a direct sum of finite cyclic groups, possibly together with infinite cyclic groups.

With such a realisation an arbitrary e-permutation π of A leaves every left coset of A_e invariant and hence every left coset of A_i for almost all $i \in \mathbb{N}$. Therefore one gets on the underlying set A a subgroup of the symmetric group, the *constricted symmetric group on A*. Actually this group has been introduced by B.A.F. Wehrfritz in [4, chapter VI]. Recall the slightly more general definition (cf. [3, S.297]) of the constricted symmetric group on an arbitrary group G:

$CS(G) = \{\pi \in S(G)|$ for some finitely generated subgroup G_π of G one has $(xG_\pi)^\pi = xG_\pi$ for all $x \in G\}$.

Notice that $CS(G)$ is isomorphic to the union of Cartesian products of the symmetric groups on left cosets with respect to all finitely generated subgroups of G. - Summing up, one gets

Theorem 3.3 *If E is homogeneous with an approximating tower $T = (t_i)_{i\in\mathbb{N}}$ then there is a countable, infinitely generated abelian group A such that:*

$$(S(E), \Omega) \simeq (CS(A), A).$$

Observe that $CS(A)$ does not depend on the group structure of A nor on the particular approximating tower. Even if the steps in the construction of the approximating tower $T = (t_i)_{i\in\mathbb{N}}$ will be permuted by σ, the resulting tower $T_\sigma = (t_{\sigma(i)})_{\sigma(i)\in\mathbb{N}}$ still yields the same group since A is abelian.

What is the incisive information distinguishing non-isomorphic constricted symmetric groups? One possibility is to search for invariants of different approximating towers, respectively approximating group sequences, yielding the same constricted symmetric group. Since A is abelian every tower can be refined if necessary by inserting for each step finitely many. Therefore assume that the approximating sequences have only steps of *step size* $|A_{i+1} : A_i|$ a power of p or infinite ones.

Moreover, a countable, infinitely abelian group $A = \bigcup_{i\in\mathbb{N}} A_i$, respectively its approximating sequence $\{A_i\}_{i\in\mathbb{N}}$ of subgroups *belongs to the same class* of the three-class-society as its corresponding approximating sequence $\{[\omega]_{t_i}\}_{t_i\in T}$ does. The countably generated free abelian group is an example for a group of the Upper Class. Clearly, countable elementary abelian (abelian locally finite) groups are examples for groups of the Lower Class. Note that, in this case, applying theorem 2.5, the affiliated constricted symmetric groups are also locally finite. If L is a non-trivial group in the Lower Class then $L \oplus \mathbb{Z}$ is in the Middle Class.

Decisive invariants for approximating sequences of subgroups are the number of different p-steps for any prime p, appearing in the affiliated index sequence $\{|A_{i+1} : A_i|\}_{i\in I\!N}$. Therefore one defines as a new measurement the *Steinitz type (supernatural number)* $St(\{A_i\}_{i\in I\!N})$ *of* A, the sequence of exponents appearing of all p-steps in a chosen approximating sequence of subgroups $\{A_i\}_{i\in I\!N}$ of A. Denote the Steinitz type of A by $St(\{A_i\}_{i\in I\!N}) := (n_\infty; n_2, n_3, n_5, \ldots, n_p, \ldots\ldots)$, whereby the coordinate n_p, for a prime p, stands for the existance of a p^{n_p}-step in the index sequence $\{|A_{i+1} : A_i|\}_{i\in I\!N}$. Further $n_\infty \in \{0, 1\}$ and $n_\infty = 1$, when ever there is at least an infinite step in the index sequence $\{|A_{i+1} : A_i|\}_{i\in I\!N}$ and $n_\infty = 0$ otherwise. We say for two groups A and B that their Steinitz types $St(\{A_i\}_{i\in I\!N})$ and $St(\{B_i\}_{i\in I\!N})$ are *equal* if their entries coincide.

As a consequence of theorem 2.7 equal Steinitz types lead to constricted symmetric groups which are permutation-equivalent and thus are isomorphic as abstract groups. Conversely, assume for groups A and B of the Lower Class (recall: as groups belonging to one class of the three-class-society they are countable, infinitely generated abelian) that $St(\{A_i\}_{i\in I\!N}) \neq St(\{B_i\}_{i\in I\!N})$. Hence there is at least one coordinate n_p in which their Steinitz types differ. Now, let n be the entry for n_p in $St(\{A_i\}_{i\in I\!N})$ and let m be the corresponding entry in $St(\{B_i\}_{i\in I\!N})$ with $n > m$, for $n, m \in I\!N$. In $CS(A)$ there is a permutation π and a subgroup A_π of A of order p^n such that π acts regularly on every left coset xA_π, for all $x \in A$. There isn't such a permutation in $CS(B)$, since $n > m$. Therefore $(CS(A), A) \not\simeq (CS(B), B)$. This proves the following

Theorem 3.4 *For any two groups A, B in the Lower Class one has:*

$$St(\{A_i\}_{i\in I\!N}) = St(\{B_i\}_{i\in I\!N}), \text{ if and only if } (CS(A), A) \simeq (CS(B), B).$$

We have shown that for groups of the Lower Class the Steinitz type characterises the constricted symmetric groups as permutation groups. However the question, whether the Steinitz type of a group of the Lower Class distinguishes the isomorphism types of constricted symmetric groups, i.e. of the $S(E)$, is left open.

For groups of the Middle Class in the three-class-society we are going to show that such a strong characterisation cannot be expected. In this case it is convenient to consider an appropriate approximating sequence of subgroups $\{A_i\}_{i\in I\!N}$ for a group A starting with an infinite step. Expand every infinite step by combining it with finitely many finite steps to a single infinite step (or shrink every infinite step in a similar way, of course). This manipulation leads to another approximating sequence of subgroups $\{A'_i\}_{i\in I\!N}$ which has still the same limit A and hence the same constricted symmetric group. But $St(\{A'_i\}_{i\in I\!N})$ differs now from $St(\{A_i\}_{i\in I\!N})$ at finitely many entries at most by finite exponent after the semicolon in the notation of the Steinitz type. Thus, Steinitz types of groups belonging to the Middle Class can be modified finitely without yielding another constricted symmetric group. Calling two Steinitz types *equivalent* if they differ finitely at most in finitely many distinct components, we define an equivalence relation on the class of all Steinitz types.

In the light of groups belonging to the Middle Class we ask, whether constricted symmetric groups which are not permutation-equivalent must have non-equivalent Steinitz types?

4 Normal subgroups

Let E be a directed and transitive set of equivalence relations on Ω. Recall from proposition 2.2 that both the group $S(\Omega)$ and $S(E)$, act n-transitively on Ω for every $n \in \mathbb{N}$. Thus, in a certain sense, they act similarly. But how similar are they? One possible approach to this question is to examine the structure of their normal systems. Do they have the same normal subgroups? The normal system of the symmetric group of countable degree is well known, since [7], and since [1] without any restriction on the cardinality of the underlying set Ω. Up to the end of this section we suppose Ω to be countable for focussing not too much on normal subgroups based only on set-theoretic properties. Again from proposition 2.2 we know that the only non-trivial normal subgroups of the symmetric group of countable degree, $S_0(\Omega)$ and $Alt(\Omega)$, the group of all finite even permutations, are also normal subgroups of $S(E)$. Is the factor ${}^{S(E)}/_{S_0(\Omega)}$ simple? The answer is: No, in general.

For $\pi \in S(E)$ and $e \in E$ we call an e-class π-*admissible* if π leaves every e-class of Ω invariant. Further denote by $\pi_{\omega,e}$ the restriction of π on the e-class $[\omega]_e$ and by $supp(\pi)$ the set of all elements in Ω moved by π. We want to call the reader's attention to the following two sets, which occur clearly as normal subgroups of $S(E)$:

Denote by $F(E)$ the set of all E-permutations acting finitely on admissible equivalence classes defined as follows: $F(E) := \{\pi \in S(E) |$ for all π-admissible e-classes one has $|supp(\pi_{\omega,e})| < \aleph_0$, for all $\omega \in \Omega\}$.

Further, let $B(E) := \{\pi \in S(E) |$ for all π-admissible e-classes there exist a $c(\pi, e) \in \mathbb{N}$ with $|supp(\pi_{\omega,e})| < c(\pi, e)$, for all $\omega \in \Omega\}$, the set of all E-permutations whose support on admissible equivalence classes is uniformly bounded by an integer c depending on the permutation π and the equivalence relation $e \in E$.

Note, that $S_0(\Omega)$ is strictly contained in $B(E)$ hence in $F(E)$ since there are permutations with infinite support in $B(E)$. As in theorem 2.5 we conclude that $B(E)$ is locally finite and $F(E)$ is locally residually finite. If E is not bounded there is a permutation in $F(E)$ with a cyclic decomposition in infinite many cycles of different finite degrees. Hence, in this case, we have $B(E) \trianglelefteq F(E)$, i.e. $B(E)$ is a *proper normal subgroup* of $F(E)$. If E is bounded we get only $B(E) = F(E)$. In the case that there is an infinite $e \in E$ one gets $F(E) \ntrianglelefteq S(E)$ since there are permutations in $S(E)$ which act regularly on every admissible equivalence class, otherwise $F(E) = S(E)$. Therefore one has the following

Proposition 4.1 *$B(E)$ is a locally finite normal subgroup and $F(E)$ is a locally residually finite normal subgroup of $S(E)$. Their relative positions depend on properties of E:*

1. *If E is bounded then $S_0(\Omega) \ntrianglelefteq B(E) = F(E) = S(E)$.*
2. *If E is finite and unbounded one has: $S_0(\Omega) \ntrianglelefteq B(E) \ntrianglelefteq F(E) = S(E)$.*
3. *If E is infinite one has: $S_0(\Omega) \ntrianglelefteq B(E) \ntrianglelefteq F(E) \ntrianglelefteq S(E)$.*

Further, we consider the set $M(E) := \{\pi \in S(E) |$ there is a $[\omega]_e$ for $\omega \in \Omega$, $e \in E$ such that $supp(\pi) \subseteq [\omega]_e\}$. Since E is transitive and directed $M(E)$ is also a normal subgroup of $S(E)$, and $M(E) = S_0(\Omega)$ if and only if E is finite. Therefore we assume from now on that E is infinite.

Furthermore, if we choose an arbitrary $\omega \in \Omega$ and fix it and assume that there is an equivalence relation $e \in E$ so that $[\omega]_e$ is infinite, $S_0(\Omega) = \bigcup_{e\in E} S_0([\omega]_e)$ and $M(E) = \bigcup_{e\in E} S([\omega]_e)$ holds since E is directed and transitive. From [7] we know that ${}^{S([\omega]_e)}/_{S_0([\omega]_e)}$ is a non-trivial simple factor and moreover for all $e \in E$ we have that $S_0([\omega]_e)$ is a maximal normal subgroup of $S([\omega]_e)$.

Let α be the canonical epimorphism mapping $S([\omega]_{e'})$ onto ${}^{S([\omega]_{e'})}/_{S_0([\omega]_{e'})}$. For every $e \in E$ with $e' \geq e$ α maps $S([\omega]_e)$ onto ${}^{S([\omega]_e)\cdot S_0([\omega]_{e'})}/_{S_0([\omega]_{e'})}$ which is isomorphic to ${}^{S([\omega]_e)}/_{S([\omega]_e)\cap S_0([\omega]_{e'})}$ by the Isomophism Theorem. Since $S_0([\omega]_e)$ is the maximal normal subgroup of $S([\omega]_e)$ we get $S_0 \cap S_0([\omega]_{e'}) = S_0([\omega]_e)$, hence

$$ {}^{S([\omega]_{e'})}/_{S_0([\omega]_{e'})} \supseteq {}^{S([\omega]_e)}/_{S_0([\omega]_e)} \text{ holds for every } e' \geq e \text{ and } e, e' \in E. $$

This yields a union of simple factor groups $\bigcup_{e\in E} {}^{S([\omega]_e)}/_{S_0([\omega]_e)} = {}^{M(E)}/_{S_0(\Omega)}$. Therefore ${}^{M(E)}/_{S_0(\Omega)}$ itself is simple. $M(E)$ is a normal subgroup of $S(E)$ minimal among the normal subgroups containing $S_0(E)$, i.e. that $M(E)$ does not show up in the normal system of the symmetric group on Ω.

Clearly, $F(E) \cap M(E) = B(E) \cap M(E) = S_0(\Omega)$, $B(E) \cdot M(E) \ntrianglelefteq F(E) \cdot M(E)$ and $F(E) \cdot M(E) \neq S(E)$, since $F(E) \not\subset S(E)$, if E is not finite. Again, from the Isomorphism Theorem we conclude the simplicity of the factors ${}^{B(E)\cdot M(E)}/_{B(E)}$ and ${}^{F(E)\cdot M(E)}/_{F(E)}$. Thus, this proves

Proposition 4.2 *$M(E)$ is a normal subgroup of $S(E)$ and the factor ${}^{M(E)}/_{S_0(\Omega)}$ is non-trivial simple if and only if E is not finite. In this case one gets:*

$$B(E) \cdot M(E) \ntrianglelefteq F(E) \cdot M(E) \ntrianglelefteq S(E).$$

The factors ${}^{B(E)\cdot M(E)}/_{B(E)}$ and ${}^{F(E)\cdot M(E)}/_{F(E)}$ are non-trivial simple.

Furthermore, let E have an approximating tower $T = (t_i)_{i\in \mathbb{N}}$. Recall that T is of the same type as E. Let $Alt(E) := \bigcup_{i\in\mathbb{N}} \prod_{\omega\in\Omega} Alt([\omega]_{t_i})$. In the case that E is finite it can be easily seen that $Alt(E)$ is a proper normal subgroup of $S(E) = F(E)$. What can be said about the cardinality of the non-trivial, elementary abelian factor group ${}^{S(E)}/_{Alt(E)}$? We shall define inductively a permutation $\pi \in S(E) \setminus Alt(E)$.

For $t_0 \in T$ we consider the partition of Ω in t_0-classes and, since Ω is countable,one has $\Omega = \bigcup_{i\in\mathbb{N}}[\omega]_{t_i}$. Suppose that every t_0-class contains at least two distinct elements ω_i, ω_i' such that a transposition $(\omega_i\omega_i')$ may be defined in $[\omega]_{t_0}$ for every $i \in \mathbb{N}$. Considering the first t_0-class $[\omega_1]_{t_0}$ let π_0 be the transposition $(\omega_1\omega_1')$. Now assume π_i has already been defined on $[\omega_1]_{t_i}$. To expand this permutation to $[\omega_1]_{t_{i+1}}$ the equivalence class of the coarser equivalence relation, suppose further without loss of generality that every t_{i+1}-class is union of at least three t_i-classes. Choose $[\omega_1]_{t_i}$, $[\omega_j]_{t_i}$, $[\omega_k]_{t_i}$ pairwise distinct t_i-classes contained in $[\omega_1]_{t_{i+1}}$. Define in $[\omega_j]_{t_i}$ and $[\omega_k]_{t_i}$ transpositions $(\omega_j\omega_j')$, respectively $(\omega_k\omega_k')$. Put π_{i+1} as ${}^{\pi_{i+1}}|_{[\omega_1]_{t_i}} = \pi_i$ and ${}^{\pi_{i+1}}|_{[\omega_1]_{t_{i+1}}\setminus[\omega_1]_{t_i}} = (\omega_j\omega_j')(\omega_k\omega_k')$. Hence π_{i+1} is odd. If one puts $\pi = \bigcup_{i\in\mathbb{N}} \pi_i$, then $\pi \in S(E)\setminus Alt(E)$ and for the cardinality of this factor one gets: $|{}^{S(E)}/_{Alt(E)}| \geq 2$. Moreover $\pi \notin S_0(\Omega) \cdot Alt(E)$ since there is an $e \in E$ such that π - by construction - is odd on infinitely many e-classes. Therefore one gets: $|{}^{S(E)}/_{Alt(E)}| \geq 4$.

Now let the set E of equivalence relations only be transitive and directed. Denote by $\tilde{S_0}(t_0) := S_0(t_0) = \prod_{\omega\in\Omega} S([\omega]_{t_0})$ and let $\tilde{S_0}(e') := \langle S_0(e'), \tilde{S_0}(e)\rangle^{S(e')}$ for $e' \geq e$. Put $\tilde{S_0}(E) := \bigcup_{e'\in E} \tilde{S_0}(e')$. $\tilde{S_0}(E)$ is a normal subgroup of $S(E)$ since for every $\pi \in S(E)$ there is an π-admissible e-class $e' \in E$ and therefore $\tilde{S_0}(e') \subseteq \tilde{S_0}(E)$ is π-invariant, for all $e' \geq e$. It can be easily seen that $\tilde{S_0}(E) = F(E)$ if E is finite. The same arguments prove that $\tilde{Alt}(E)$ is also a normal subgroup, independent of E being bounded, finite or infinite, for $\tilde{Alt}(E) := \bigcup_{e'\in E} \tilde{Alt}(e')$ with $\tilde{Alt}(t_0) := Alt(t_0) = \prod_{\omega\in\Omega} Alt([\omega]_{t_0})$ and $\tilde{Alt}(e') := \langle Alt(e'), \tilde{Alt}_0(e)\rangle^{Alt(e')}$ for $e' \geq e$. One gets the following

Proposition 4.3 *If the set E of equivalence relations is transitive and directed then the sets $\tilde{S_0}(E)$ and $\tilde{Alt}(E)$ are normal subgroups of $S(E)$. Further, $F(E)$ and $\tilde{S_0}(E)$, respectively $\tilde{Alt}(E)$ and $Alt(E)$ coincide if E is finite. Hence ${}^{S(E)}/_{S_0(\Omega)}$ is not simple.*

References

[1] Reinhold Baer: Die Kompositionsreihe der Gruppen aller eineindeutigen Abbildungen einer unendlichen Menge auf sich, *Studia Mathematica* **5** *(1934), 15-17.*

[2] Graham Higman/Bernhard H. Neumann/Hanna Neumann: Embedding theorems for groups, *J. London Math. Soc.* **24** *(1949), 247-254.* **MR** 11 # 322.

[3] Otto H. Kegel: Examples of highly transitive permutation groups, *Rend. Sem. Mat. Univ. Padova* **63** *(1980), 295-300.* **MR** 82f:20042.

[4] Otto H. Kegel/Bertram A.F. Wehrfritz: *Locally Finite Groups*, North-Holland, Amsterdam 1973. **MR** 57 # 9848.

[5] Nina V. Kroshko/Vitalii I. Sushchansky: Direct Limits of Symmetric and Alternating Groups with Strictly Diagonal Embeddings, *Arch. Math*, to appear.

[6] Bernhard H. Neumann: Identical relations in groups. I., *Math. Ann.* **114** *(1937), 506-525.*

[7] J.Schreier/Stanislaw M. Ulam: Über die Permutationsgruppe der natürlichen Zahlenfolge, *Studia Mathematica* **4** *(1933), 134-144.*

E-mail: ohk@ibm2.mathematik.uni-freiburg.de

Index

MIX
Papier aus verantwortungsvollen Quellen
Paper from responsible sources
FSC® C105338

If you have any concerns about our products,
you can contact us on
ProductSafety@springernature.com

In case Publisher is established outside the EU,
the EU authorized representative is:
Springer Nature Customer Service Center GmbH
Europaplatz 3, 69115 Heidelberg, Germany

Printed by Libri Plureos GmbH
in Hamburg, Germany